Advanced R

Second Edition

Chapman & Hall/CRC
The R Series

Series Editors

John M. Chambers, Department of Statistics, Stanford University, California, USA
Torsten Hothorn, Division of Biostatistics, University of Zurich, Switzerland
Duncan Temple Lang, Department of Statistics, University of California, Davis, USA
Hadley Wickham, RStudio, Boston, Massachusetts, USA

Recently Published Titles

For more information about this series, please visit: https://www.crcpress.com/go/the-r-series

Advanced R

Second Edition

Hadley Wickham

CRC Press

Taylor & Francis Group

Boca Raton London New York

CRC Press is an imprint of the
Taylor & Francis Group, an **informa** business

A CHAPMAN & HALL BOOK

CRC Press
Taylor & Francis Group
6000 Broken Sound Parkway NW, Suite 300
Boca Raton, FL 33487-2742

Printed on acid-free paper
Version Date: 20190422

International Standard Book Number-13: 978-0-367-25537-4 (Hardback)
International Standard Book Number-13: 978-0-815-38457-1 (Paperback)

Visit the Taylor & Francis Web site at
http://www.taylorandfrancis.com

and the CRC Press Web site at
http://www.crcpress.com

To Mina, the best book writing companion.
We miss you.

Contents

Preface

Welcome to the second edition of *Advanced R*. I had three main goals for this edition:

- Improve coverage of important concepts that I fully understood only after the publication of the first edition.

- Reduce coverage of topics time has shown to be less useful, or that I think are really exciting but turn out not to be that practical.

- Generally make the material easier to understand with better text, clearer code, and many more diagrams.

If you're familiar with the first edition, this preface describes the major changes so that you can focus your reading on the new areas. If you're reading a printed version of this book you'll notice one big change very quickly: *Advanced R* is now in colour! This has considerably improved the syntax highlighting of code chunks, and made it much easier to create helpful diagrams. I have taken advantage of this and included over 100 new diagrams throughout the book.

Another big change in this version is the use of new packages, particularly rlang (http://rlang.r-lib.org), which provides a clean interface to low-level data structures and operations. The first edition used base R functions almost exclusively, which created some pedagogical challenges because many functions evolved independently over multiple years, making it hard to see the big underlying ideas hidden amongst the incidental variations in function names and arguments. I continue to show base equivalents in sidebars, footnotes, and where needed, in individual sections, but if you want to see the purest base R expression of the ideas in this book, I recommend reading the first edition, which you can find online at http://adv-r.had.co.nz.

The foundations of R have not changed in the five years since the first edition, but my understanding of them certainly has. Thus, the overall structure of "Foundations" has remained roughly the same, but many of the individual chapters have been considerably improved:

- Chapter 2, "Names and values", is a brand new chapter that helps you understand the difference between objects and names of objects. This helps you more accurately predict when R will make a copy of a data structure, and lays important groundwork to understand functional programming.

- Chapter 3, "Vectors" (previously called data structures), has been rewritten to focus on vector types like integers, factors, and data frames. It contains more details of important S3 vectors (like dates and date-times), discusses the data frame variation provided by the tibble package [Müller and Wickham, 2018], and generally reflects my improved understanding of vector data types.

- Chapter 4, "Subsetting", now distinguishes between [and [[by their intention: [extracts many values and [[extracts a single value (previously they were characterised by whether they "simplified" or "preserved"). Section 4.3 draws the "train" to help you understand how [[works with lists, and introduces new functions that provide more consistent behaviour for out-of-bounds indices.

- Chapter 5, "Control flow", is a new chapter: somehow I previously forgot about important tools like if statements and for loops!

- Chapter 6, "Functions", has an improved ordering, introduces the pipe (%>%) as a third way to compose functions (Section 6.3), and has considerably improved coverage of function forms (Section 6.8).

- Chapter 7, "Environments", has a reorganised treatment of special environments (Section 7.4), and a much improved discussion of the call stack (Section 7.5).

- Chapter 8, "Conditions", contains material previously in "Exceptions and debugging", and much new content on how R's condition system works. It also shows you how to create your own custom condition classes (Section 8.5).

The chapters following Part I, Foundations, have been re-organised around the three most important programming paradigms in R: functional programming, object-oriented programming, and metaprogramming.

- Functional programming is now more cleanly divided into the three main techniques: "Functionals" (Chapter 9), "Function factories" (Chapter 10), and "Function operators" (Chapter 11). I've focussed in on ideas that have practical applications in data science and reduced the amount of pure theory.

  These chapters now use functions provided by the purrr package [Henry and Wickham, 2018a], which allow me to focus more on the underlying ideas and less on the incidental details. This led to a considerable simplification of the function operators chapter since a major use was to work around the absence of ellipses (. . .) in base functionals.

- Object-oriented programming (OOP) now forms a major section of the book with completely new chapters on base types (Chapter 12), S3 (Chapter 13), S4 (Chapter 15), R6 (Chapter 14), and the tradeoffs between the systems (Chapter 16).

These chapters focus on how the different object systems work, not how to use them effectively. This is unfortunate, but necessary, because many of the technical details are not described elsewhere, and effective use of OOP needs a whole book of its own.

- Metaprogramming (previously called "computing on the language") describes the suite of tools that you can use to generate code with code. Compared to the first edition this material has been substantially expanded and now focusses on "tidy evaluation", a set of ideas and theory that make metaprogramming safe, well-principled, and accessible to many more R programmers. Chapter 17, "Big picture" coarsely lays out how all the pieces fit together; Chapter 18, "Expressions", describes the underlying data structures; Chapter 19, "Quasiquotation", covers quoting and unquoting; Chapter 20, "Evaluation", explains evaluation of code in special environments; and Chapter 21, "Translations", pulls all the themes together to show how you might translate from one (programming) language to another.

The final section of the book pulls together the chapters on programming techniques: profiling, measuring and improving performance, and Rcpp. The contents are very similar to the first edition, although the organisation is a little different. I have made light updates throughout these chapters particularly to use newer packges (microbenchmark -> bench, lineprof -> profvis), but the majority of the text is the same.

While the second edition has mostly expanded coverage of existing material, five chapters have been removed:

- The vocabulary chapter has been removed because it was always a bit of an odd duck, and there are more effective ways to present vocabulary lists than in a book chapter.

- The style chapter has been replaced with an online style guide, http://style.tidyverse.org/. The style guide is paired with the new styler package [Müller and Walthert, 2018] which can automatically apply many of the rules.

- The C chapter has been moved to https://github.com/hadley/r-internals, which, over time, will provide a guide to writing C code that works with R's data structures.

- The memory chapter has been removed. Much of the material has been integrated into Chapter 2 and the remainder felt excessively technical and not that important to understand.

- The chapter on R's performance as a language was removed. It delivered few actionable insights, and became dated as R changed.

1

Introduction

I have now been programming in R for over 15 years, and have been doing it full-time for the last five years. This has given me the luxury of time to examine how the language works. This book is my attempt to pass on what I've learned so that you can understand the intricacies of R as quickly and painlessly as possible. Reading it will help you avoid the mistakes I've made and dead ends I've gone down, and will teach you useful tools, techniques, and idioms that can help you to attack many types of problems. In the process, I hope to show that, despite its sometimes frustrating quirks, R is, at its heart, an elegant and beautiful language, well tailored for data science.

1.1 Why R?

If you are new to R, you might wonder what makes learning such a quirky language worthwhile. To me, some of the best features are:

- It's free, open source, and available on every major platform. As a result, if you do your analysis in R, anyone can easily replicate it, regardless of where they live or how much money they earn.

- R has a diverse and welcoming community, both online (e.g. the #rstats twitter community (https://twitter.com/search?q=%23rstats)) and in person (like the many R meetups (https://www.meetup.com/topics/r-programming-language/)). Two particularly inspiring community groups are rweekly newsletter (https://rweekly.org) which makes it easy to keep up to date with R, and R-Ladies (http://r-ladies.org) which has made a wonderfully welcoming community for women and other minority genders.

- A massive set of packages for statistical modelling, machine learning, visualisation, and importing and manipulating data. Whatever model or graphic you're trying to do, chances are that someone has already tried to do it and you can learn from their efforts.

- Powerful tools for communicating your results. RMarkdown (https://rmarkdown.rstudio.com) makes it easy to turn your results into HTML files, PDFs, Word documents, PowerPoint presentations, dashboards and more. Shiny (http://shiny.rstudio.com) allows you to make beautiful interactive apps without any knowledge of HTML or javascript.

- RStudio, the IDE (http://www.rstudio.com/ide/), provides an integrated development environment, tailored to the needs of data science, interactive data analysis, and statistical programming.

- Cutting edge tools. Researchers in statistics and machine learning will often publish an R package to accompany their articles. This means immediate access to the very latest statistical techniques and implementations.

- Deep-seated language support for data analysis. This includes features like missing values, data frames, and vectorisation.

- A strong foundation of functional programming. The ideas of functional programming are well suited to the challenges of data science, and the R language is functional at heart, and provides many primitives needed for effective functional programming.

- RStudio, the company (https://www.rstudio.com), which makes money by selling professional products to teams of R users, and turns around and invests much of that money back into the open source community (over 50% of software engineers at RStudio work on open source projects). I work for RStudio because I fundamentally believe in its mission.

- Powerful metaprogramming facilities. R's metaprogramming capabilities allow you to write magically succinct and concise functions and provide an excellent environment for designing domain-specific languages like ggplot2, dplyr, data.table, and more.

- The ease with which R can connect to high-performance programming languages like C, Fortran, and C++.

Of course, R is not perfect. R's biggest challenge (and opportunity!) is that most R users are not programmers. This means that:

- Much of the R code you'll see in the wild is written in haste to solve a pressing problem. As a result, code is not very elegant, fast, or easy to understand. Most users do not revise their code to address these shortcomings.

- Compared to other programming languages, the R community is more focussed on results than processes. Knowledge of software engineering best practices is patchy. For example, not enough R programmers use source code control or automated testing.

- Metaprogramming is a double-edged sword. Too many R functions use tricks to reduce the amount of typing at the cost of making code that is hard to understand and that can fail in unexpected ways.

- Inconsistency is rife across contributed packages, and even within base R. You are confronted with over 25 years of evolution every time you use R, and this can make learning R tough because there are so many special cases to remember.

- R is not a particularly fast programming language, and poorly written R code can be terribly slow. R is also a profligate user of memory.

Personally, I think these challenges create a great opportunity for experienced programmers to have a profound positive impact on R and the R community. R users do care about writing high quality code, particularly for reproducible research, but they don't yet have the skills to do so. I hope this book will not only help more R users to become R programmers, but also encourage programmers from other languages to contribute to R.

1.2 Who should read this book

This book is aimed at two complementary audiences:

- Intermediate R programmers who want to dive deeper into R, understand how the language works, and learn new strategies for solving diverse problems.

- Programmers from other languages who are learning R and want to understand why R works the way it does.

To get the most out of this book, you'll need to have written a decent amount of code in R or another programming language. You should be familiar with the basics of data analysis (i.e. data import, manipulation, and visualisation), have written a number of functions, and be familiar with the installation and use of CRAN packages.

This book walks the narrow line between being a reference book (primarily used for lookup), and being linearly readable. This involves some tradeoffs, because it's difficult to linearise material while still keeping related materials together, and some concepts are much easier to explain if you're already familiar with specific technical vocabulary. I've tried to use footnotes and cross-references to make sure you can still make sense even if you just dip your toes in a chapter.

1.3   What you will get out of this book

This book delivers the knowledge that I think an advanced R programmer should possess: a deep understanding of the fundamentals coupled with a broad vocabulary that means that you can tactically learn more about a topic when needed.

After reading this book, you will:

- Be familiar with the foundations of R. You will understand complex data types and the best ways to perform operations on them. You will have a deep understanding of how functions work, you'll know what environments are, and how to make use of the condition system.

- Understand what functional programming means, and why it is a useful tool for data science. You'll be able to quickly learn how to use existing tools, and have the knowledge to create your own functional tools when needed.

- Know about R's rich variety of object-oriented systems. You'll be most familiar with S3, but you'll know of S4 and R6 and where to look for more information when needed.

- Appreciate the double-edged sword of metaprogramming. You'll be able to create functions that use tidy evaluation, saving typing and creating elegant code to express important operations. You'll also understand the dangers and when to avoid it.

- Have a good intuition for which operations in R are slow or use a lot of memory. You'll know how to use profiling to pinpoint performance bottlenecks, and you'll know enough C++ to convert slow R functions to fast C++ equivalents.

1.4   What you will not learn

This book is about R the programming language, not R the data analysis tool. If you are looking to improve your data science skills, I instead recommend that you learn about the tidyverse (https://www.tidyverse.org/), a collection of consistent packages developed by me and my colleagues. In this book you'll learn the techniques used to develop the tidyverse packages; if you want to instead learn how to use them, I recommend *R for Data Science* (http://r4ds.had.co.nz/).

If you want to share your R code with others, you will need to make an R package. This allows you to bundle code along with documentation and unit tests, and easily distribute it via CRAN. In my opinion, the easiest way to develop packages is with devtools (`http://devtools.r-lib.org`), roxygen2 (`http://klutometis.github.io/roxygen/`), testthat (`http://testthat.r-lib.org`), and usethis (`http://usethis.r-lib.org`). You can learn about using these packages to make your own package in *R packages* (`http://r-pkgs.had.co.nz/`).

1.5 Meta-techniques

There are two meta-techniques that are tremendously helpful for improving your skills as an R programmer: reading source code and adopting a scientific mindset.

Reading source code is important because it will help you write better code. A great place to start developing this skill is to look at the source code of the functions and packages you use most often. You'll find things that are worth emulating in your own code and you'll develop a sense of taste for what makes good R code. You will also see things that you don't like, either because its virtues are not obvious or it offends your sensibilities. Such code is nonetheless valuable, because it helps make concrete your opinions on good and bad code.

A scientific mindset is extremely helpful when learning R. If you don't understand how something works, you should develop a hypothesis, design some experiments, run them, and record the results. This exercise is extremely useful since if you can't figure something out and need to get help, you can easily show others what you tried. Also, when you learn the right answer, you'll be mentally prepared to update your world view.

1.6 Recommended reading

Because the R community mostly consists of data scientists, not computer scientists, there are relatively few books that go deep in the technical underpinnings of R. In my personal journey to understand R, I've found it particularly helpful to use resources from other programming languages. R has aspects of both functional and object-oriented (OO) programming languages. Learning how these concepts are expressed in R will help you leverage your existing knowledge of other programming languages, and will help you identify areas where you can improve.

To understand why R's object systems work the way they do, I found *The Structure and Interpretation of Computer Programs*[1] [Abelson et al., 1996] (SICP) to be particularly helpful. It's a concise but deep book, and after reading it, I felt for the first time that I could actually design my own object-oriented system. The book was my first introduction to the encapsulated paradigm of object-oriented programming found in R, and it helped me understand the strengths and weaknesses of this system. SICP also teaches the functional mindset where you create functions that are simple individually, and which become powerful when composed together.

To understand the trade-offs that R has made compared to other programming languages, I found *Concepts, Techniques and Models of Computer Programming* [Van-Roy and Haridi, 2004] extremely helpful. It helped me understand that R's copy-on-modify semantics make it substantially easier to reason about code, and that while its current implementation is not particularly efficient, it is a solvable problem.

If you want to learn to be a better programmer, there's no place better to turn than *The Pragmatic Programmer* [Hunt and Thomas, 1990]. This book is language agnostic, and provides great advice for how to be a better programmer.

1.7   Getting help

Currently, there are three main venues to get help when you're stuck and can't figure out what's causing the problem: RStudio Community (`https://community.rstudio.com/`), StackOverflow (`http://stackoverflow.com`) and the R-help mailing list (`https://stat.ethz.ch/mailman/listinfo/r-help`). You can get fantastic help in each venue, but they do have their own cultures and expectations. It's usually a good idea to spend a little time lurking, learning about community expectations, before you put up your first post.

Some good general advice:

- Make sure you have the latest version of R and of the package (or packages) you are having problems with. It may be that your problem is the result of a recently fixed bug.

- Spend some time creating a **repr**oducible **ex**ample, or reprex. This will help others help you, and often leads to a solution without asking others, because in the course of making the problem reproducible you often figure

---

[1]You can read it online for free at `https://mitpress.mit.edu/sites/default/files/sicp/full-text/book/book.html`

out the root cause. I highly recommend learning and using the reprex (`https://reprex.tidyverse.org/`) package.

If you are looking for specific help solving the exercises in this book, solutions from Malte Grosser and Henning Bumann are available at `https://advanced-r-solutions.rbind.io`.

1.8   Acknowledgments

I would like to thank the many contributors to R-devel and R-help and, more recently, Stack Overflow and RStudio Community. There are too many to name individually, but I'd particularly like to thank Luke Tierney, John Chambers, JJ Allaire, and Brian Ripley for generously giving their time and correcting my countless misunderstandings.

This book was written in the open (`https://github.com/hadley/adv-r/`), and chapters were advertised on twitter (`https://twitter.com/hadleywickham`) when complete. It is truly a community effort: many people read drafts, fixed typos, suggested improvements, and contributed content. Without those contributors, the book wouldn't be nearly as good as it is, and I'm deeply grateful for their help. Special thanks go to Jeff Hammerbacher, Peter Li, Duncan Murdoch, and Greg Wilson, who all read the book from cover-to-cover and provided many fixes and suggestions.

A big thank you to all 386 contributors (in alphabetical order by username): Aaron Wolen (@aaronwolen), @absolutelyNoWarranty, Adam Hunt (@adamphunt), @agrabovsky, Alexander Grueneberg (@agrueneberg), Anthony Damico (@ajdamico), James Manton (@ajdm), Aaron Schumacher (@ajschumacher), Alan Dipert (@alandipert), Alex Brown (@alexbbrown), @alexperrone, Alex Whitworth (@alexWhitworth), Alexandros Kokkalis (@alko989), @amarchin, Amelia McNamara (@AmeliaMN), Bryce Mecum (@amoeba), Andrew Laucius (@andrewla), Andrew Bray (@andrewpbray), Andrie de Vries (@andrie), Angela Li (@angela-li), @aranlunzer, Ari Lamstein (@arilamstein), @asnr, Andy Teucher (@ateucher), Albert Vilella (@avilella), baptiste (@baptiste), Brian G. Barkley (@BarkleyBG), Mara Averick (@batpigandme), Byron (@bcjaeger), Brandon Greenwell (@bgreenwell), Brandon Hurr (@bhive01), Jason Knight (@binarybana), Brett Klamer (@bklamer), Jesse Anderson (@blindjesse), Brian Mayer (@blmayer), Benjamin L. Moore (@blmoore), Brian Diggs (@BrianDiggs), Brian S. Yandell (@byandell), @carey1024, Chip Hogg (@chiphogg), Chris Muir (@ChrisMuir), Christopher Gandrud (@christophergandrud), Clay Ford (@clayford), Colin Fay (@ColinFay), @cortinah, Cameron Plouffe (@cplouffe), Carson Sievert (@cpsievert), Craig Citro (@craigcitro), Craig Grabowski (@craiggrabowski),

Christopher Roach (@croach), Peter Meilstrup (@crowding), Crt Ahlin (@crtahlin), Carlos Scheidegger (@cscheid), Colin Gillespie (@csgillespie), Christopher Brown (@ctbrown), Davor Cubranic (@cubranic), Darren Cusanovich (@cusanovich), Christian G. Warden (@cwarden), Charlotte Wickham (@cwickham), Dean Attali (@daattali), Dan Sullivan (@dan87134), Daniel Barnett (@daniel-barnett), Daniel (@danielruc91), Kenny Darrell (@darrkj), Tracy Nance (@datapixie), Dave Childers (@davechilders), David Vukovic (@david-vukovic), David Rubinger (@davidrubinger), David Chudzicki (@dchudz), Deependra Dhakal (@DeependraD), Daisuke ICHIKAWA (@dichika), david kahle (@dkahle), David LeBauer (@dlebauer), David Schweizer (@dlschweizer), David Montaner (@dmontaner), @dmurdoch, Zhuoer Dong (@dongzhuoer), Doug Mitarotonda (@dougmitarotonda), Dragoș Moldovan-Grünfeld (@dragosmg), Jonathan Hill (@Dripdrop12), @drtjc, Julian During (@duju211), @duncanwadsworth, @eaurele, Dirk Eddelbuettel (@eddelbuettel), @EdFineOKL, Eduard Szöcs (@EDiLD), Edwin Thoen (@EdwinTh), Ethan Heinzen (@eheinzen), @eijoac, Joel Schwartz (@eipi10), Eric Ronald Legrand (@elegrand), Elio Campitelli (@eliocamp), Ellis Valentiner (@ellisvalentiner), Emil Hvitfeldt (@EmilHvitfeldt), Emil Rehnberg (@EmilRehnberg), Daniel Lee (@erget), Eric C. Anderson (@eriqande), Enrico Spinielli (@espinielli), @etb, David Hajage (@eusebe), Fabian Scheipl (@fabian-s), @flammy0530, François Michonneau (@fmichonneau), Francois Pepin (@fpepin), Frank Farach (@frankfarach), @freezby, Frans van Dunné (@FvD), @fyears, @gagnagaman, Garrett Grolemund (@garrettgman), Gavin Simpson (@gavinsimpson), Brooke Anderson (@geanders), @gezakiss7, @gggtest, Gökçen Eraslan (@gokceneraslan), Josh Goldberg (@GoldbergData), Georg Russ (@gr650), @grasshoppermouse, Gregor Thomas (@gregorp), Garrett See (@gsee), Ari Friedman (@gsk3), Gunnlaugur Thor Briem (@gthb), Greg Wilson (@gvwilson), Hamed (@hamedbh), Jeff Hammerbacher (@hammer), Harley Day (@harleyday), @hassaad85, @helmingstay, Henning (@henningsway), Henrik Bengtsson (@HenrikBengtsson), Ching Boon (@hoscb), @hplieninger, Hörmet Yiltiz (@hyiltiz), Iain Dillingham (@iaindillingham), @IanKopacka, Ian Lyttle (@ijlyttle), Ilan Man (@ilanman), Imanuel Costigan (@imanuelcostigan), Thomas Bürli (@initdch), Os Keyes (@Ironholds), @irudnyts, i (@isomorphisms), Irene Steves (@isteves), Jan Gleixner (@jan-glx), Jannes Muenchow (@jannes-m), Jason Asher (@jasonasher), Jason Davies (@jasondavies), Chris (@jastingo), jcborras (@jcborras), Joe Cheng (@jcheng5), John Blischak (@jdblischak), @jeharmse, Lukas Burk (@jemus42), Jennifer (Jenny) Bryan (@jennybc), Justin Jent (@jentjr), Jeston (@JestonBlu), Josh Cook (@jhrcook), Jim Hester (@jimhester), @JimInNashville, @jimmyliu2017, Jim Vine (@jimvine), Jinlong Yang (@jinlong25), J.J. Allaire (@jjallaire), @JMHay, Jochen Van de Velde (@jochenvdv), Johann Hibschman (@johannh), John Baumgartner (@johnbaums), John Horton (@johnjosephhorton), @johnthomas12, Jon Calder (@jonmcalder), Jon Harmon (@jonthegeek), Julia Gustavsen (@jooolia), JorneBiccler (@JorneBiccler), Jeffrey Arnold (@jrnold), Joyce Robbins (@jtr13), Juan Manuel Truppia (@juancentro), @juangomezduaso, Kevin Markham (@justmarkham), john verzani (@jverzani), Michael Kane (@kaneplusplus), Bart Kastermans (@kasterma), Kevin D'Auria (@kdauria), Karandeep Singh (@kdpsingh), Ken Williams (@kenahoo), Kendon Bell (@kendonB), Kent Johnson (@kent37), Kevin Ushey (@kevinushey), 电线杆 (@kfeng123), Karl Forner (@kforner), Kirill Sevastyanenko (@kirillseva), Brian Knaus (@knausb), Kirill Müller (@krlmlr), Kriti Sen Sharma (@ksens), Kai Tang (唐恺) (@ktang), Kevin Wright (@kwstat), suo.lawrence.liu@gmail.com (mailto:suo.lawrence.liu@gmail.com) (@Lawrence-Liu), @ldfmrails, Kevin Kainan Li (@legendre6891), Rachel Severson (@leighseverson), Laurent Gatto (@lgatto), C. Jason Liang (@liangcj), Steve Lianoglou (@lianos), Yongfu Liao (@liao961120), Likan (@likanzhan), @lindbrook, Lingbing Feng (@Lingbing), Marcel Ramos (@LiNk-NY), Zhongpeng Lin (@linzhp), Lionel Henry (@lionel-), Lluís (@llrs), myq (@lrcg), Luke W Johnston (@lwjohnst86), Kevin Lynagh (@lynaghk), @MajoroMask, Malcolm Barrett (@malcolmbarrett), @mannyishere, @mascaretti, Matt (@mattbaggott), Matthew Grogan (@mattgrogan), @matthewhillary, Matthieu Gomez (@matthieugomez), Matt Malin (@mattmalin), Mauro Lepore (@maurolepore), Max Ghenis (@MaxGhenis), Maximilian Held (@maxheld83), Michal Bojanowski (@mbojan), Mark Rosenstein (@mbrmbr), Michael Sumner (@mdsumner), Jun Mei (@meijun), merkliopas (@merkliopas), mfrasco (@mfrasco), Michael Bach (@michaelbach), Michael Bishop (@MichaelMBishop), Michael Buckley (@michaelmikebuckley), Michael Quinn (@michaelquinn32), @miguelmorin, Michael (@mikekaminsky), Mine Cetinkaya-Rundel (@mine-cetinkaya-rundel), @mjsduncan, Mamoun Benghezal (@MoBeng), Matt Pettis (@mpettis), Martin Morgan (@mtmorgan), Guy Dawson (@Mullefa), Nacho Caballero (@nachocab), Natalya Rapstine (@natalya-patrikeeva), Nick Carchedi (@ncarchedi), Pascal Burkhard (@Nenuial), Noah Greifer (@ngreifer), Nicholas Vasile (@nickv9), Nikos Ignatiadis (@nignatiadis), Nina Munkholt Jakobsen (@nmjakobsen), Xavier Laviron (@norival), Nick Pullen (@nstjhp), Oge Nnadi (@ogennadi), Oliver Paisley (@oliverpaisley), Pariksheet Nanda (@omsai), Øystein Sørensen (@osorensen), Paul (@otepoti), Otho Mantegazza (@othomantegazza), Dewey Dunnington (@paleolimbot), Paola Corrales (@paocorrales), Parker Abercrombie (@parkerabercrombie), Patrick Hausmann (@patperu), Patrick Miller (@patr1ckm), Patrick Werkmeister (@Patrick01), @paulponcet, @pdb61, Tom Crockett (@pelotom), @pengyu, Jeremiah (@perryjer1), Peter Hickey (@PeteHaitch), Phil Chalmers (@philchalmers), Jose Antonio Magaña Mesa (@picarus), Pierre Casadebaig (@picasa), Antonio Piccolboni (@piccolbo), Pierre Roudier (@pierreroudier), Poor Yorick (@pooryorick), Marie-Helene Burle (@prosoitos), Peter Schulam (@pschulam), John (@quantbo), Quyu Kong (@qykong), Ramiro Magno (@ramiromagno), Ramnath Vaidyanathan (@ramnathv), Kun Ren (@renkun-ken), Richard Reeve (@richardreeve), Richard Cotton (@richierocks), Robert M Flight (@rmflight), R. Mark Sharp (@rmsharp), Robert Krzyzanowski (@robertzk), @robiRagan, Romain François (@romainfrancois), Ross Holmberg (@rossholmberg), Ricardo Pietrobon (@rpietro), @rrunner, Ryan Walker (@rtwalker), @rubenfcasal, Rob Weyant (@rweyant), Rumen Zarev (@rzarev), Nan Wang (@sailingwave), Samuel Perreault (@samperochkin), @sbgraves237, Scott Kostyshak (@scottkosty), Scott Leishman (@scttl), Sean Hughes (@seaaan), Sean Anderson (@seananderson), Sean Carmody (@seancarmody), Sebastian (@sebastian-c), Matthew Sedaghatfar (@sedaghatfar), @see24, Sven E. Templer (@setempler), @sflippl, @shabbybanks, Steven Pav (@shabbychef), Shannon Rush (@shannonrush), S'busiso Mkhondwane (@sibusiso16), Sigfried Gold (@Sigfried), Simon O'Hanlon (@simonohanlon101), Simon Potter (@sjp), Leo Razoumov (@slonik-az), Richard M. Smith (@Smudgerville), Steve (@SplashDance), Scott Ritchie (@sritchie73), Tim Cole (@statist7), @ste-fan, @stephens999, Steve Walker (@stevencarlislewalker), Stefan Widgren (@stewid), Homer Strong (@strongh), Suman Khanal (@sumanstats), Dirk (@surmann), Sebastien Vigneau (@svigneau), Steven Nydick (@swnydick), Taekyun Kim (@taekyunk), Tal Galili (@talgalili), @Tazinho, Tyler Bradley (@tbradley1013), Tom B (@tbuckl), @tdenes, @thomasherbig, Thomas (@thomaskern), Thomas Lin Pedersen (@thomasp85), Thomas Zumbrunn (@thomaszumbrunn), Tim Waterhouse (@timwaterhouse), TJ Mahr (@tjmahr), Thomas Nagler (@tnagler), Anton Antonov (@tonytonov), Ben Torvaney (@Torvaney), Jeff Allen (@trestletech), Tyler Rinker (@trinker), Chitu Okoli (@Tripartio), Kirill Tsukanov (@tskir), Terence Teo (@tteo), Tim Triche, Jr. (@ttriche), @tyhenkaline, Tyler Ritchie (@tylerritchie), Tyler Littlefield (@tyluRp), Varun Agrawal (@varun729), Vijay Barve (@vijaybarve), Victor (@vkryukov), Vaidotas Zemlys-Balevičius (@vzemlys), Winston Chang (@wch), Linda Chin (@wchi144), Welliton Souza (@Welliton309), Gregg Whitworth (@whitwort), Will Beasley (@wibeasley), William R Bauer (@WilCrofter), William Doane (@WilDoane), Sean Wilkinson (@wilkinson), Christof Winter (@winterschlaefer), Jake Thompson (@wjakethompson), Bill Carver (@wmc3), Wolfgang Huber (@wolfganghuber), Krishna Sankar (@xsankar), Yihui Xie (@yihui), yang (@yiluheihei), Yoni Ben-Meshulam (@yoni), @yuchouchen, Yuqi Liao (@yuqiliao), Hiroaki Yutani (@yutannihilation), Zachary Foster (@zachary-foster), @zachcp, @zackham, Sergio Oller (@zeehio), Edward Cho (@zerokarmaleft), Albert Zhao (@zxzb).

1.9 Conventions

Throughout this book I use f() to refer to functions, g to refer to variables and function parameters, and h/ to paths.

Larger code blocks intermingle input and output. Output is commented (#>) so that if you have an electronic version of the book, e.g., https://adv-r. hadley.nz/, you can easily copy and paste examples into R.

Many examples use random numbers. These are made reproducible by `set.seed(1014)`, which is executed automatically at the start of each chapter.

1.10   Colophon

This book was written in bookdown (`http://bookdown.org/`) inside RStudio (`http://www.rstudio.com/ide/`). The website (`https://adv-r.hadley.nz/`) is hosted with netlify (`http://netlify.com/`), and automatically updated after every commit by travis-ci (`https://travis-ci.org/`). The complete source is available from GitHub (`https://github.com/hadley/adv-r`). Code in the printed book is set in inconsolata (`http://levien.com/type/myfonts/inconsolata.html`). Emoji images in the printed book come from the open-licensed Twitter Emoji (`https://github.com/twitter/twemoji`).

This version of the book was built with R version 3.5.2 (2018-12-20) and the following packages.

package	version	source
bench	1.0.1	Github (r-lib/bench@97844d5)
bookdown	0.9	CRAN (R 3.5.0)
dbplyr	1.3.0.9000	local
desc	1.2.0	Github (r-lib/desc@42b9578)
emo	0.0.0.9000	Github (hadley/emo@02a5206)
ggbeeswarm	0.6.0	CRAN (R 3.5.0)
ggplot2	3.0.0	CRAN (R 3.5.0)
knitr	1.20	standard (@1.20)
lobstr	1.0.1	CRAN (R 3.5.1)
memoise	1.1.0.9000	Github (hadley/memoise@1650ad7)
png	0.1-7	CRAN (R 3.5.0)
profvis	0.3.5	CRAN (R 3.5.1)
Rcpp	1.0.0.1	Github (RcppCore/Rcpp@0c9f683)
rlang	0.3.1.9000	Github (r-lib/rlang@7243c6d)
rmarkdown	1.11	CRAN (R 3.5.0)
RSQLite	2.1.1.9002	Github (r-dbi/RSQLite@0db36af)
scales	1.0.0	CRAN (R 3.5.0)
sessioninfo	1.1.1	CRAN (R 3.5.1)
sloop	1.0.0.9000	local
testthat	2.0.1.9000	local
tidyr	0.8.3.9000	local
vctrs	0.1.0.9002	Github (r-lib/vctrs@098154c)
zeallot	0.1.0	CRAN (R 3.5.0)

Part I

Foundations

Introduction

To start your journey in mastering R, the following six chapters will help you learn the foundational components of R. I expect that you've already seen many of these pieces before, but you probably have not studied them deeply. To help check your existing knowledge, each chapter starts with a quiz; if you get all the questions right, feel free to skip to the next chapter!

1. Chapter 2 teaches you about an important distinction that you probably haven't thought deeply about: the difference between an object and its name. Improving your mental model here will help you make better predictions about when R copies data and hence which basic operations are cheap and which are expensive.

2. Chapter 3 dives into the details of vectors, helping you learn how the different types of vector fit together. You'll also learn about attributes, which allow you to store arbitrary metadata, and form the basis for two of R's object-oriented programming toolkits.

3. Chapter 4 describes how to use subsetting to write clear, concise, and efficient R code. Understanding the fundamental components will allow you to solve new problems by combining the building blocks in novel ways.

4. Chapter 5 presents tools of control flow that allow you to only execute code under certain conditions, or to repeatedly execute code with changing inputs. These include the important if and for constructs, as well as related tools like switch() and while.

5. Chapter 6 deals with functions, the most important building blocks of R code. You'll learn exactly how they work, including the scoping rules, which govern how R looks up values from names. You'll also learn more of the details behind lazy evaluation, and how you can control what happens when you exit a function.

6. Chapter 7 describes a data structure that is crucial for understanding how R works, but quite unimportant for data analysis: the environment. Environments are the data structure that binds names to values, and they power important tools like package namespaces. Unlike most programming languages, environments in R are "first

class" which means that you can manipulate them just like other objects.

7. Chapter 8 concludes the foundations of R with an exploration of "conditions", the umbrella term used to describe errors, warnings, and messages. You've certainly encountered these before, so in this chapter you learn how to signal them appropriately in your own functions, and how to handle them when signalled elsewhere.

2

Names and values

2.1 Introduction

In R, it is important to understand the distinction between an object and its name. Doing so will help you:

- More accurately predict the performance and memory usage of your code.
- Write faster code by avoiding accidental copies, a major source of slow code.
- Better understand R's functional programming tools.

The goal of this chapter is to help you understand the distinction between names and values, and when R will copy an object.

Quiz

Answer the following questions to see if you can safely skip this chapter. You can find the answers at the end of the chapter in Section 2.7.

1. Given the following data frame, how do I create a new column called "3" that contains the sum of 1 and 2? You may only use $, not [[. What makes 1, 2, and 3 challenging as variable names?

   ```
   df <- data.frame(runif(3), runif(3))
   names(df) <- c(1, 2)
   ```

2. In the following code, how much memory does y occupy?

   ```
   x <- runif(1e6)
   y <- list(x, x, x)
   ```

3. On which line does a get copied in the following example?

```
a <- c(1, 5, 3, 2)
b <- a
b[[1]] <- 10
```

Outline

- Section 2.2 introduces you to the distinction between names and values, and discusses how <- creates a binding, or reference, between a name and a value.

- Section 2.3 describes when R makes a copy: whenever you modify a vector, you're almost certainly creating a new, modified vector. You'll learn how to use tracemem() to figure out when a copy actually occurs. Then you'll explore the implications as they apply to function calls, lists, data frames, and character vectors.

- Section 2.4 explores the implications of the previous two sections on how much memory an object occupies. Since your intuition may be profoundly wrong and since utils::object.size() is unfortunately inaccurate, you'll learn how to use lobstr::obj_size().

- Section 2.5 describes the two important exceptions to copy-on-modify: with environments and values with a single name, objects are actually modified in place.

- Section 2.6 concludes the chapter with a discussion of the garbage collector, which frees up the memory used by objects no longer referenced by a name.

Prerequisites

We'll use the lobstr (https://github.com/r-lib/lobstr) package to dig into the internal representation of R objects.

```
library(lobstr)
```

Sources

The details of R's memory management are not documented in a single place. Much of the information in this chapter was gleaned from a close reading of the documentation (particularly ?Memory and ?gc), the memory profiling (http://cran.r-project.org/doc/manuals/R-exts.html#Profiling-R-code-for-memory-use) section of *Writing R extensions* [R Core Team, 2018a], and the SEXPs (http://cran.r-project.org/doc/manuals/R-ints.html#SEXPs) section of *R internals* [R Core Team, 2018b]. The rest I figured out by reading the C source code, performing small experiments, and asking questions on R-devel. Any mistakes are entirely mine.

2.2   Binding basics

Consider this code:

```
x <- c(1, 2, 3)
```

It's easy to read it as: "create an object named 'x', containing the values 1, 2, and 3". Unfortunately, that's a simplification that will lead to inaccurate predictions about what R is actually doing behind the scenes. It's more accurate to say that this code is doing two things:

- It's creating an object, a vector of values, c(1, 2, 3).
- And it's binding that object to a name, x.

In other words, the object, or value, doesn't have a name; it's actually the name that has a value.

To further clarify this distinction, I'll draw diagrams like this:

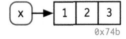

The name, x, is drawn with a rounded rectangle. It has an arrow that points to (or binds or references) the value, the vector c(1, 2, 3). The arrow points in opposite direction to the assignment arrow: <- creates a binding from the name on the left-hand side to the object on the right-hand side.

Thus, you can think of a name as a reference to a value. For example, if you run this code, you don't get another copy of the value c(1, 2, 3), you get another binding to the existing object:

```
y <- x
```

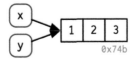

You might have noticed that the value c(1, 2, 3) has a label: 0x74b. While the vector doesn't have a name, I'll occasionally need to refer to an object independent of its bindings. To make that possible, I'll label values with a unique identifier. These identifiers have a special form that looks like the object's memory "address", i.e. the location in memory where the object is stored. But because the actual memory addresses changes every time the code is run, we use these identifiers instead.

You can access an object's identifier with lobstr::obj_addr(). Doing so allows you to see that both x and y point to the same identifier:

```
obj_addr(x)
#> [1] "0x7f8e16850e58"
obj_addr(y)
#> [1] "0x7f8e16850e58"
```

These identifiers are long, and change every time you restart R.

It can take some time to get your head around the distinction between names and values, but understanding this is really helpful in functional programming where functions can have different names in different contexts.

2.2.1 Non-syntactic names

R has strict rules about what constitutes a valid name. A **syntactic** name must consist of letters[1], digits, . and _ but can't begin with _ or a digit. Additionally, you can't use any of the **reserved words** like TRUE, NULL, if, and function (see the complete list in ?Reserved). A name that doesn't follow these rules is a **non-syntactic** name; if you try to use them, you'll get an error:

```
_abc <- 1
#> Error: unexpected input in "_"

if <- 10
#> Error: unexpected assignment in "if <-"
```

It's possible to override these rules and use any name, i.e., any sequence of characters, by surrounding it with backticks:

---

[1]Surprisingly, precisely what constitutes a letter is determined by your current locale. That means that the syntax of R code can actually differ from computer to computer, and that it's possible for a file that works on one computer to not even parse on another! Avoid this problem by sticking to ASCII characters (i.e. A-Z) as much as possible.

```
`_abc` <- 1
`_abc`
#> [1] 1

`if` <- 10
`if`
#> [1] 10
```

While it's unlikely you'd deliberately create such crazy names, you need to understand how these crazy names work because you'll come across them, most commonly when you load data that has been created outside of R.

You *can* also create non-syntactic bindings using single or double quotes (e.g. "_abc" <- 1) instead of backticks, but you shouldn't, because you'll have to use a different syntax to retrieve the values. The ability to use strings on the left hand side of the assignment arrow is an historical artefact, used before R supported backticks.

2.2.2 Exercises

1. Explain the relationship between a, b, c and d in the following code:

   ```
   a <- 1:10
   b <- a
   c <- b
   d <- 1:10
   ```

2. The following code accesses the mean function in multiple ways. Do they all point to the same underlying function object? Verify this with lobstr::obj_addr().

   ```
   mean
   base::mean
   get("mean")
   evalq(mean)
   match.fun("mean")
   ```

3. By default, base R data import functions, like read.csv(), will automatically convert non-syntactic names to syntactic ones. Why

might this be problematic? What option allows you to suppress this behaviour

4. What rules does make.names() use to convert non-syntactic names into syntactic ones?

5. I slightly simplified the rules that govern syntactic names. Why is .123e1 not a syntactic name? Read ?make.names for the full details.

2.3 Copy-on-modify

Consider the following code. It binds x and y to the same underlying value, then modifies y^2.

```
x <- c(1, 2, 3)
y <- x

y[[3]] <- 4
x
#> [1] 1 2 3
```

Modifying y clearly didn't modify x. So what happened to the shared binding? While the value associated with y changed, the original object did not. Instead, R created a new object, 0xcd2, a copy of 0x74b with one value changed, then rebound y to that object.

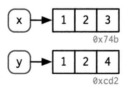

This behaviour is called **copy-on-modify**. Understanding it will radically improve your intuition about the performance of R code. A related way to describe this behaviour is to say that R objects are unchangeable, or **immutable**. However, I'll generally avoid that term because there are a couple of important exceptions to copy-on-modify that you'll learn about in Section 2.5.

---

[2]You may be surprised to see [[used to subset a numeric vector. We'll come back to this in Section 4.3, but in brief, I think you should always use [[when you are getting or setting a single element.

When exploring copy-on-modify behaviour interactively, be aware that you'll get different results inside of RStudio. That's because the environment pane must make a reference to each object in order to display information about it. This distorts your interactive exploration but doesn't affect code inside of functions, and so doesn't affect performance during data analysis. For experimentation, I recommend either running R directly from the terminal, or using RMarkdown (like this book).

2.3.1 `tracemem()`

You can see when an object gets copied with the help of `base::tracemem()`. Once you call that function with an object, you'll get the object's current address:

```
x <- c(1, 2, 3)
cat(tracemem(x), "\n")
#> <0x7f80c0e0ffc8>
```

From then on, whenever that object is copied, `tracemem()` will print a message telling you which object was copied, its new address, and the sequence of calls that led to the copy:

```
y <- x
y[[3]] <- 4L
#> tracemem[0x7f80c0e0ffc8 -> 0x7f80c4427f40]:
```

If you modify y again, it won't get copied. That's because the new object now only has a single name bound to it, so R applies modify-in-place optimisation. We'll come back to this in Section 2.5.

```
y[[3]] <- 5L

untracemem(y)
```

`untracemem()` is the opposite of `tracemem()`; it turns tracing off.

2.3.2 Function calls

The same rules for copying also apply to function calls. Take this code:

```
f <- function(a) {
  a
}

x <- c(1, 2, 3)
cat(tracemem(x), "\n")
#> <0x7f8e10fd5d38>

z <- f(x)
# there's no copy here!

untracemem(x)
```

While f() is running, the a inside the function points to the same value as the x does outside the function:

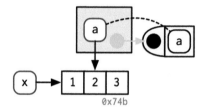

You'll learn more about the conventions used in this diagram in Section 7.4.4. In brief: the function f() is depicted by the yellow object on the right. It has a formal argument, a, which becomes a binding (indicated by dotted black line) in the execution environment (the gray box) when the function is run.

Once f() completes, x and z will point to the same object. 0x74b never gets copied because it never gets modified. If f() did modify x, R would create a new copy, and then z would bind that object.

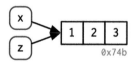

2.3.3   Lists

It's not just names (i.e. variables) that point to values; elements of lists do too. Consider this list, which is superficially very similar to the numeric vector above:

```r
l1 <- list(1, 2, 3)
```

This list is more complex because instead of storing the values itself, it stores references to them:

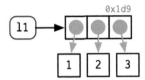

This is particularly important when we modify a list:

```r
l2 <- l1
```

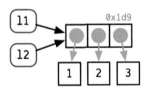

```r
l2[[3]] <- 4
```

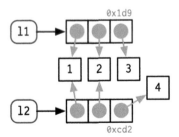

Like vectors, lists use copy-on-modify behaviour; the original list is left unchanged, and R creates a modified copy. This, however, is a **shallow** copy: the list object and its bindings are copied, but the values pointed to by the bindings are not. The opposite of a shallow copy is a deep copy where the contents of every reference are copied. Prior to R 3.1.0, copies were always deep copies.

To see values that are shared across lists, use `lobstr::ref()`. `ref()` prints the memory address of each object, along with a local ID so that you can easily cross-reference shared components.

```
ref(l1, l2)
#>  ▋ [1:0x7f8e161c9198] <list>
#>  ├─[2:0x7f8e1689c178] <dbl>
#>  ├─[3:0x7f8e1689c140] <dbl>
#>  └─[4:0x7f8e1689c108] <dbl>
#>
#>  ▋ [5:0x7f8e162d0c98] <list>
#>  ├─[2:0x7f8e1689c178]
#>  ├─[3:0x7f8e1689c140]
#>  └─[6:0x7f8e163423a0] <dbl>
```

2.3.4   Data frames

Data frames are lists of vectors, so copy-on-modify has important consequences when you modify a data frame. Take this data frame as an example:

```
d1 <- data.frame(x = c(1, 5, 6), y = c(2, 4, 3))
```

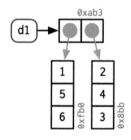

If you modify a column, only *that* column needs to be modified; the others will still point to their original references:

```
d2 <- d1
d2[, 2] <- d2[, 2] * 2
```

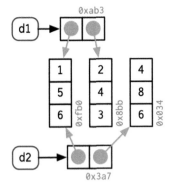

However, if you modify a row, every column is modified, which means every column must be copied:

```
d3 <- d1
d3[1, ] <- d3[1, ] * 3
```

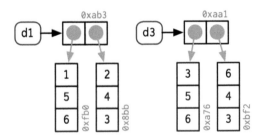

2.3.5 Character vectors

The final place that R uses references is with character vectors[3]. I usually draw character vectors like this:

```
x <- c("a", "a", "abc", "d")
```

But this is a polite fiction. R actually uses a **global string pool** where each element of a character vector is a pointer to a unique string in the pool:

---

[3]Confusingly, a character vector is a vector of strings, not individual characters.

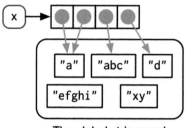

The global string pool

You can request that ref() show these references by setting the character argument to TRUE:

```
ref(x, character = TRUE)
#> █ [1:0x7f8e149ef008] <chr>
#> ├─[2:0x7f8e1094d598] <string: "a">
#> ├─[2:0x7f8e1094d598]
#> ├─[3:0x7f8e152e21c8] <string: "abc">
#> └─[4:0x7f8e10b0e6b0] <string: "d">
```

This has a profound impact on the amount of memory a character vector uses but is otherwise generally unimportant, so elsewhere in the book I'll draw character vectors as if the strings lived inside a vector.

2.3.6 Exercises

1. Why is tracemem(1:10) not useful?

2. Explain why tracemem() shows two copies when you run this code. Hint: carefully look at the difference between this code and the code shown earlier in the section.

   ```
   x <- c(1L, 2L, 3L)
   tracemem(x)

   x[[3]] <- 4
   ```

3. Sketch out the relationship between the following objects:

   ```
   a <- 1:10
   b <- list(a, a)
   c <- list(b, a, 1:10)
   ```

4. What happens when you run this code?

```
x <- list(1:10)
x[[2]] <- x
```

Draw a picture.

2.4 Object size

You can find out how much memory an object takes with `lobstr::obj_size()`[4]:

```
obj_size(letters)
#> 1,712 B
obj_size(ggplot2::diamonds)
#> 3,456,344 B
```

Since the elements of lists are references to values, the size of a list might be much smaller than you expect:

```
x <- runif(1e6)
obj_size(x)
#> 8,000,048 B

y <- list(x, x, x)
obj_size(y)
#> 8,000,128 B
```

y is only 80 bytes[5] bigger than x. That's the size of an empty list with three elements:

```
obj_size(list(NULL, NULL, NULL))
#> 80 B
```

Similarly, because R uses a global string pool character vectors take up less memory than you might expect: repeating a string 1000 times does not make it take up 1000 times as much memory.

---

[4]Beware of the `utils::object.size()` function. It does not correctly account for shared references and will return sizes that are too large.

[5]If you're running 32-bit R, you'll see slightly different sizes.

```
banana <- "bananas bananas bananas"
obj_size(banana)
#> 136 B
obj_size(rep(banana, 100))
#> 928 B
```

References also make it challenging to think about the size of individual objects. obj_size(x) + obj_size(y) will only equal obj_size(x, y) if there are no shared values. Here, the combined size of x and y is the same as the size of y:

```
obj_size(x, y)
#> 8,000,128 B
```

Finally, R 3.5.0 and later versions have a feature that might lead to surprises: ALTREP, short for **alternative representation**. This allows R to represent certain types of vectors very compactly. The place you are most likely to see this is with : because instead of storing every single number in the sequence, R just stores the first and last number. This means that every sequence, no matter how large, is the same size:

```
obj_size(1:3)
#> 680 B
obj_size(1:1e3)
#> 680 B
obj_size(1:1e6)
#> 680 B
obj_size(1:1e9)
#> 680 B
```

2.4.1   Exercises

1. In the following example, why are object.size(y) and obj_size(y) so radically different? Consult the documentation of object.size().

   ```
   y <- rep(list(runif(1e4)), 100)

   object.size(y)
   #> 8005648 bytes
   obj_size(y)
   #> 80,896 B
   ```

2. Take the following list. Why is its size somewhat misleading?

```
funs <- list(mean, sd, var)
obj_size(funs)
#> 17,608 B
```

3. Predict the output of the following code:

```
a <- runif(1e6)
obj_size(a)

b <- list(a, a)
obj_size(b)
obj_size(a, b)

b[[1]][[1]] <- 10
obj_size(b)
obj_size(a, b)

b[[2]][[1]] <- 10
obj_size(b)
obj_size(a, b)
```

2.5 Modify-in-place

As we've seen above, modifying an R object usually creates a copy. There are two exceptions:

- Objects with a single binding get a special performance optimisation.
- Environments, a special type of object, are always modified in place.

2.5.1 Objects with a single binding

If an object has a single name bound to it, R will modify it in place:

```
v <- c(1, 2, 3)
```

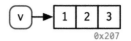

```
v[[3]] <- 4
```

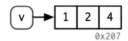

(Note the object IDs here: v continues to bind to the same object, 0x207.)

Two complications make predicting exactly when R applies this optimisation challenging:

- When it comes to bindings, R can currently[6] only count 0, 1, or many. That means that if an object has two bindings, and one goes away, the reference count does not go back to 1: one less than many is still many. In turn, this means that R will make copies when it sometimes doesn't need to.

- Whenever you call the vast majority of functions, it makes a reference to the object. The only exception are specially written "primitive" C functions. These can only be written by R-core and occur mostly in the base package.

Together, these two complications make it hard to predict whether or not a copy will occur. Instead, it's better to determine it empirically with trace-mem().

Let's explore the subtleties with a case study using for loops. For loops have a reputation for being slow in R, but often that slowness is caused by every iteration of the loop creating a copy. Consider the following code. It subtracts the median from each column of a large data frame:

```
x <- data.frame(matrix(runif(5 * 1e4), ncol = 5))
medians <- vapply(x, median, numeric(1))

for (i in seq_along(medians)) {
  x[[i]] <- x[[i]] - medians[[i]]
}
```

This loop is surprisingly slow because each iteration of the loop copies the data frame. You can see this by using tracemem():

---

[6]By the time you read this, this may have changed, as plans are afoot to improve reference counting: https://developer.r-project.org/Refcnt.html

```
cat(tracemem(x), "\n")
#> <0x7f80c429e020>

for (i in 1:5) {
  x[[i]] <- x[[i]] - medians[[i]]
}
#> tracemem[0x7f80c429e020 -> 0x7f80c0c144d8]:
#> tracemem[0x7f80c0c144d8 -> 0x7f80c0c14540]: [[<-.data.frame [[<-
#> tracemem[0x7f80c0c14540 -> 0x7f80c0c145a8]: [[<-.data.frame [[<-
#> tracemem[0x7f80c0c145a8 -> 0x7f80c0c14610]:
#> tracemem[0x7f80c0c14610 -> 0x7f80c0c14678]: [[<-.data.frame [[<-
#> tracemem[0x7f80c0c14678 -> 0x7f80c0c146e0]: [[<-.data.frame [[<-
#> tracemem[0x7f80c0c146e0 -> 0x7f80c0c14748]:
#> tracemem[0x7f80c0c14748 -> 0x7f80c0c147b0]: [[<-.data.frame [[<-
#> tracemem[0x7f80c0c147b0 -> 0x7f80c0c14818]: [[<-.data.frame [[<-
#> tracemem[0x7f80c0c14818 -> 0x7f80c0c14880]:
#> tracemem[0x7f80c0c14880 -> 0x7f80c0c148e8]: [[<-.data.frame [[<-
#> tracemem[0x7f80c0c148e8 -> 0x7f80c0c14950]: [[<-.data.frame [[<-
#> tracemem[0x7f80c0c14950 -> 0x7f80c0c149b8]:
#> tracemem[0x7f80c0c149b8 -> 0x7f80c0c14a20]: [[<-.data.frame [[<-
#> tracemem[0x7f80c0c14a20 -> 0x7f80c0c14a88]: [[<-.data.frame [[<-

untracemem(x)
```

In fact, each iteration copies the data frame not once, not twice, but three times! Two copies are made by [[.data.frame, and a further copy[7] is made because [[.data.frame is a regular function that increments the reference count of x.

We can reduce the number of copies by using a list instead of a data frame. Modifying a list uses internal C code, so the references are not incremented and only a single copy is made:

```
y <- as.list(x)
cat(tracemem(y), "\n")
#> <0x7f80c5c3de20>

for (i in 1:5) {
  y[[i]] <- y[[i]] - medians[[i]]
}
#> tracemem[0x7f80c5c3de20 -> 0x7f80c48de210]:
```

---

[7]These copies are shallow: they only copy the reference to each individual column, not the contents of the columns. This means the performance isn't terrible, but it's obviously not as good as it could be.

While it's not hard to determine when a copy is made, it is hard to prevent it. If you find yourself resorting to exotic tricks to avoid copies, it may be time to rewrite your function in C++, as described in Chapter 25.

2.5.2 Environments

You'll learn more about environments in Chapter 7, but it's important to mention them here because their behaviour is different from that of other objects: environments are always modified in place. This property is sometimes described as **reference semantics** because when you modify an environment all existing bindings to that environment continue to have the same reference.

Take this environment, which we bind to e1 and e2:

```
e1 <- rlang::env(a = 1, b = 2, c = 3)
e2 <- e1
```

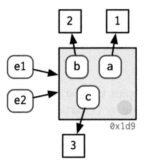

If we change a binding, the environment is modified in place:

```
e1$c <- 4
e2$c
#> [1] 4
```

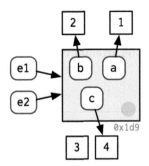

This basic idea can be used to create functions that "remember" their previous state. See Section 10.2.4 for more details. This property is also used to implement the R6 object-oriented programming system, the topic of Chapter 14.

One consequence of this is that environments can contain themselves:

```
e <- rlang::env()
e$self <- e

ref(e)
#>  [1:0x7f8e18918480] <env>
#> └─self = [1:0x7f8e18918480]
```

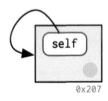

This is a unique property of environments!

2.5.3 Exercises

1. Explain why the following code doesn't create a circular list.

   ```
   x <- list()
   x[[1]] <- x
   ```

2. Wrap the two methods for subtracting medians into two functions, then use the 'bench' package [Hester, 2018] to carefully com-

pare their speeds. How does performance change as the number of columns increase?

3. What happens if you attempt to use `tracemem()` on an environment?

2.6 Unbinding and the garbage collector

Consider this code:

```
x <- 1:3
```

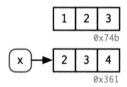

```
x <- 2:4
```

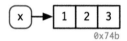

```
rm(x)
```

We created two objects, but by the time the code finishes, neither object is bound to a name. How do these objects get deleted? That's the job of the **garbage collector**, or GC for short. The GC frees up memory by deleting R objects that are no longer used, and by requesting more memory from the operating system if needed.

R uses a **tracing** GC. This means it traces every object that's reachable from the global[8] environment, and all objects that are, in turn, reachable

---

[8]And every environment in the current call stack.

from those objects (i.e. the references in lists and environments are searched recursively). The garbage collector does not use the modify-in-place reference count described above. While these two ideas are closely related, the internal data structures are optimised for different use cases.

The garbage collector (GC) runs automatically whenever R needs more memory to create a new object. Looking from the outside, it's basically impossible to predict when the GC will run. In fact, you shouldn't even try. If you want to find out when the GC runs, call `gcinfo(TRUE)` and GC will print a message to the console every time it runs.

You can force garbage collection by calling `gc()`. But despite what you might have read elsewhere, there's never any *need* to call `gc()` yourself. The only reasons you might *want* to call `gc()` is to ask R to return memory to your operating system so other programs can use it, or for the side-effect that tells you how much memory is currently being used:

```
gc()
#>            used (Mb) gc trigger (Mb) limit (Mb) max used (Mb)
#> Ncells   720514 38.5    1329171   71         NA  1329171   71
#> Vcells  5455049 41.7   19041363  145      16384 15603142  119
```

`lobstr::mem_used()` is a wrapper around `gc()` that prints the total number of bytes used:

```
mem_used()
#> 83,989,184 B
```

This number won't agree with the amount of memory reported by your operating system. There are three reasons:

1. It includes objects created by R but not by the R interpreter.

2. Both R and the operating system are lazy: they won't reclaim memory until it's actually needed. R might be holding on to memory because the OS hasn't yet asked for it back.

3. R counts the memory occupied by objects but there may be empty gaps due to deleted objects. This problem is known as memory fragmentation.

2.7   Quiz answers

1. You must quote non-syntactic names with backticks: `: for example,
   the variables 1, 2, and 3.

   ```
   df <- data.frame(runif(3), runif(3))
   names(df) <- c(1, 2)

   df$`3` <- df$`1` + df$`2`
   ```

2. It occupies about 8 MB.

   ```
   x <- runif(1e6)
   y <- list(x, x, x)
   obj_size(y)
   #> 8,000,128 B
   ```

3. a is copied when b is modified, b[[1]] <- 10.

3

Vectors

3.1 Introduction

This chapter discusses the most important family of data types in base R: vectors[1]. While you've probably already used many (if not all) of the different types of vectors, you may not have thought deeply about how they're inter-related. In this chapter, I won't cover individual vectors types in too much detail, but I will show you how all the types fit together as a whole. If you need more details, you can find them in R's documentation.

Vectors come in two flavours: atomic vectors and lists[2]. They differ in terms of their elements' types: for atomic vectors, all elements must have the same type; for lists, elements can have different types. While not a vector, NULL is closely related to vectors and often serves the role of a generic zero length vector. This diagram, which we'll be expanding on throughout this chapter, illustrates the basic relationships:

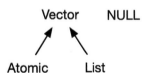

Every vector can also have **attributes**, which you can think of as a named list of arbitrary metadata. Two attributes are particularly important. The **dimension** attribute turns vectors into matrices and arrays and the **class** attribute powers the S3 object system. While you'll learn how to use S3 in Chapter 13, here you'll learn about some of the most important S3 vectors: factors, date and times, data frames, and tibbles. And while 2D structures

---

[1]Collectively, all the other data types are known as "node" types, which include things like functions and environments. You're most likely to come across this highly technical term when using gc(): the "N" in Ncells stands for nodes and the "V" in Vcells stands for vectors.

[2]A few places in R's documentation call lists generic vectors to emphasise their difference from atomic vectors.

like matrices and data frames are not necessarily what come to mind when you think of vectors, you'll also learn why R considers them to be vectors.

Quiz

Take this short quiz to determine if you need to read this chapter. If the answers quickly come to mind, you can comfortably skip this chapter. You can check your answers in Section 3.8.

1. What are the four common types of atomic vectors? What are the two rare types?

2. What are attributes? How do you get them and set them?

3. How is a list different from an atomic vector? How is a matrix different from a data frame?

4. Can you have a list that is a matrix? Can a data frame have a column that is a matrix?

5. How do tibbles behave differently from data frames?

Outline

- Section 3.2 introduces you to the atomic vectors: logical, integer, double, and character. These are R's simplest data structures.

- Section 3.3 takes a small detour to discuss attributes, R's flexible metadata specification. The most important attributes are names, dimensions, and class.

- Section 3.4 discusses the important vector types that are built by combining atomic vectors with special attributes. These include factors, dates, datetimes, and durations.

- Section 3.5 dives into lists. Lists are very similar to atomic vectors, but have one key difference: an element of a list can be any data type, including another list. This makes them suitable for representing hierarchical data.

- Section 3.6 teaches you about data frames and tibbles, which are used to represent rectangular data. They combine the behaviour of lists and matrices to make a structure ideally suited for the needs of statistical data.

3.2 Atomic vectors

There are four primary types of atomic vectors: logical, integer, double, and character (which contains strings). Collectively integer and double vectors are known as numeric vectors[3]. There are two rare types: complex and raw. I won't discuss them further because complex numbers are rarely needed in statistics, and raw vectors are a special type that's only needed when handling binary data.

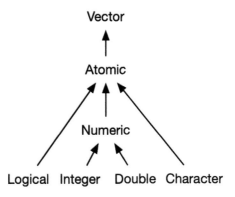

3.2.1 Scalars

Each of the four primary types has a special syntax to create an individual value, AKA a **scalar**[4]:

- Logicals can be written in full (TRUE or FALSE), or abbreviated (T or F).

- Doubles can be specified in decimal (0.1234), scientific (1.23e4), or hexadecimal (0xcafe) form. There are three special values unique to doubles: Inf, -Inf, and NaN (not a number). These are special values defined by the floating point standard.

- Integers are written similarly to doubles but must be followed by L[5] (1234L, 1e4L, or 0xcafeL), and can not contain fractional values.

---

[3]This is a slight simplification as R does not use "numeric" consistently, which we'll come back to in Section 12.3.1.

[4]Technically, the R language does not possess scalars. Everything that looks like a scalar is actually a vector of length one. This is mostly a theoretical distinction, but it does mean that expressions like 1[1] work.

[5]L is not intuitive, and you might wonder where it comes from. At the time L was added to R, R's integer type was equivalent to a long integer in C, and C code could use a suffix of l or L to force a number to be a long integer. It was decided that l was too visually similar to i (used for complex numbers in R), leaving L.

- Strings are surrounded by " ("hi") or ' ('bye'). Special characters are escaped with \; see ?Quotes for full details.

3.2.2 Making longer vectors with c()

To create longer vectors from shorter ones, use c(), short for combine:

```
lgl_var <- c(TRUE, FALSE)
int_var <- c(1L, 6L, 10L)
dbl_var <- c(1, 2.5, 4.5)
chr_var <- c("these are", "some strings")
```

When the inputs are atomic vectors, c() always creates another atomic vector; i.e. it flattens:

```
c(c(1, 2), c(3, 4))
#> [1] 1 2 3 4
```

In diagrams, I'll depict vectors as connected rectangles, so the above code could be drawn as follows:

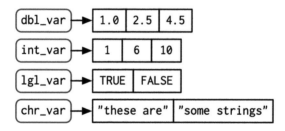

You can determine the type of a vector with typeof()[6] and its length with length().

```
typeof(lgl_var)
#> [1] "logical"
typeof(int_var)
#> [1] "integer"
typeof(dbl_var)
#> [1] "double"
typeof(chr_var)
#> [1] "character"
```

---

[6]You may have heard of the related mode() and storage.mode() functions. Do not use them: they exist only for compatibility with S.

3.2.3 Missing values

R represents missing, or unknown values, with special sentinel value: NA (short for not applicable). Missing values tend to be infectious: most computations involving a missing value will return another missing value.

```
NA > 5
#> [1] NA
10 * NA
#> [1] NA
!NA
#> [1] NA
```

There are only a few exceptions to this rule. These occur when some identity holds for all possible inputs:

```
NA ^ 0
#> [1] 1
NA | TRUE
#> [1] TRUE
NA & FALSE
#> [1] FALSE
```

Propagation of missingness leads to a common mistake when determining which values in a vector are missing:

```
x <- c(NA, 5, NA, 10)
x == NA
#> [1] NA NA NA NA
```

This result is correct (if a little surprising) because there's no reason to believe that one missing value has the same value as another. Instead, use is.na() to test for the presence of missingness:

```
is.na(x)
#> [1]  TRUE FALSE  TRUE FALSE
```

NB: Technically there are four missing values, one for each of the atomic types: NA (logical), NA_integer_ (integer), NA_real_ (double), and NA_character_ (character). This distinction is usually unimportant because NA will be automatically coerced to the correct type when needed.

3.2.4   Testing and coercion

Generally, you can **test** if a vector is of a given type with an `is.*()` function, but these functions need to be used with care. `is.logical()`, `is.integer()`, `is.double()`, and `is.character()` do what you might expect: they test if a vector is a character, double, integer, or logical. Avoid `is.vector()`, `is.atomic()`, and `is.numeric()`: they don't test if you have a vector, atomic vector, or numeric vector; you'll need to carefully read the documentation to figure out what they actually do.

For atomic vectors, type is a property of the entire vector: all elements must be the same type. When you attempt to combine different types they will be **coerced** in a fixed order: character → double → integer → logical. For example, combining a character and an integer yields a character:

```
str(c("a", 1))
#>  chr [1:2] "a" "1"
```

Coercion often happens automatically. Most mathematical functions (`+`, `log`, `abs`, etc.) will coerce to numeric. This coercion is particularly useful for logical vectors because `TRUE` becomes 1 and `FALSE` becomes 0.

```
x <- c(FALSE, FALSE, TRUE)
as.numeric(x)
#> [1] 0 0 1

# Total number of TRUEs
sum(x)
#> [1] 1

# Proportion that are TRUE
mean(x)
#> [1] 0.333
```

Generally, you can deliberately coerce by using an `as.*()` function, like `as.logical()`, `as.integer()`, `as.double()`, or `as.character()`. Failed coercion of strings generates a warning and a missing value:

```
as.integer(c("1", "1.5", "a"))
#> Warning: NAs introduced by coercion
#> [1]  1  1 NA
```

3.2.5 Exercises

1. How do you create raw and complex scalars? (See ?raw and ?complex.)

2. Test your knowledge of the vector coercion rules by predicting the output of the following uses of c():

```
c(1, FALSE)
c("a", 1)
c(TRUE, 1L)
```

3. Why is 1 == "1" true? Why is -1 < FALSE true? Why is "one" < 2 false?

4. Why is the default missing value, NA, a logical vector? What's special about logical vectors? (Hint: think about c(FALSE, NA_character_).)

5. Precisely what do is.atomic(), is.numeric(), and is.vector() test for?

3.3 Attributes

You might have noticed that the set of atomic vectors does not include a number of important data structures like matrices, arrays, factors, or datetimes. These types are built on top of atomic vectors by adding attributes. In this section, you'll learn the basics of attributes, and how the dim attribute makes matrices and arrays. In the next section you'll learn how the class attribute is used to create S3 vectors, including factors, dates, and date-times.

3.3.1 Getting and setting

You can think of attributes as name-value pairs[7] that attach metadata to an object. Individual attributes can be retrieved and modified with attr(), or retrieved en masse with attributes(), and set en masse with structure().

---

[7]Attributes behave like named lists, but are actually pairlists. Pairlists are functionally indistinguishable from lists, but are profoundly different under the hood. You'll learn more about them in Section 18.6.1.

```
a <- 1:3
attr(a, "x") <- "abcdef"
attr(a, "x")
#> [1] "abcdef"

attr(a, "y") <- 4:6
str(attributes(a))
#> List of 2
#>  $ x: chr "abcdef"
#>  $ y: int [1:3] 4 5 6

# Or equivalently
a <- structure(
  1:3,
  x = "abcdef",
  y = 4:6
)
str(attributes(a))
#> List of 2
#>  $ x: chr "abcdef"
#>  $ y: int [1:3] 4 5 6
```

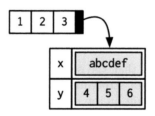

Attributes should generally be thought of as ephemeral. For example, most attributes are lost by most operations:

```
attributes(a[1])
#> NULL
attributes(sum(a))
#> NULL
```

There are only two attributes that are routinely preserved:

- **names**, a character vector giving each element a name.
- **dim**, short for dimensions, an integer vector, used to turn vectors into matrices or arrays.

To preserve other attributes, you'll need to create your own S3 class, the topic of Chapter 13.

3.3.2 Names

You can name a vector in three ways:

```
# When creating it:
x <- c(a = 1, b = 2, c = 3)

# By assigning a character vector to names()
x <- 1:3
names(x) <- c("a", "b", "c")

# Inline, with setNames():
x <- setNames(1:3, c("a", "b", "c"))
```

Avoid using `attr(x, "names")` as it requires more typing and is less readable than `names(x)`. You can remove names from a vector by using `unname(x)` or `names(x) <- NULL`.

To be technically correct, when drawing the named vector x, I should draw it like so:

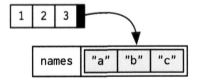

However, names are so special and so important, that unless I'm trying specifically to draw attention to the attributes data structure, I'll use them to label the vector directly:

To be useful with character subsetting (e.g. Section 4.5.1) names should be unique, and non-missing, but this is not enforced by R. Depending on how the names are set, missing names may be either "" or `NA_character_`. If all names are missing, `names()` will return `NULL`.

3.3.3 Dimensions

Adding a dim attribute to a vector allows it to behave like a 2-dimensional **matrix** or a multi-dimensional **array**. Matrices and arrays are primarily mathematical and statistical tools, not programming tools, so they'll be used infrequently and only covered briefly in this book. Their most important feature is multidimensional subsetting, which is covered in Section 4.2.3.

You can create matrices and arrays with matrix() and array(), or by using the assignment form of dim():

```
# Two scalar arguments specify row and column sizes
a <- matrix(1:6, nrow = 2, ncol = 3)
a
#>      [,1] [,2] [,3]
#> [1,]    1    3    5
#> [2,]    2    4    6

# One vector argument to describe all dimensions
b <- array(1:12, c(2, 3, 2))
b
#> , , 1
#>
#>      [,1] [,2] [,3]
#> [1,]    1    3    5
#> [2,]    2    4    6
#>
#> , , 2
#>
#>      [,1] [,2] [,3]
#> [1,]    7    9   11
#> [2,]    8   10   12

# You can also modify an object in place by setting dim()
c <- 1:6
dim(c) <- c(3, 2)
c
#>      [,1] [,2]
#> [1,]    1    4
#> [2,]    2    5
#> [3,]    3    6
```

Many of the functions for working with vectors have generalisations for matrices and arrays:

Vector	Matrix	Array
names()	rownames(), colnames()	dimnames()
length()	nrow(), ncol()	dim()
c()	rbind(), cbind()	abind::abind()
—	t()	aperm()
is.null(dim(x))	is.matrix()	is.array()

A vector without a dim attribute set is often thought of as 1-dimensional, but actually has NULL dimensions. You also can have matrices with a single row or single column, or arrays with a single dimension. They may print similarly, but will behave differently. The differences aren't too important, but it's useful to know they exist in case you get strange output from a function (tapply() is a frequent offender). As always, use str() to reveal the differences.

```
str(1:3)                     # 1d vector
#>   int [1:3] 1 2 3
str(matrix(1:3, ncol = 1))   # column vector
#>   int [1:3, 1] 1 2 3
str(matrix(1:3, nrow = 1))   # row vector
#>   int [1, 1:3] 1 2 3
str(array(1:3, 3))           # "array" vector
#>   int [1:3(1d)] 1 2 3
```

3.3.4 Exercises

1. How is setNames() implemented? How is unname() implemented? Read the source code.

2. What does dim() return when applied to a 1-dimensional vector? When might you use NROW() or NCOL()?

3. How would you describe the following three objects? What makes them different from 1:5?

```
x1 <- array(1:5, c(1, 1, 5))
x2 <- array(1:5, c(1, 5, 1))
x3 <- array(1:5, c(5, 1, 1))
```

4. An early draft used this code to illustrate structure():

```
structure(1:5, comment = "my attribute")
#> [1] 1 2 3 4 5
```

But when you print that object you don't see the comment attribute. Why? Is the attribute missing, or is there something else
special about it? (Hint: try using help.)

3.4 S3 atomic vectors

One of the most important vector attributes is class, which underlies the S3
object system. Having a class attribute turns an object into an **S3 object**,
which means it will behave differently from a regular vector when passed to
a **generic** function. Every S3 object is built on top of a base type, and often
stores additional information in other attributes. You'll learn the details of
the S3 object system, and how to create your own S3 classes, in Chapter 13.

In this section, we'll discuss four important S3 vectors used in base R:

- Categorical data, where values come from a fixed set of levels recorded in
  factor vectors.

- Dates (with day resolution), which are recorded in **Date** vectors.

- Date-times (with second or sub-second resolution), which are stored in
  POSIXct vectors.

- Durations, which are stored in **difftime** vectors.

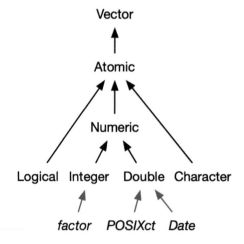

3.4.1 Factors

A factor is a vector that can contain only predefined values. It is used to store categorical data. Factors are built on top of an integer vector with two attributes: a class, "factor", which makes it behave differently from regular integer vectors, and levels, which defines the set of allowed values.

```
x <- factor(c("a", "b", "b", "a"))
x
#> [1] a b b a
#> Levels: a b

typeof(x)
#> [1] "integer"
attributes(x)
#> $levels
#> [1] "a" "b"
#>
#> $class
#> [1] "factor"
```

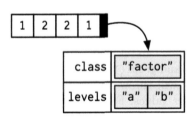

Factors are useful when you know the set of possible values but they're not all present in a given dataset. In contrast to a character vector, when you tabulate a factor you'll get counts of all categories, even unobserved ones:

```
sex_char <- c("m", "m", "m")
sex_factor <- factor(sex_char, levels = c("m", "f"))

table(sex_char)
#> sex_char
#> m
#> 3
table(sex_factor)
#> sex_factor
#> m f
#> 3 0
```

Ordered factors are a minor variation of factors. In general, they behave like regular factors, but the order of the levels is meaningful (low, medium, high) (a property that is automatically leveraged by some modelling and visualisation functions).

```
grade <- ordered(c("b", "b", "a", "c"), levels = c("c", "b", "a"))
grade
#> [1] b b a c
#> Levels: c < b < a
```

In base R[8] you tend to encounter factors very frequently because many base R functions (like `read.csv()` and `data.frame()`) automatically convert character vectors to factors. This is suboptimal because there's no way for those functions to know the set of all possible levels or their correct order: the levels are a property of theory or experimental design, not of the data. Instead, use the argument `stringsAsFactors = FALSE` to suppress this behaviour, and then manually convert character vectors to factors using your knowledge of the "theoretical" data. To learn about the historical context of this behaviour, I recommend *stringsAsFactors: An unauthorized biography* (http://simplystatistics.org/2015/07/24/stringsasfactors-an-unauthorized-biography/) by Roger Peng, and *stringsAsFactors = <sigh>* (http://notstatschat.tumblr.com/post/124987394001/stringsasfactors-sigh) by Thomas Lumley.

While factors look like (and often behave like) character vectors, they are built on top of integers. So be careful when treating them like strings. Some string methods (like `gsub()` and `grepl()`) will automatically coerce factors to strings, others (like `nchar()`) will throw an error, and still others will (like `c()`) use the underlying integer values. For this reason, it's usually best to explicitly convert factors to character vectors if you need string-like behaviour.

3.4.2 Dates

Date vectors are built on top of double vectors. They have class "Date" and no other attributes:

```
today <- Sys.Date()

typeof(today)
#> [1] "double"
attributes(today)
```

---

[8]The tidyverse never automatically coerces characters to factors, and provides the forcats [Wickham, 2018] package specifically for working with factors.

```
#> $class
#> [1] "Date"
```

The value of the double (which can be seen by stripping the class), represents the number of days since 1970-01-01[9]:

```
date <- as.Date("1970-02-01")
unclass(date)
#> [1] 31
```

3.4.3 Date-times

Base R[10] provides two ways of storing date-time information, POSIXct, and POSIXlt. These are admittedly odd names: "POSIX" is short for Portable Operating System Interface, which is a family of cross-platform standards. "ct" standards for calendar time (the `time_t` type in C), and "lt" for local time (the `struct tm` type in C). Here we'll focus on POSIXct, because it's the simplest, is built on top of an atomic vector, and is most appropriate for use in data frames. POSIXct vectors are built on top of double vectors, where the value represents the number of seconds since 1970-01-01.

```
now_ct <- as.POSIXct("2018-08-01 22:00", tz = "UTC")
now_ct
#> [1] "2018-08-01 22:00:00 UTC"

typeof(now_ct)
#> [1] "double"
attributes(now_ct)
#> $class
#> [1] "POSIXct" "POSIXt"
#>
#> $tzone
#> [1] "UTC"
```

The `tzone` attribute controls only how the date-time is formatted; it does not control the instant of time represented by the vector. Note that the time is not printed if it is midnight.

---

[9]This special date is known as the Unix Epoch.

[10]The tidyverse provides the lubridate [Grolemund and Wickham, 2011] package for working with date-times. It provides a number of convenient helpers that work with the base POSIXct type.

```
structure(now_ct, tzone = "Asia/Tokyo")
#> [1] "2018-08-02 07:00:00 JST"
structure(now_ct, tzone = "America/New_York")
#> [1] "2018-08-01 18:00:00 EDT"
structure(now_ct, tzone = "Australia/Lord_Howe")
#> [1] "2018-08-02 08:30:00 +1030"
structure(now_ct, tzone = "Europe/Paris")
#> [1] "2018-08-02 CEST"
```

3.4.4 Durations

Durations, which represent the amount of time between pairs of dates or date-times, are stored in difftimes. Difftimes are built on top of doubles, and have a units attribute that determines how the integer should be interpreted:

```
one_week_1 <- as.difftime(1, units = "weeks")
one_week_1
#> Time difference of 1 weeks

typeof(one_week_1)
#> [1] "double"
attributes(one_week_1)
#> $class
#> [1] "difftime"
#>
#> $units
#> [1] "weeks"

one_week_2 <- as.difftime(7, units = "days")
one_week_2
#> Time difference of 7 days

typeof(one_week_2)
#> [1] "double"
attributes(one_week_2)
#> $class
#> [1] "difftime"
#>
#> $units
#> [1] "days"
```

3.4.5 Exercises

1. What sort of object does `table()` return? What is its type? What attributes does it have? How does the dimensionality change as you tabulate more variables?

2. What happens to a factor when you modify its levels?

```
f1 <- factor(letters)
levels(f1) <- rev(levels(f1))
```

3. What does this code do? How do f2 and f3 differ from f1?

```
f2 <- rev(factor(letters))

f3 <- factor(letters, levels = rev(letters))
```

3.5 Lists

Lists are a step up in complexity from atomic vectors: each element can be any type, not just vectors. Technically speaking, each element of a list is actually the same type because, as you saw in Section 2.3.3, each element is really a *reference* to another object, which can be any type.

3.5.1 Creating

You construct lists with `list()`:

```
l1 <- list(
  1:3,
  "a",
  c(TRUE, FALSE, TRUE),
  c(2.3, 5.9)
)

typeof(l1)
#> [1] "list"
```

```
str(l1)
#> List of 4
#>  $ : int [1:3] 1 2 3
#>  $ : chr "a"
#>  $ : logi [1:3] TRUE FALSE TRUE
#>  $ : num [1:2] 2.3 5.9
```

Because the elements of a list are references, creating a list does not involve copying the components into the list. For this reason, the total size of a list might be smaller than you might expect.

```
lobstr::obj_size(mtcars)
#> 7,208 B
```

```
l2 <- list(mtcars, mtcars, mtcars, mtcars)
lobstr::obj_size(l2)
#> 7,288 B
```

Lists can contain complex objects so it's not possible to pick a single visual style that works for every list. Generally I'll draw lists like vectors, using colour to remind you of the hierarchy.

Lists are sometimes called **recursive** vectors because a list can contain other lists. This makes them fundamentally different from atomic vectors.

```
l3 <- list(list(list(1)))
str(l3)
#> List of 1
#>  $ :List of 1
#>   ..$ :List of 1
#>   .. ..$ : num 1
```

c() will combine several lists into one. If given a combination of atomic vectors and lists, c() will coerce the vectors to lists before combining them. Compare the results of list() and c():

```
14 <- list(list(1, 2), c(3, 4))
15 <- c(list(1, 2), c(3, 4))
str(14)
#> List of 2
#>  $ :List of 2
#>   ..$ : num 1
#>   ..$ : num 2
#>  $ : num [1:2] 3 4
str(15)
#> List of 4
#>  $ : num 1
#>  $ : num 2
#>  $ : num 3
#>  $ : num 4
```

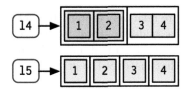

3.5.2 Testing and coercion

The `typeof()` a list is `list`. You can test for a list with `is.list()`, and coerce to a list with `as.list()`.

```
list(1:3)
#> [[1]]
#> [1] 1 2 3
as.list(1:3)
#> [[1]]
#> [1] 1
#>
#> [[2]]
#> [1] 2
#>
#> [[3]]
#> [1] 3
```

You can turn a list into an atomic vector with `unlist()`. The rules for the resulting type are complex, not well documented, and not always equivalent to what you'd get with `c()`.

3.5.3   Matrices and arrays

With atomic vectors, the dimension attribute is commonly used to create matrices. With lists, the dimension attribute can be used to create list-matrices or list-arrays:

```
l <- list(1:3, "a", TRUE, 1.0)
dim(l) <- c(2, 2)
l
#>      [,1]      [,2]
#> [1,] Integer,3 TRUE
#> [2,] "a"       1

l[[1, 1]]
#> [1] 1 2 3
```

These data structures are relatively esoteric but they can be useful if you want to arrange objects in a grid-like structure. For example, if you're running models on a spatio-temporal grid, it might be more intuitive to store the models in a 3D array that matches the grid structure.

3.5.4   Exercises

1. List all the ways that a list differs from an atomic vector.

2. Why do you need to use `unlist()` to convert a list to an atomic vector? Why doesn't `as.vector()` work?

3. Compare and contrast `c()` and `unlist()` when combining a date and date-time into a single vector.

3.6   Data frames and tibbles

The two most important S3 vectors built on top of lists are data frames and tibbles.

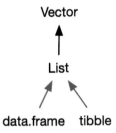

If you do data analysis in R, you're going to be using data frames. A data frame is a named list of vectors with attributes for (column) names, row.names[11], and its class, "data.frame":

```
df1 <- data.frame(x = 1:3, y = letters[1:3])
typeof(df1)
#> [1] "list"

attributes(df1)
#> $names
#> [1] "x" "y"
#>
#> $class
#> [1] "data.frame"
#>
#> $row.names
#> [1] 1 2 3
```

In contrast to a regular list, a data frame has an additional constraint: the length of each of its vectors must be the same. This gives data frames their rectangular structure and explains why they share the properties of both matrices and lists:

- A data frame has rownames()[12] and colnames(). The names() of a data frame are the column names.

- A data frame has nrow() rows and ncol() columns. The length() of a data frame gives the number of columns.

---

[11]Row names are one of the most surprisingly complex data structures in R. They've also been a persistent source of performance issues over the years. The most straightforward implementation is a character or integer vector, with one element for each row. But there's also a compact representation for "automatic" row names (consecutive integers), created by .set_row_names(). R 3.5 has a special way of deferring integer to character conversion that is specifically designed to speed up lm(); see https://svn.r-project.org/R/branches/ALTREP/ALTREP.html#deferred_string_conversions for details.

[12]Technically, you are encouraged to use row.names(), not rownames() with data frames, but this distinction is rarely important.

Data frames are one of the biggest and most important ideas in R, and one of the things that makes R different from other programming languages. However, in the over 20 years since their creation, the ways that people use R have changed, and some of the design decisions that made sense at the time data frames were created now cause frustration.

This frustration lead to the creation of the tibble [Müller and Wickham, 2018], a modern reimagining of the data frame. Tibbles are designed to be (as much as possible) drop-in replacements for data frames that fix those frustrations. A concise, and fun, way to summarise the main differences is that tibbles are lazy and surly: they do less and complain more. You'll see what that means as you work through this section.

Tibbles are provided by the tibble package and share the same structure as data frames. The only difference is that the class vector is longer, and includes tbl_df. This allows tibbles to behave differently in the key ways which we'll discuss below.

```
library(tibble)

df2 <- tibble(x = 1:3, y = letters[1:3])
typeof(df2)
#> [1] "list"

attributes(df2)
#> $names
#> [1] "x" "y"
#>
#> $row.names
#> [1] 1 2 3
#>
#> $class
#> [1] "tbl_df"      "tbl"          "data.frame"
```

3.6.1 Creating

You create a data frame by supplying name-vector pairs to data.frame():

```
df <- data.frame(
  x = 1:3,
  y = c("a", "b", "c")
)
str(df)
#> 'data.frame':     3 obs. of  2 variables:
```

```
#>   $ x: int  1 2 3
#>   $ y: Factor w/ 3 levels "a","b","c": 1 2 3
```

Beware of the default conversion of strings to factors. Use `stringsAsFactors` = `FALSE` to suppress this and keep character vectors as character vectors:

```
df1 <- data.frame(
  x = 1:3,
  y = c("a", "b", "c"),
  stringsAsFactors = FALSE
)
str(df1)
#> 'data.frame':    3 obs. of  2 variables:
#>   $ x: int  1 2 3
#>   $ y: chr  "a" "b" "c"
```

Creating a tibble is similar to creating a data frame. The difference between the two is that tibbles never coerce their input (this is one feature that makes them lazy):

```
df2 <- tibble(
  x = 1:3,
  y = c("a", "b", "c")
)
str(df2)
#> Classes 'tbl_df', 'tbl' and 'data.frame':    3 obs. of  2 variables:
#>   $ x: int  1 2 3
#>   $ y: chr  "a" "b" "c"
```

Additionally, while data frames automatically transform non-syntactic names (unless `check.names` = `FALSE`), tibbles do not (although they do print non-syntactic names surrounded by `).

```
names(data.frame(`1` = 1))
#> [1] "X1"

names(tibble(`1` = 1))
#> [1] "1"
```

While every element of a data frame (or tibble) must have the same length, both `data.frame()` and `tibble()` will recycle shorter inputs. However, while data frames automatically recycle columns that are an integer multiple of the longest column, tibbles will only recycle vectors of length one.

```
data.frame(x = 1:4, y = 1:2)
#>   x y
#> 1 1 1
#> 2 2 2
#> 3 3 1
#> 4 4 2
data.frame(x = 1:4, y = 1:3)
#> Error in data.frame(x = 1:4, y = 1:3): arguments imply differing
#> number of rows: 4, 3

tibble(x = 1:4, y = 1)
#> # A tibble: 4 x 2
#>       x     y
#>   <int> <dbl>
#> 1     1     1
#> 2     2     1
#> 3     3     1
#> 4     4     1
tibble(x = 1:4, y = 1:2)
#> Error: Tibble columns must have consistent lengths, only values of
#> length one are recycled:
#> * Length 2: Column `y`
#> * Length 4: Column `x`
```

There is one final difference: `tibble()` allows you to refer to variables created during construction:

```
tibble(
  x = 1:3,
  y = x * 2
)
#> # A tibble: 3 x 2
#>       x     y
#>   <int> <dbl>
#> 1     1     2
#> 2     2     4
#> 3     3     6
```

(Inputs are evaluated left-to-right.)

When drawing data frames and tibbles, rather than focussing on the implementation details, i.e. the attributes:

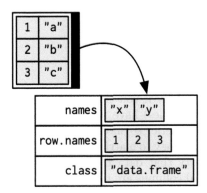

I'll draw them the same way as a named list, but arrange them to emphasise their columnar structure.

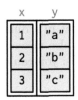

3.6.2 Row names

Data frames allow you to label each row with a name, a character vector containing only unique values:

```
df3 <- data.frame(
  age = c(35, 27, 18),
  hair = c("blond", "brown", "black"),
  row.names = c("Bob", "Susan", "Sam")
)
df3
#>       age  hair
#> Bob    35 blond
#> Susan  27 brown
#> Sam    18 black
```

You can get and set row names with `rownames()`, and you can use them to subset rows:

```
rownames(df3)
#> [1] "Bob"   "Susan" "Sam"
```

```
df3["Bob", ]
#>     age hair
#> Bob  35 blond
```

Row names arise naturally if you think of data frames as 2D structures like matrices: columns (variables) have names so rows (observations) should too. Most matrices are numeric, so having a place to store character labels is important. But this analogy to matrices is misleading because matrices possess an important property that data frames do not: they are transposable. In matrices the rows and columns are interchangeable, and transposing a matrix gives you another matrix (transposing again gives you the original matrix). With data frames, however, the rows and columns are not interchangeable: the transpose of a data frame is not a data frame.

There are three reasons why row names are undesirable:

- Metadata is data, so storing it in a different way to the rest of the data is fundamentally a bad idea. It also means that you need to learn a new set of tools to work with row names; you can't use what you already know about manipulating columns.

- Row names are a poor abstraction for labelling rows because they only work when a row can be identified by a single string. This fails in many cases, for example when you want to identify a row by a non-character vector (e.g. a time point), or with multiple vectors (e.g. position, encoded by latitude and longitude).

- Row names must be unique, so any duplication of rows (e.g. from bootstrapping) will create new row names. If you want to match rows from before and after the transformation, you'll need to perform complicated string surgery.

```
df3[c(1, 1, 1), ]
#>       age hair
#> Bob    35 blond
#> Bob.1  35 blond
#> Bob.2  35 blond
```

For these reasons, tibbles do not support row names. Instead the tibble package provides tools to easily convert row names into a regular column with either `rownames_to_column()`, or the `rownames` argument in `as_tibble()`:

```
as_tibble(df3, rownames = "name")
#> # A tibble: 3 x 3
#>   name   age hair
#>   <chr> <dbl> <fct>
```

```
#> 1 Bob      35 blond
#> 2 Susan    27 brown
#> 3 Sam      18 black
```

3.6.3 Printing

One of the most obvious differences between tibbles and data frames is how they print. I assume that you're already familiar with how data frames are printed, so here I'll highlight some of the biggest differences using an example dataset included in the dplyr package:

```
dplyr::starwars
#> # A tibble: 87 x 13
#>    name  height  mass hair_color skin_color  eye_color birth_year
#>    <chr>  <int> <dbl> <chr>      <chr>       <chr>          <dbl>
#>  1 Luke...   172    77 blond      fair        blue              19
#>  2 C-3PO     167    75 <NA>       gold        yellow           112
#>  3 R2-D2      96    32 <NA>       white, bl... red               33
#>  4 Dart...   202   136 none       white       yellow          41.9
#>  5 Leia...   150    49 brown      light       brown             19
#>  6 Owen...   178   120 brown, gr... light      blue              52
#>  7 Beru...   165    75 brown      light       blue              47
#>  8 R5-D4      97    32 <NA>       white, red  red               NA
#>  9 Bigg...   183    84 black      light       brown             24
#> 10 Obi-...   182    77 auburn, w... fair       blue-gray         57
#> # ... with 77 more rows, and 6 more variables: gender <chr>,
#> #   homeworld <chr>, species <chr>, films <list>, vehicles <list>,
#> #   starships <list>
```

- Tibbles only show the first 10 rows and all the columns that will fit on screen. Additional columns are shown at the bottom.

- Each column is labelled with its type, abbreviated to three or four letters.

- Wide columns are truncated to avoid having a single long string occupy an entire row. (This is still a work in progress: it's a tricky tradeoff between showing as many columns as possible and showing columns in their entirety.)

- When used in console environments that support it, colour is used judiciously to highlight important information, and de-emphasise supplemental details.

3.6.4 Subsetting

As you will learn in Chapter 4, you can subset a data frame or a tibble like a 1D structure (where it behaves like a list), or a 2D structure (where it behaves like a matrix).

In my opinion, data frames have two undesirable subsetting behaviours:

• When you subset columns with df[, vars], you will get a vector if vars selects one variable, otherwise you'll get a data frame. This is a frequent source of bugs when using [in a function, unless you always remember to use df[, vars, drop = FALSE].

• When you attempt to extract a single column with df$x and there is no column x, a data frame will instead select any variable that starts with x. If no variable starts with x, df$x will return NULL. This makes it easy to select the wrong variable or to select a variable that doesn't exist.

Tibbles tweak these behaviours so that a [always returns a tibble, and a $ doesn't do partial matching and warns if it can't find a variable (this is what makes tibbles surly).

```
df1 <- data.frame(xyz = "a")
df2 <- tibble(xyz = "a")

str(df1$x)
#>   Factor w/ 1 level "a": 1
str(df2$x)
#> Warning: Unknown or uninitialised column: 'x'.
#>   NULL
```

A tibble's insistence on returning a data frame from [can cause problems with legacy code, which often uses df[, "col"] to extract a single column. If you want a single column, I recommend using df[["col"]]. This clearly communicates your intent, and works with both data frames and tibbles.

3.6.5 Testing and coercing

To check if an object is a data frame or tibble, use is.data.frame():

```
is.data.frame(df1)
#> [1] TRUE
is.data.frame(df2)
#> [1] TRUE
```

Typically, it should not matter if you have a tibble or data frame, but if you need to be certain, use `is_tibble()`:

```
is_tibble(df1)
#> [1] FALSE
is_tibble(df2)
#> [1] TRUE
```

You can coerce an object to a data frame with `as.data.frame()` or to a tibble with `as_tibble()`.

3.6.6 List columns

Since a data frame is a list of vectors, it is possible for a data frame to have a column that is a list. This is very useful because a list can contain any other object: this means you can put any object in a data frame. This allows you to keep related objects together in a row, no matter how complex the individual objects are. You can see an application of this in the "Many Models" chapter of *R for Data Science*, `http://r4ds.had.co.nz/many-models.html`.

List-columns are allowed in data frames but you have to do a little extra work by either adding the list-column after creation or wrapping the list in `I()`[13].

```
df <- data.frame(x = 1:3)
df$y <- list(1:2, 1:3, 1:4)

data.frame(
  x = 1:3,
  y = I(list(1:2, 1:3, 1:4))
)
#>   x           y
#> 1 1        1, 2
#> 2 2     1, 2, 3
#> 3 3  1, 2, 3, 4
```

---

[13] `I()` is short for identity and is often used to indicate that an input should be left as is, and not automatically transformed.

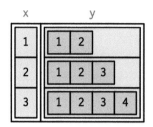

List columns are easier to use with tibbles because they can be directly included inside tibble() and they will be printed tidily:

```
tibble(
  x = 1:3,
  y = list(1:2, 1:3, 1:4)
)
#> # A tibble: 3 x 2
#>       x y
#>   <int> <list>
#> 1     1 <int [2]>
#> 2     2 <int [3]>
#> 3     3 <int [4]>
```

3.6.7   Matrix and data frame columns

As long as the number of rows matches the data frame, it's also possible to have a matrix or array as a column of a data frame. (This requires a slight extension to our definition of a data frame: it's not the length() of each column that must be equal, but the NROW().) As for list-columns, you must either add it after creation, or wrap it in I().

```
dfm <- data.frame(
  x = 1:3 * 10
)
dfm$y <- matrix(1:9, nrow = 3)
dfm$z <- data.frame(a = 3:1, b = letters[1:3], stringsAsFactors = FALSE)

str(dfm)
#> 'data.frame':    3 obs. of  3 variables:
#>  $ x: num  10 20 30
#>  $ y: int [1:3, 1:3] 1 2 3 4 5 6 7 8 9
#>  $ z:'data.frame':    3 obs. of  2 variables:
```

```
#>    ..$ a: int  3 2 1
#>    ..$ b: chr  "a" "b" "c"
```

Matrix and data frame columns require a little caution. Many functions that work with data frames assume that all columns are vectors. Also, the printed display can be confusing.

```
dfm[1, ]
#>    x y.1 y.2 y.3 z.a z.b
#> 1 10   1   4   7   3   a
```

3.6.8 Exercises

1. Can you have a data frame with zero rows? What about zero columns?

2. What happens if you attempt to set rownames that are not unique?

3. If df is a data frame, what can you say about t(df), and t(t(df))? Perform some experiments, making sure to try different column types.

4. What does as.matrix() do when applied to a data frame with columns of different types? How does it differ from data.matrix()?

3.7 NULL

To finish up this chapter, I want to talk about one final important data structure that's closely related to vectors: NULL. NULL is special because it has a unique type, is always length zero, and can't have any attributes:

```
typeof(NULL)
#> [1] "NULL"

length(NULL)
#> [1] 0

x <- NULL
attr(x, "y") <- 1
#> Error in attr(x, "y") <- 1: attempt to set an attribute on NULL
```

You can test for NULLs with is.null():

```
is.null(NULL)
#> [1] TRUE
```

There are two common uses of NULL:

- To represent an empty vector (a vector of length zero) of arbitrary type. For example, if you use c() but don't include any arguments, you get NULL, and concatenating NULL to a vector will leave it unchanged:

```
c()
#> NULL
```

- To represent an absent vector. For example, NULL is often used as a default function argument, when the argument is optional but the default value requires some computation (see Section 6.5.3 for more on this). Contrast this with NA which is used to indicate that an *element* of a vector is absent.

If you're familiar with SQL, you'll know about relational NULL and might expect it to be the same as R's. However, the database NULL is actually equivalent to R's NA.

3.8   Quiz answers

1. The four common types of atomic vector are logical, integer, double and character. The two rarer types are complex and raw.

2. Attributes allow you to associate arbitrary additional metadata to any object. You can get and set individual attributes with attr(x, "y") and attr(x, "y") <- value; or you can get and set all attributes at once with attributes().

3. The elements of a list can be any type (even a list); the elements of an atomic vector are all of the same type. Similarly, every element of a matrix must be the same type; in a data frame, different columns can have different types.

4. You can make a list-array by assigning dimensions to a list. You can make a matrix a column of a data frame with `df$x <- matrix()`, or by using `I()` when creating a new data frame `data.frame(x = I(matrix()))`.

5. Tibbles have an enhanced print method, which never coerces strings to factors, and provide stricter subsetting methods.

4

Subsetting

4.1 Introduction

R's subsetting operators are fast and powerful. Mastering them allows you to succinctly perform complex operations in a way that few other languages can match. Subsetting in R is easy to learn but hard to master because you need to internalise a number of interrelated concepts:

- There are six ways to subset atomic vectors.

- There are three subsetting operators, `[[`, `[`, and `$`.

- Subsetting operators interact differently with different vector types (e.g., atomic vectors, lists, factors, matrices, and data frames).

- Subsetting can be combined with assignment.

Subsetting is a natural complement to `str()`. While `str()` shows you all the pieces of any object (its structure), subsetting allows you to pull out the pieces that you're interested in. For large, complex objects, I highly recommend using the interactive RStudio Viewer, which you can activate with `View(my_object)`.

Quiz

Take this short quiz to determine if you need to read this chapter. If the answers quickly come to mind, you can comfortably skip this chapter. Check your answers in Section 4.6.

1. What is the result of subsetting a vector with positive integers, negative integers, a logical vector, or a character vector?

2. What's the difference between `[`, `[[`, and `$` when applied to a list?

3. When should you use `drop = FALSE`?

4. If x is a matrix, what does `x[] <- 0` do? How is it different from `x <- 0`?

5. How can you use a named vector to relabel categorical variables?

Outline

- Section 4.2 starts by teaching you about [. You'll learn the six ways to subset atomic vectors. You'll then learn how those six ways act when used to subset lists, matrices, and data frames.

- Section 4.3 expands your knowledge of subsetting operators to include [[and $ and focuses on the important principles of simplifying versus preserving.

- In Section 4.4 you'll learn the art of subassignment, which combines subsetting and assignment to modify parts of an object.

- Section 4.5 leads you through eight important, but not obvious, applications of subsetting to solve problems that you often encounter in data analysis.

4.2 Selecting multiple elements

Use [to select any number of elements from a vector. To illustrate, I'll apply [to 1D atomic vectors, and then show how this generalises to more complex objects and more dimensions.

4.2.1 Atomic vectors

Let's explore the different types of subsetting with a simple vector, x.

```
x <- c(2.1, 4.2, 3.3, 5.4)
```

Note that the number after the decimal point represents the original position in the vector.

There are six things that you can use to subset a vector:

- **Positive integers** return elements at the specified positions:

```
x[c(3, 1)]
#> [1] 3.3 2.1
x[order(x)]
#> [1] 2.1 3.3 4.2 5.4

# Duplicate indices will duplicate values
x[c(1, 1)]
#> [1] 2.1 2.1
```

```
# Real numbers are silently truncated to integers
x[c(2.1, 2.9)]
#> [1] 4.2 4.2
```

- **Negative integers** exclude elements at the specified positions:

```
x[-c(3, 1)]
#> [1] 4.2 5.4
```

Note that you can't mix positive and negative integers in a single subset:

```
x[c(-1, 2)]
#> Error in x[c(-1, 2)]: only 0's may be mixed with negative subscripts
```

- **Logical vectors** select elements where the corresponding logical value is TRUE. This is probably the most useful type of subsetting because you can write an expression that uses a logical vector:

```
x[c(TRUE, TRUE, FALSE, FALSE)]
#> [1] 2.1 4.2
x[x > 3]
#> [1] 4.2 3.3 5.4
```

In x[y], what happens if x and y are different lengths? The behaviour is controlled by the **recycling rules** where the shorter of the two is recycled to the length of the longer. This is convenient and easy to understand when one of x and y is length one, but I recommend avoiding recycling for other lengths because the rules are inconsistently applied throughout base R.

```
x[c(TRUE, FALSE)]
#> [1] 2.1 3.3
# Equivalent to
x[c(TRUE, FALSE, TRUE, FALSE)]
#> [1] 2.1 3.3
```

Note that a missing value in the index always yields a missing value in the output:

```
x[c(TRUE, TRUE, NA, FALSE)]
#> [1] 2.1 4.2  NA
```

- **Nothing** returns the original vector. This is not useful for 1D vectors, but, as you'll see shortly, is very useful for matrices, data frames, and arrays. It can also be useful in conjunction with assignment.

```
x[]
#> [1] 2.1 4.2 3.3 5.4
```

- **Zero** returns a zero-length vector. This is not something you usually do on purpose, but it can be helpful for generating test data.

```
x[0]
#> numeric(0)
```

- If the vector is named, you can also use **character vectors** to return elements with matching names.

```
(y <- setNames(x, letters[1:4]))
#>   a   b   c   d
#> 2.1 4.2 3.3 5.4
y[c("d", "c", "a")]
#>   d   c   a
#> 5.4 3.3 2.1

# Like integer indices, you can repeat indices
y[c("a", "a", "a")]
#>   a   a   a
#> 2.1 2.1 2.1

# When subsetting with [, names are always matched exactly
z <- c(abc = 1, def = 2)
z[c("a", "d")]
#> <NA> <NA>
#>   NA   NA
```

NB: Factors are not treated specially when subsetting. This means that subsetting will use the underlying integer vector, not the character levels. This is typically unexpected, so you should avoid subsetting with factors:

```
y[factor("b")]
#>   a
#> 2.1
```

4.2.2   Lists

Subsetting a list works in the same way as subsetting an atomic vector. Using [always returns a list; [[and $, as described in Section 4.3, lets you pull out elements of a list.

4.2.3   Matrices and arrays

You can subset higher-dimensional structures in three ways:

- With multiple vectors.
- With a single vector.
- With a matrix.

The most common way of subsetting matrices (2D) and arrays (>2D) is a simple generalisation of 1D subsetting: supply a 1D index for each dimension, separated by a comma. Blank subsetting is now useful because it lets you keep all rows or all columns.

```
a <- matrix(1:9, nrow = 3)
colnames(a) <- c("A", "B", "C")
a[1:2, ]
#>      A B C
#> [1,] 1 4 7
#> [2,] 2 5 8
a[c(TRUE, FALSE, TRUE), c("B", "A")]
#>      B A
#> [1,] 4 1
#> [2,] 6 3
a[0, -2]
#>      A C
```

By default, [simplifies the results to the lowest possible dimensionality. For example, both of the following expressions return 1D vectors. You'll learn how to avoid "dropping" dimensions in Section 4.2.5:

```
a[1, ]
#> A B C
#> 1 4 7
a[1, 1]
#> A
#> 1
```

Because both matrices and arrays are just vectors with special attributes, you
can subset them with a single vector, as if they were a 1D vector. Note that
arrays in R are stored in column-major order:

```
vals <- outer(1:5, 1:5, FUN = "paste", sep = ",")
vals
#>      [,1]  [,2]  [,3]  [,4]  [,5]
#> [1,] "1,1" "1,2" "1,3" "1,4" "1,5"
#> [2,] "2,1" "2,2" "2,3" "2,4" "2,5"
#> [3,] "3,1" "3,2" "3,3" "3,4" "3,5"
#> [4,] "4,1" "4,2" "4,3" "4,4" "4,5"
#> [5,] "5,1" "5,2" "5,3" "5,4" "5,5"

vals[c(4, 15)]
#> [1] "4,1" "5,3"
```

You can also subset higher-dimensional data structures with an integer matrix
(or, if named, a character matrix). Each row in the matrix specifies the location
of one value, and each column corresponds to a dimension in the array. This
means that you can use a 2 column matrix to subset a matrix, a 3 column
matrix to subset a 3D array, and so on. The result is a vector of values:

```
select <- matrix(ncol = 2, byrow = TRUE, c(
  1, 1,
  3, 1,
  2, 4
))
vals[select]
#> [1] "1,1" "3,1" "2,4"
```

4.2.4 Data frames and tibbles

Data frames have the characteristics of both lists and matrices:

- When subsetting with a single index, they behave like lists and index the
  columns, so df[1:2] selects the first two columns.

- When subsetting with two indices, they behave like matrices, so df[1:3,]
  selects the first three *rows* (and all the columns)[1].

---

[1]If you're coming from Python this is likely to be confusing, as you'd probably expect
df[1:3, 1:2] to select three columns and two rows. Generally, R "thinks" about dimensions
in terms of rows and columns while Python does so in terms of columns and rows.

```r
df <- data.frame(x = 1:3, y = 3:1, z = letters[1:3])

df[df$x == 2, ]
#>   x y z
#> 2 2 2 b
df[c(1, 3), ]
#>   x y z
#> 1 1 3 a
#> 3 3 1 c

# There are two ways to select columns from a data frame
# Like a list
df[c("x", "z")]
#>   x z
#> 1 1 a
#> 2 2 b
#> 3 3 c
# Like a matrix
df[, c("x", "z")]
#>   x z
#> 1 1 a
#> 2 2 b
#> 3 3 c

# There's an important difference if you select a single
# column: matrix subsetting simplifies by default, list
# subsetting does not.
str(df["x"])
#> 'data.frame':    3 obs. of  1 variable:
#>  $ x: int  1 2 3
str(df[, "x"])
#>  int [1:3] 1 2 3
```

Subsetting a tibble with [always returns a tibble:

```r
df <- tibble::tibble(x = 1:3, y = 3:1, z = letters[1:3])

str(df["x"])
#> Classes 'tbl_df', 'tbl' and 'data.frame':    3 obs. of  1 variable:
#>  $ x: int  1 2 3
str(df[, "x"])
#> Classes 'tbl_df', 'tbl' and 'data.frame':    3 obs. of  1 variable:
#>  $ x: int  1 2 3
```

4.2.5   Preserving dimensionality

By default, subsetting a matrix or data frame with a single number, a single
name, or a logical vector containing a single TRUE, will simplify the returned
output, i.e. it will return an object with lower dimensionality. To preserve the
original dimensionality, you must use drop = FALSE.

- For matrices and arrays, any dimensions with length 1 will be dropped:

```
a <- matrix(1:4, nrow = 2)
str(a[1, ])
#>  int [1:2] 1 3

str(a[1, , drop = FALSE])
#>  int [1, 1:2] 1 3
```

- Data frames with a single column will return just that column:

```
df <- data.frame(a = 1:2, b = 1:2)
str(df[, "a"])
#>  int [1:2] 1 2

str(df[, "a", drop = FALSE])
#> 'data.frame':    2 obs. of  1 variable:
#>  $ a: int  1 2
```

The default drop = TRUE behaviour is a common source of bugs in functions:
you check your code with a data frame or matrix with multiple columns, and
it works. Six months later, you (or someone else) uses it with a single column
data frame and it fails with a mystifying error. When writing functions, get in
the habit of always using drop = FALSE when subsetting a 2D object. For this
reason, tibbles default to drop = FALSE, and [always returns another tibble.

Factor subsetting also has a drop argument, but its meaning is rather different.
It controls whether or not levels (rather than dimensions) are preserved, and
it defaults to FALSE. If you find you're using drop = TRUE a lot it's often a sign
that you should be using a character vector instead of a factor.

```
z <- factor(c("a", "b"))
z[1]
#> [1] a
#> Levels: a b
z[1, drop = TRUE]
#> [1] a
#> Levels: a
```

4.2.6 Exercises

1. Fix each of the following common data frame subsetting errors:

```
mtcars[mtcars$cyl = 4, ]
mtcars[-1:4, ]
mtcars[mtcars$cyl <= 5]
mtcars[mtcars$cyl == 4 | 6, ]
```

2. Why does the following code yield five missing values? (Hint: why is it different from x[NA_real_]?)

```
x <- 1:5
x[NA]
#> [1] NA NA NA NA NA
```

3. What does upper.tri() return? How does subsetting a matrix with it work? Do we need any additional subsetting rules to describe its behaviour?

```
x <- outer(1:5, 1:5, FUN = "*")
x[upper.tri(x)]
```

4. Why does mtcars[1:20] return an error? How does it differ from the similar mtcars[1:20,]?

5. Implement your own function that extracts the diagonal entries from a matrix (it should behave like diag(x) where x is a matrix).

6. What does df[is.na(df)] <- 0 do? How does it work?

4.3 Selecting a single element

There are two other subsetting operators: [[and $. [[is used for extracting single items, while x$y is a useful shorthand for x[["y"]].

4.3.1 [[

[[is most important when working with lists because subsetting a list with [always returns a smaller list. To help make this easier to understand we can use a metaphor:

> *If list* x *is a train carrying objects, then* x[[5]] *is the object in car 5;* x[4:6] *is a train of cars 4-6.*
>
> — *@RLangTip,*
> https://twitter.com/RLangTip/status/268375867468681216

Let's use this metaphor to make a simple list:

```
x <- list(1:3, "a", 4:6)
```

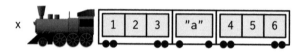

When extracting a single element, you have two options: you can create a smaller train, i.e., fewer carriages, or you can extract the contents of a particular carriage. This is the difference between [and [[:

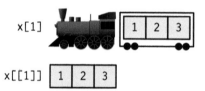

When extracting multiple (or even zero!) elements, you have to make a smaller train:

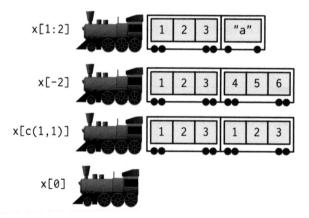

Because [[can return only a single item, you must use it with either a single positive integer or a single string. If you use a vector with [[, it will subset recursively, i.e. x[[c(1, 2)]] is equivalent to x[[1]][[2]]. This is a quirky feature that few know about, so I recommend avoiding it in favour of purrr::pluck(), which you'll learn about in Section 4.3.3.

While you must use [[when working with lists, I'd also recommend using it with atomic vectors whenever you want to extract a single value. For example, instead of writing:

```
for (i in 2:length(x)) {
  out[i] <- fun(x[i], out[i - 1])
}
```

It's better to write:

```
for (i in 2:length(x)) {
  out[[i]] <- fun(x[[i]], out[[i - 1]])
}
```

Doing so reinforces the expectation that you are getting and setting individual values.

4.3.2 $

$ is a shorthand operator: x$y is roughly equivalent to x[["y"]]. It's often used to access variables in a data frame, as in mtcars$cyl or diamonds$carat. One common mistake with $ is to use it when you have the name of a column stored in a variable:

```
var <- "cyl"
# Doesn't work - mtcars$var translated to mtcars[["var"]]
mtcars$var
#> NULL

# Instead use [[
mtcars[[var]]
#>  [1] 6 6 4 6 8 6 8 4 4 6 6 8 8 8 8 8 8 4 4 4 4 8 8 8 8 4 4 4 8 6 8 4
```

The one important difference between $ and [[is that $ does (left-to-right) partial matching:

```
x <- list(abc = 1)
x$a
#> [1] 1
x[["a"]]
#> NULL
```

To help avoid this behaviour I highly recommend setting the global option warnPartialMatchDollar to TRUE:

```
options(warnPartialMatchDollar = TRUE)
x$a
#> Warning in x$a: partial match of 'a' to 'abc'
#> [1] 1
```

(For data frames, you can also avoid this problem by using tibbles, which never do partial matching.)

4.3.3 Missing and out-of-bounds indices

It's useful to understand what happens with [[when you use an "invalid" index. The following table summarise what happens when you subset a logical vector, list, and NULL with a zero-length object (like NULL or logical()), out-of-bounds values (OOB), or a missing value (e.g. NA_integer_) with [[. Each cell shows the result of subsetting the data structure named in the row by the type of index described in the column. I've only shown the results for logical vectors, but other atomic vectors behave similarly, returning elements of the same type (NB: int = integer; chr = character).

row[[col]]	Zero-length	OOB (int)	OOB (chr)	Missing
Atomic	Error	Error	Error	Error
List	Error	Error	NULL	NULL
NULL	NULL	NULL	NULL	NULL

If the vector being indexed is named, then the names of OOB, missing, or NULL components will be <NA>.

The inconsistencies in the table above led to the development of purrr::pluck() and purrr::chuck(). When the element is missing, pluck() always returns NULL (or the value of the .default argument) and chuck() always throws an error. The behaviour of pluck() makes it well suited for indexing into deeply nested data structures where the component you want may not exist (as is common when working with JSON data from web APIs).

`pluck()` also allows you to mix integer and character indices, and provides an alternative default value if an item does not exist:

```
x <- list(
  a = list(1, 2, 3),
  b = list(3, 4, 5)
)

purrr::pluck(x, "a", 1)
#> [1] 1

purrr::pluck(x, "c", 1)
#> NULL

purrr::pluck(x, "c", 1, .default = NA)
#> [1] NA
```

4.3.4 `@` and `slot()`

There are two additional subsetting operators, which are needed for S4 objects: `@` (equivalent to `$`), and `slot()` (equivalent to `[[`). `@` is more restrictive than `$` in that it will return an error if the slot does not exist. These are described in more detail in Chapter 15.

4.3.5 Exercises

1. Brainstorm as many ways as possible to extract the third value from the `cyl` variable in the `mtcars` dataset.

2. Given a linear model, e.g., `mod <- lm(mpg ~ wt, data = mtcars)`, extract the residual degrees of freedom. Then extract the R squared from the model summary (`summary(mod)`)

4.4 Subsetting and assignment

All subsetting operators can be combined with assignment to modify selected values of an input vector: this is called subassignment. The basic form is `x[i] <- value`:

```
x <- 1:5
x[c(1, 2)] <- c(101, 102)
x
#> [1] 101 102   3   4   5
```

I recommend that you should make sure that length(value) is the same as length(x[i]), and that i is unique. This is because, while R will recycle if needed, those rules are complex (particularly if i contains missing or duplicated values) and may cause problems.

With lists, you can use x[[i]] <- NULL to remove a component. To add a literal NULL, use x[i] <- list(NULL):

```
x <- list(a = 1, b = 2)
x[["b"]] <- NULL
str(x)
#> List of 1
#>  $ a: num 1

y <- list(a = 1, b = 2)
y["b"] <- list(NULL)
str(y)
#> List of 2
#>  $ a: num 1
#>  $ b: NULL
```

Subsetting with nothing can be useful with assignment because it preserves the structure of the original object. Compare the following two expressions. In the first, mtcars remains a data frame because you are only changing the contents of mtcars, not mtcars itself. In the second, mtcars becomes a list because you are changing the object it is bound to.

```
mtcars[] <- lapply(mtcars, as.integer)
is.data.frame(mtcars)
#> [1] TRUE

mtcars <- lapply(mtcars, as.integer)
is.data.frame(mtcars)
#> [1] FALSE
```

4.5 Applications

The principles described above have a wide variety of useful applications. Some of the most important are described below. While many of the basic principles of subsetting have already been incorporated into functions like subset(), merge(), dplyr::arrange(), a deeper understanding of how those principles have been implemented will be valuable when you run into situations where the functions you need don't exist.

4.5.1 Lookup tables (character subsetting)

Character matching is a powerful way to create lookup tables. Say you want to convert abbreviations:

```
x <- c("m", "f", "u", "f", "f", "m", "m")
lookup <- c(m = "Male", f = "Female", u = NA)
lookup[x]
#>       m        f        u        f        f        m        m
#>  "Male" "Female"       NA "Female" "Female"  "Male"   "Male"
```

Note that if you don't want names in the result, use unname() to remove them.

```
unname(lookup[x])
#> [1] "Male"   "Female" NA       "Female" "Female" "Male"   "Male"
```

4.5.2 Matching and merging by hand (integer subsetting)

You can also have more complicated lookup tables with multiple columns of information. For example, suppose we have a vector of integer grades, and a table that describes their properties:

```
grades <- c(1, 2, 2, 3, 1)

info <- data.frame(
  grade = 3:1,
  desc = c("Excellent", "Good", "Poor"),
  fail = c(F, F, T)
)
```

Then, let's say we want to duplicate the `info` table so that we have a row for each value in `grades`. An elegant way to do this is by combining `match()` and integer subsetting (`match(needles, haystack)` returns the position where each needle is found in the `haystack`).

```
id <- match(grades, info$grade)
id
#> [1] 3 2 2 1 3
info[id, ]
#>      grade        desc  fail
#> 3        1        Poor  TRUE
#> 2        2        Good FALSE
#> 2.1      2        Good FALSE
#> 1        3   Excellent FALSE
#> 3.1      1        Poor  TRUE
```

If you're matching on multiple columns, you'll need to first collapse them into a single column (with e.g. `interaction()`). Typically, however, you're better off switching to a function designed specifically for joining multiple tables like `merge()`, or `dplyr::left_join()`.

4.5.3   Random samples and bootstraps (integer subsetting)

You can use integer indices to randomly sample or bootstrap a vector or data frame. Just use `sample(n)` to generate a random permutation of `1:n`, and then use the results to subset the values:

```
df <- data.frame(x = c(1, 2, 3, 1, 2), y = 5:1, z = letters[1:5])

# Randomly reorder
df[sample(nrow(df)), ]
#>   x y z
#> 1 1 5 a
#> 4 1 2 d
#> 2 2 4 b
#> 5 2 1 e
#> 3 3 3 c

# Select 3 random rows
df[sample(nrow(df), 3), ]
#>   x y z
#> 3 3 3 c
#> 2 2 4 b
```

```
#> 1  1 5 a
```

```
# Select 6 bootstrap replicates
df[sample(nrow(df), 6, replace = TRUE), ]
#>     x y z
#> 4    1 2 d
#> 4.1 1 2 d
#> 5    2 1 e
#> 1    1 5 a
#> 1.1 1 5 a
#> 2    2 4 b
```

The arguments of `sample()` control the number of samples to extract, and also whether sampling is done with or without replacement.

4.5.4   Ordering (integer subsetting)

`order()` takes a vector as its input and returns an integer vector describing how to order the subsetted vector[2]:

```
x <- c("b", "c", "a")
order(x)
#> [1] 3 1 2
x[order(x)]
#> [1] "a" "b" "c"
```

To break ties, you can supply additional variables to `order()`. You can also change the order from ascending to descending by using `decreasing = TRUE`. By default, any missing values will be put at the end of the vector; however, you can remove them with `na.last = NA` or put them at the front with `na.last = FALSE`.

For two or more dimensions, `order()` and integer subsetting makes it easy to order either the rows or columns of an object:

```
# Randomly reorder df
df2 <- df[sample(nrow(df)), 3:1]
df2
#>   z y x
#> 3 c 3 3
```

---

[2]These are "pull" indices, i.e., `order(x)[i]` is an index of where each `x[i]` is located. It is not an index of where `x[i]` should be sent.

```
#> 1 a 5 1
#> 2 b 4 2
#> 4 d 2 1
#> 5 e 1 2

df2[order(df2$x), ]
#>    z y x
#> 1 a 5 1
#> 4 d 2 1
#> 2 b 4 2
#> 5 e 1 2
#> 3 c 3 3
df2[, order(names(df2))]
#>    x y z
#> 3 3 3 c
#> 1 1 5 a
#> 2 2 4 b
#> 4 1 2 d
#> 5 2 1 e
```

You can sort vectors directly with `sort()`, or similarly `dplyr::arrange()`, to sort a data frame.

4.5.5 Expanding aggregated counts (integer subsetting)

Sometimes you get a data frame where identical rows have been collapsed into one and a count column has been added. `rep()` and integer subsetting make it easy to uncollapse, because we can take advantage of `rep()`s vectorisation: `rep(x, y)` repeats `x[i]` `y[i]` times.

```
df <- data.frame(x = c(2, 4, 1), y = c(9, 11, 6), n = c(3, 5, 1))
rep(1:nrow(df), df$n)
#> [1] 1 1 1 2 2 2 2 2 3

df[rep(1:nrow(df), df$n), ]
#>      x  y n
#> 1    2  9 3
#> 1.1  2  9 3
#> 1.2  2  9 3
#> 2    4 11 5
#> 2.1  4 11 5
#> 2.2  4 11 5
#> 2.3  4 11 5
```

```
#> 2.4 4 11 5
#> 3    1  6 1
```

4.5.6 Removing columns from data frames (character subsetting)

There are two ways to remove columns from a data frame. You can set individual columns to `NULL`:

```
df <- data.frame(x = 1:3, y = 3:1, z = letters[1:3])
df$z <- NULL
```

Or you can subset to return only the columns you want:

```
df <- data.frame(x = 1:3, y = 3:1, z = letters[1:3])
df[c("x", "y")]
#>   x y
#> 1 1 3
#> 2 2 2
#> 3 3 1
```

If you only know the columns you don't want, use set operations to work out which columns to keep:

```
df[setdiff(names(df), "z")]
#>   x y
#> 1 1 3
#> 2 2 2
#> 3 3 1
```

4.5.7 Selecting rows based on a condition (logical subsetting)

Because logical subsetting allows you to easily combine conditions from multiple columns, it's probably the most commonly used technique for extracting rows out of a data frame.

```
mtcars[mtcars$gear == 5, ]
#>                mpg cyl disp  hp drat   wt qsec vs am gear carb
#> Porsche 914-2 26.0   4 120.3  91 4.43 2.14 16.7  0  1    5    2
#> Lotus Europa  30.4   4  95.1 113 3.77 1.51 16.9  1  1    5    2
```

```
#> Ford Pantera L 15.8   8 351.0 264 4.22 3.17 14.5 0  1   5   4
#> Ferrari Dino   19.7   6 145.0 175 3.62 2.77 15.5 0  1   5   6
#> Maserati Bora  15.0   8 301.0 335 3.54 3.57 14.6 0  1   5   8

mtcars[mtcars$gear == 5 & mtcars$cyl == 4, ]
#>               mpg cyl  disp  hp drat   wt qsec vs am gear carb
#> Porsche 914-2 26.0   4 120.3  91 4.43 2.14 16.7 0  1   5   2
#> Lotus Europa  30.4   4  95.1 113 3.77 1.51 16.9 1  1   5   2
```

Remember to use the vector boolean operators & and |, not the short-circuiting scalar operators && and ||, which are more useful inside if statements. And don't forget De Morgan's laws (http://en.wikipedia.org/wiki/De_Morgan's_laws), which can be useful to simplify negations:

- !(X & Y) is the same as !X | !Y
- !(X | Y) is the same as !X & !Y

For example, !(X & !(Y | Z)) simplifies to !X | !!(Y|Z), and then to !X | Y | Z.

4.5.8 Boolean algebra versus sets (logical and integer subsetting)

It's useful to be aware of the natural equivalence between set operations (integer subsetting) and Boolean algebra (logical subsetting). Using set operations is more effective when:

- You want to find the first (or last) TRUE.

- You have very few TRUEs and very many FALSEs; a set representation may be faster and require less storage.

which() allows you to convert a Boolean representation to an integer representation. There's no reverse operation in base R but we can easily create one:

```
x <- sample(10) < 4
which(x)
#> [1] 2 5 8

unwhich <- function(x, n) {
  out <- rep_len(FALSE, n)
  out[x] <- TRUE
  out
}
```

```
unwhich(which(x), 10)
#>  [1] FALSE  TRUE FALSE FALSE  TRUE FALSE FALSE  TRUE FALSE FALSE
```

Let's create two logical vectors and their integer equivalents, and then explore the relationship between Boolean and set operations.

```
(x1 <- 1:10 %% 2 == 0)
#>  [1] FALSE  TRUE FALSE  TRUE FALSE  TRUE FALSE  TRUE FALSE  TRUE
(x2 <- which(x1))
#> [1]  2  4  6  8 10
(y1 <- 1:10 %% 5 == 0)
#>  [1] FALSE FALSE FALSE FALSE  TRUE FALSE FALSE FALSE FALSE  TRUE
(y2 <- which(y1))
#> [1]  5 10

# X & Y <-> intersect(x, y)
x1 & y1
#>  [1] FALSE FALSE FALSE FALSE FALSE FALSE FALSE FALSE FALSE  TRUE
intersect(x2, y2)
#> [1] 10

# X | Y <-> union(x, y)
x1 | y1
#>  [1] FALSE  TRUE FALSE  TRUE  TRUE  TRUE FALSE  TRUE FALSE  TRUE
union(x2, y2)
#> [1]  2  4  6  8 10  5

# X & !Y <-> setdiff(x, y)
x1 & !y1
#>  [1] FALSE  TRUE FALSE  TRUE FALSE  TRUE FALSE  TRUE FALSE FALSE
setdiff(x2, y2)
#> [1] 2 4 6 8

# xor(X, Y) <-> setdiff(union(x, y), intersect(x, y))
xor(x1, y1)
#>  [1] FALSE  TRUE FALSE  TRUE  TRUE  TRUE FALSE  TRUE FALSE FALSE
setdiff(union(x2, y2), intersect(x2, y2))
#> [1] 2 4 6 8 5
```

When first learning subsetting, a common mistake is to use x[which(y)] instead of x[y]. Here the which() achieves nothing: it switches from logical to integer subsetting but the result is exactly the same. In more general cases, there are two important differences.

- When the logical vector contains NA, logical subsetting replaces these values with NA while which() simply drops these values. It's not uncommon to use which() for this side-effect, but I don't recommend it: nothing about the name "which" implies the removal of missing values.

- x[-which(y)] is **not** equivalent to x[!y]: if y is all FALSE, which(y) will be integer(0) and -integer(0) is still integer(0), so you'll get no values, instead of all values.

In general, avoid switching from logical to integer subsetting unless you want, for example, the first or last TRUE value.

4.5.9 Exercises

1. How would you randomly permute the columns of a data frame? (This is an important technique in random forests.) Can you simultaneously permute the rows and columns in one step?

2. How would you select a random sample of m rows from a data frame? What if the sample had to be contiguous (i.e., with an initial row, a final row, and every row in between)?

3. How could you put the columns in a data frame in alphabetical order?

4.6 Quiz answers

1. Positive integers select elements at specific positions, negative integers drop elements; logical vectors keep elements at positions corresponding to TRUE; character vectors select elements with matching names.

2. [selects sub-lists: it always returns a list. If you use it with a single positive integer, it returns a list of length one. [[selects an element within a list. $ is a convenient shorthand: x$y is equivalent to x[["y"]].

3. Use drop = FALSE if you are subsetting a matrix, array, or data frame and you want to preserve the original dimensions. You should almost always use it when subsetting inside a function.

4. If x is a matrix, x[] <- 0 will replace every element with 0, keeping the same number of rows and columns. In contrast, x <- 0 completely replaces the matrix with the value 0.

5. A named character vector can act as a simple lookup table: `c(x = 1, y = 2, z = 3)[c("y", "z", "x")]`

5

Control flow

5.1 Introduction

There are two primary tools of control flow: choices and loops. Choices, like if statements and `switch()` calls, allow you to run different code depending on the input. Loops, like for and while, allow you to repeatedly run code, typically with changing options. I'd expect that you're already familiar with the basics of these functions so I'll briefly cover some technical details and then introduce some useful, but lesser known, features.

The condition system (messages, warnings, and errors), which you'll learn about in Chapter 8, also provides non-local control flow.

Quiz

Want to skip this chapter? Go for it, if you can answer the questions below. Find the answers at the end of the chapter in Section 5.4.

- What is the different between if and ifelse()?

- In the following code, what will the value of y be if x is TRUE? What if x is FALSE? What if x is NA?

```
y <- if (x) 3
```

- What does switch("x", x = , y = 2, z = 3) return?

Outline

- Section 5.2 dives into the details of if, then discusses the close relatives ifelse() and switch().

- Section 5.3 starts off by reminding you of the basic structure of the for loop in R, discusses some common pitfalls, and then talks about the related while and repeat statements.

5.2   Choices

The basic form of an if statement in R is as follows:

```
if (condition) true_action
if (condition) true_action else false_action
```

If condition is TRUE, true_action is evaluated; if condition is FALSE, the optional false_action is evaluated.

Typically the actions are compound statements contained within {:

```
grade <- function(x) {
  if (x > 90) {
    "A"
  } else if (x > 80) {
    "B"
  } else if (x > 50) {
    "C"
  } else {
    "F"
  }
}
```

if returns a value so that you can assign the results:

```
x1 <- if (TRUE) 1 else 2
x2 <- if (FALSE) 1 else 2

c(x1, x2)
#> [1] 1 2
```

(I recommend assigning the results of an if statement only when the entire expression fits on one line; otherwise it tends to be hard to read.)

When you use the single argument form without an else statement, if invisibly (Section 6.7.2) returns NULL if the condition is FALSE. Since functions like c() and paste() drop NULL inputs, this allows for a compact expression of certain idioms:

```
greet <- function(name, birthday = FALSE) {
  paste0(
```

```
    "Hi ", name,
    if (birthday) " and HAPPY BIRTHDAY"
  )
}
greet("Maria", FALSE)
#> [1] "Hi Maria"
greet("Jaime", TRUE)
#> [1] "Hi Jaime and HAPPY BIRTHDAY"
```

5.2.1   Invalid inputs

The `condition` should evaluate to a single `TRUE` or `FALSE`. Most other inputs
will generate an error:

```
if ("x") 1
#> Error in if ("x") 1: argument is not interpretable as logical
if (logical()) 1
#> Error in if (logical()) 1: argument is of length zero
if (NA) 1
#> Error in if (NA) 1: missing value where TRUE/FALSE needed
```

The exception is a logical vector of length greater than 1, which generates a
warning:

```
if (c(TRUE, FALSE)) 1
#> Warning in if (c(TRUE, FALSE)) 1: the condition has length > 1 and
#> only the first element will be used
#> [1] 1
```

In R 3.5.0 and greater, thanks to Henrik Bengtsson (https://github.com/
HenrikBengtsson/Wishlist-for-R/issues/38), you can turn this into an error
by setting an environment variable:

```
Sys.setenv("_R_CHECK_LENGTH_1_CONDITION_" = "true")
if (c(TRUE, FALSE)) 1
#> Error in if (c(TRUE, FALSE)) 1: the condition has length > 1
```

I think this is good practice as it reveals a clear mistake that you might
otherwise miss if it were only shown as a warning.

5.2.2 Vectorised if

Given that if only works with a single TRUE or FALSE, you might wonder what to do if you have a vector of logical values. Handling vectors of values is the job of ifelse(): a vectorised function with test, yes, and no vectors (that will be recycled to the same length):

```
x <- 1:10
ifelse(x %% 5 == 0, "XXX", as.character(x))
#>  [1] "1"   "2"   "3"   "4"   "XXX" "6"   "7"   "8"   "9"   "XXX"

ifelse(x %% 2 == 0, "even", "odd")
#>  [1] "odd"  "even" "odd"  "even" "odd"  "even" "odd"  "even" "odd"
#> [10] "even"
```

Note that missing values will be propagated into the output.

I recommend using ifelse() only when the yes and no vectors are the same type as it is otherwise hard to predict the output type. See about https://vctrs.r-lib.org/articles/stability.html#ifelse for additional discussion.

Another vectorised equivalent is the more general dplyr::case_when(). It uses a special syntax to allow any number of condition-vector pairs:

```
dplyr::case_when(
  x %% 35 == 0 ~ "fizz buzz",
  x %% 5 == 0 ~ "fizz",
  x %% 7 == 0 ~ "buzz",
  is.na(x) ~ "???",
  TRUE ~ as.character(x)
)
#>  [1] "1"    "2"    "3"    "4"    "fizz" "6"    "buzz" "8"    "9"
#> [10] "fizz"
```

5.2.3 switch() statement

Closely related to if is the switch()-statement. It's a compact, special purpose equivalent that lets you replace code like:

```
x_option <- function(x) {
  if (x == "a") {
    "option 1"
  } else if (x == "b") {
```

```
      "option 2"
    } else if (x == "c") {
      "option 3"
    } else {
      stop("Invalid `x` value")
    }
}
```

with the more succinct:

```
x_option <- function(x) {
  switch(x,
    a = "option 1",
    b = "option 2",
    c = "option 3",
    stop("Invalid `x` value")
  )
}
```

The last component of a `switch()` should always throw an error, otherwise unmatched inputs will invisibly return `NULL`:

```
(switch("c", a = 1, b = 2))
#> NULL
```

If multiple inputs have the same output, you can leave the right hand side of = empty and the input will "fall through" to the next value. This mimics the behaviour of C's `switch` statement:

```
legs <- function(x) {
  switch(x,
    cow = ,
    horse = ,
    dog = 4,
    human = ,
    chicken = 2,
    plant = 0,
    stop("Unknown input")
  )
}
legs("cow")
#> [1] 4
```

```
legs("dog")
#> [1] 4
```

It is also possible to use switch() with a numeric x, but is harder to read, and has undesirable failure modes if x is a not a whole number. I recommend using switch() only with character inputs.

5.2.4   Exercises

1. What type of vector does each of the following calls to ifelse() return?

   ```
   ifelse(TRUE, 1, "no")
   ifelse(FALSE, 1, "no")
   ifelse(NA, 1, "no")
   ```

   Read the documentation and write down the rules in your own words.

2. Why does the following code work?

   ```
   x <- 1:10
   if (length(x)) "not empty" else "empty"
   #> [1] "not empty"

   x <- numeric()
   if (length(x)) "not empty" else "empty"
   #> [1] "empty"
   ```

5.3   Loops

For loops are used to iterate over items in a vector. They have the following basic form:

```
for (item in vector) perform_action
```

For each item in `vector`, `perform_action` is called once; updating the value of `item` each time.

```
for (i in 1:3) {
  print(i)
}
#> [1] 1
#> [1] 2
#> [1] 3
```

(When iterating over a vector of indices, it's conventional to use very short variable names like `i`, `j`, or `k`.)

N.B.: `for` assigns the `item` to the current environment, overwriting any existing variable with the same name:

```
i <- 100
for (i in 1:3) {}
i
#> [1] 3
```

There are two ways to terminate a for loop early:

- `next` exits the current iteration.
- `break` exits the entire for loop.

```
for (i in 1:10) {
  if (i < 3)
    next

  print(i)

  if (i >= 5)
    break
}
#> [1] 3
#> [1] 4
#> [1] 5
```

5.3.1 Common pitfalls

There are three common pitfalls to watch out for when using `for`. First, if you're generating data, make sure to preallocate the output container. Otherwise the loop will be very slow; see Sections 23.2.2 and 24.6 for more details. The vector() function is helpful here.

```
means <- c(1, 50, 20)
out <- vector("list", length(means))
for (i in 1:length(means)) {
  out[[i]] <- rnorm(10, means[[i]])
}
```

Next, beware of iterating over 1:length(x), which will fail in unhelpful ways if x has length 0:

```
means <- c()
out <- vector("list", length(means))
for (i in 1:length(means)) {
  out[[i]] <- rnorm(10, means[[i]])
}
#> Error in rnorm(10, means[[i]]): invalid arguments
```

This occurs because : works with both increasing and decreasing sequences:

```
1:length(means)
#> [1] 1 0
```

Use seq_along(x) instead. It always returns a value the same length as x:

```
seq_along(means)
#> integer(0)

out <- vector("list", length(means))
for (i in seq_along(means)) {
  out[[i]] <- rnorm(10, means[[i]])
}
```

Finally, you might encounter problems when iterating over S3 vectors, as loops typically strip the attributes:

```
xs <- as.Date(c("2020-01-01", "2010-01-01"))
for (x in xs) {
  print(x)
}
#> [1] 18262
#> [1] 14610
```

Work around this by calling `[[` yourself:

```
for (i in seq_along(xs)) {
  print(xs[[i]])
}
#> [1] "2020-01-01"
#> [1] "2010-01-01"
```

5.3.2 Related tools

`for` loops are useful if you know in advance the set of values that you want to iterate over. If you don't know, there are two related tools with more flexible specifications:

- `while(condition) action`: performs `action` while `condition` is `TRUE`.

- `repeat(action)`: repeats `action` forever (i.e. until it encounters `break`).

R does not have an equivalent to the `do {action} while (condition)` syntax found in other languages.

You can rewrite any `for` loop to use `while` instead, and you can rewrite any `while` loop to use `repeat`, but the converses are not true. That means `while` is more flexible than `for`, and `repeat` is more flexible than `while`. It's good practice, however, to use the least-flexible solution to a problem, so you should use `for` wherever possible.

Generally speaking you shouldn't need to use for loops for data analysis tasks, as `map()` and `apply()` already provide less flexible solutions to most problems. You'll learn more in Chapter 9.

5.3.3 Exercises

1. Why does this code succeed without errors or warnings?

   ```
   x <- numeric()
   out <- vector("list", length(x))
   for (i in 1:length(x)) {
     out[i] <- x[i] ^ 2
   }
   out
   ```

2. When the following code is evaluated, what can you say about the vector being iterated?

```
xs <- c(1, 2, 3)
for (x in xs) {
  xs <- c(xs, x * 2)
}
xs
#> [1] 1 2 3 2 4 6
```

3. What does the following code tell you about when the index is updated?

```
for (i in 1:3) {
  i <- i * 2
  print(i)
}
#> [1] 2
#> [1] 4
#> [1] 6
```

5.4   Quiz answers

- `if` works with scalars; `ifelse()` works with vectors.

- When `x` is `TRUE`, `y` will be 3; when `FALSE`, `y` will be `NULL`; when `NA` the `if` statement will throw an error.

- This `switch()` statement makes use of fall-through so it will return 2. See details in Section 5.2.3.

6

Functions

6.1 Introduction

If you're reading this book, you've probably already created many R functions and know how to use them to reduce duplication in your code. In this chapter, you'll learn how to turn that informal, working knowledge into more rigorous, theoretical understanding. And while you'll see some interesting tricks and techniques along the way, keep in mind that what you'll learn here will be important for understanding the more advanced topics discussed later in the book.

Quiz

Answer the following questions to see if you can safely skip this chapter. You can find the answers in Section 6.9.

1. What are the three components of a function?

2. What does the following code return?

```
x <- 10
f1 <- function(x) {
  function() {
    x + 10
  }
}
f1()()
```

3. How would you usually write this code?

```
`+`(1, `*`(2, 3))
```

4. How could you make this call easier to read?

```
mean(, TRUE, x = c(1:10, NA))
```

5. Does the following code throw an error when executed? Why or why not?

```
f2 <- function(a, b) {
  a * 10
}
f2(10, stop("This is an error!"))
```

6. What is an infix function? How do you write it? What's a replacement function? How do you write it?

7. How do you ensure that cleanup action occurs regardless of how a function exits?

Outline

- Section 6.2 describes the basics of creating a function, the three main components of a function, and the exception to many function rules: primitive functions (which are implemented in C, not R).

- Section 6.3 discusses the strengths and weaknesses of the three forms of function composition commonly used in R code.

- Section 6.4 shows you how R finds the value associated with a given name, i.e. the rules of lexical scoping.

- Section 6.5 is devoted to an important property of function arguments: they are only evaluated when used for the first time.

- Section 6.6 discusses the special ... argument, which allows you to pass on extra arguments to another function.

- Section 6.7 discusses the two primary ways that a function can exit, and how to define an exit handler, code that is run on exit, regardless of what triggers it.

- Section 6.8 shows you the various ways in which R disguises ordinary function calls, and how you can use the standard prefix form to better understand what's going on.

6.2 Function fundamentals

To understand functions in R you need to internalise two important ideas:

- Functions can be broken down into three components: arguments, body, and environment.

There are exceptions to every rule, and in this case, there is a small selection of "primitive" base functions that are implemented purely in C.

- Functions are objects, just as vectors are objects.

6.2.1 Function components

A function has three parts:

- The formals(), the list of arguments that control how you call the function.
- The body(), the code inside the function.
- The environment(), the data structure that determines how the function finds the values associated with the names.

While the formals and body are specified explicitly when you create a function, the environment is specified implicitly, based on *where* you defined the function. The function environment always exists, but it is only printed when the function isn't defined in the global environment.

```
f02 <- function(x, y) {
  # A comment
  x + y
}

formals(f02)
#> $x
#>
#>
#> $y

body(f02)
#> {
#>     x + y
#> }
```

```
environment(f02)
#> <environment: R_GlobalEnv>
```

I'll draw functions as in the following diagram. The black dot on the left is the environment. The two blocks to the right are the function arguments. I won't draw the body, because it's usually large, and doesn't help you understand the shape of the function.

Like all objects in R, functions can also possess any number of additional attributes(). One attribute used by base R is srcref, short for source reference. It points to the source code used to create the function. The srcref is used for printing because, unlike body(), it contains code comments and other formatting.

```
attr(f02, "srcref")
#> function(x, y) {
#>    # A comment
#>    x + y
#> }
```

6.2.2 Primitive functions

There is one exception to the rule that a function has three components. Primitive functions, like sum() and [, call C code directly.

```
sum
#> function (..., na.rm = FALSE)  .Primitive("sum")
`[`
#> .Primitive("[")
```

They have either type builtin or type special.

```
typeof(sum)
#> [1] "builtin"
typeof(`[`)
#> [1] "special"
```

These functions exist primarily in C, not R, so their `formals()`, `body()`, and `environment()` are all `NULL`:

```
formals(sum)
#> NULL
body(sum)
#> NULL
environment(sum)
#> NULL
```

Primitive functions are only found in the base package. While they have certain performance advantages, this benefit comes at a price: they are harder to write. For this reason, R-core generally avoids creating them unless there is no other option.

6.2.3 First-class functions

It's very important to understand that R functions are objects in their own right, a language property often called "first-class functions". Unlike in many other languages, there is no special syntax for defining and naming a function: you simply create a function object (with `function`) and bind it to a name with `<-`:

```
f01 <- function(x) {
  sin(1 / x ^ 2)
}
```

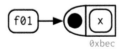

While you almost always create a function and then bind it to a name, the binding step is not compulsory. If you choose not to give a function a name, you get an **anonymous function**. This is useful when it's not worth the effort to figure out a name:

```
lapply(mtcars, function(x) length(unique(x)))
Filter(function(x) !is.numeric(x), mtcars)
integrate(function(x) sin(x) ^ 2, 0, pi)
```

A final option is to put functions in a list:

```
funs <- list(
  half = function(x) x / 2,
  double = function(x) x * 2
)

funs$double(10)
#> [1] 20
```

In R, you'll often see functions called **closures**. This name reflects the fact that R functions capture, or enclose, their environments, which you'll learn more about in Section 7.4.2.

6.2.4 Invoking a function

You normally call a function by placing with arguments wrapped in parentheses: `mean(1:10, na.rm = TRUE)`. But what happens if you have the arguments already in a data structure?

```
args <- list(1:10, na.rm = TRUE)
```

You can instead use `do.call()`: it has two arguments. The function to call, and a list containing the function arguments:

```
do.call(mean, args)
#> [1] 5.5
```

We'll come back to this idea in Section 19.6.

6.2.5 Exercises

1. Given a name, like `"mean"`, `match.fun()` lets you find a function. Given a function, can you find its name? Why doesn't that make sense in R?

2. It's possible (although typically not useful) to call an anonymous function. Which of the two approaches below is correct? Why?

    ```
    function(x) 3()
    #> function(x) 3()
    (function(x) 3)()
    #> [1] 3
    ```

3. A good rule of thumb is that an anonymous function should fit on one line and shouldn't need to use {}. Review your code. Where could you have used an anonymous function instead of a named function? Where should you have used a named function instead of an anonymous function?

4. What function allows you to tell if an object is a function? What function allows you to tell if a function is a primitive function?

5. This code makes a list of all functions in the base package.

```
objs <- mget(ls("package:base", all = TRUE), inherits = TRUE)
funs <- Filter(is.function, objs)
```

Use it to answer the following questions:

a. Which base function has the most arguments?

b. How many base functions have no arguments? What's special about those functions?

c. How could you adapt the code to find all primitive functions?

6. What are the three important components of a function?

7. When does printing a function not show the environment it was created in?

6.3 Function composition

Base R provides two ways to compose multiple function calls. For example, imagine you want to compute the population standard deviation using sqrt() and mean() as building blocks:

```
square <- function(x) x^2
deviation <- function(x) x - mean(x)
```

You either nest the function calls:

```
x <- runif(100)

sqrt(mean(square(deviation(x))))
#> [1] 0.274
```

Or you save the intermediate results as variables:

```
out <- deviation(x)
out <- square(out)
out <- mean(out)
out <- sqrt(out)
out
#> [1] 0.274
```

The magrittr package [Bache and Wickham, 2014] provides a third option:
the binary operator %>%, which is called the pipe and is pronounced as "and
then".

```
library(magrittr)

x %>%
  deviation() %>%
  square() %>%
  mean() %>%
  sqrt()
#> [1] 0.274
```

x %>% f() is equivalent to f(x); x %>% f(y) is equivalent to f(x, y). The pipe
allows you to focus on the high-level composition of functions rather than the
low-level flow of data; the focus is on what's being done (the verbs), rather
than on what's being modified (the nouns). This style is common in Haskell
and F#, the main inspiration for magrittr, and is the default style in stack
based programming languages like Forth and Factor.

Each of the three options has its own strengths and weaknesses:

- Nesting, f(g(x)), is concise, and well suited for short sequences. But longer
  sequences are hard to read because they are read inside out and right to
  left. As a result, arguments can get spread out over long distances creating
  the Dagwood sandwich (https://en.wikipedia.org/wiki/Dagwood_sandwich)
  problem.

- Intermediate objects, y <- f(x); g(y), requires you to name intermediate
  objects. This is a strength when objects are important, but a weakness when
  values are truly intermediate.

- Piping, x %>% f() %>% g(), allows you to read code in straightforward left-to-
  right fashion and doesn't require you to name intermediate objects. But you
  can only use it with linear sequences of transformations of a single object. It
  also requires an additional third party package and assumes that the reader
  understands piping.

Most code will use a combination of all three styles. Piping is more common in data analysis code, as much of an analysis consists of a sequence of transformations of an object (like a data frame or plot). I tend to use piping infrequently in packages; not because it is a bad idea, but because it's often a less natural fit.

6.4 Lexical scoping

In Chapter 2, we discussed assignment, the act of binding a name to a value. Here we'll discuss **scoping**, the act of finding the value associated with a name.

The basic rules of scoping are quite intuitive, and you've probably already internalised them, even if you never explicitly studied them. For example, what will the following code return, 10 or 20?[1]

```r
x <- 10
g01 <- function() {
  x <- 20
  x
}

g01()
```

In this section, you'll learn the formal rules of scoping as well as some of its more subtle details. A deeper understanding of scoping will help you to use more advanced functional programming tools, and eventually, even to write tools that translate R code into other languages.

R uses **lexical scoping**[2]: it looks up the values of names based on how a function is defined, not how it is called. "Lexical" here is not the English adjective that means relating to words or a vocabulary. It's a technical CS term that tells us that the scoping rules use a parse-time, rather than a runtime structure.

R's lexical scoping follows four primary rules:

- Name masking

---

[1]I'll "hide" the answers to these challenges in the footnotes. Try solving them before looking at the answer; this will help you to better remember the correct answer. In this case, g01() will return 20.

[2]Functions that automatically quote one or more arguments can override the default scoping rules to implement other varieties of scoping. You'll learn more about that in Chapter 20.

- Functions versus variables
- A fresh start
- Dynamic lookup

6.4.1   Name masking

The basic principle of lexical scoping is that names defined inside a function mask names defined outside a function. This is illustrated in the following example.

```
x <- 10
y <- 20
g02 <- function() {
  x <- 1
  y <- 2
  c(x, y)
}
g02()
#> [1] 1 2
```

If a name isn't defined inside a function, R looks one level up.

```
x <- 2
g03 <- function() {
  y <- 1
  c(x, y)
}
g03()
#> [1] 2 1

# And this doesn't change the previous value of y
y
#> [1] 20
```

The same rules apply if a function is defined inside another function. First, R looks inside the current function. Then, it looks where that function was defined (and so on, all the way up to the global environment). Finally, it looks in other loaded packages.

Run the following code in your head, then confirm the result by running the code.[3]

---

[3]`g04()` returns `c(1, 2, 3)`.

```
x <- 1
g04 <- function() {
  y <- 2
  i <- function() {
    z <- 3
    c(x, y, z)
  }
  i()
}
g04()
```

The same rules also apply to functions created by other functions, which I call manufactured functions, the topic of Chapter 10.

6.4.2 Functions versus variables

In R, functions are ordinary objects. This means the scoping rules described above also apply to functions:

```
g07 <- function(x) x + 1
g08 <- function() {
  g07 <- function(x) x + 100
  g07(10)
}
g08()
#> [1] 110
```

However, when a function and a non-function share the same name (they must, of course, reside in different environments), applying these rules gets a little more complicated. When you use a name in a function call, R ignores non-function objects when looking for that value. For example, in the code below, g09 takes on two different values:

```
g09 <- function(x) x + 100
g10 <- function() {
  g09 <- 10
  g09(g09)
}
g10()
#> [1] 110
```

For the record, using the same name for different things is confusing and best avoided!

6.4.3 A fresh start

What happens to values between invocations of a function? Consider the example below. What will happen the first time you run this function? What will happen the second time?[4] (If you haven't seen exists() before, it returns TRUE if there's a variable with that name and returns FALSE if not.)

```
g11 <- function() {
  if (!exists("a")) {
    a <- 1
  } else {
    a <- a + 1
  }
  a
}

g11()
g11()
```

You might be surprised that g11() always returns the same value. This happens because every time a function is called a new environment is created to host its execution. This means that a function has no way to tell what happened the last time it was run; each invocation is completely independent. We'll see some ways to get around this in Section 10.2.4.

6.4.4 Dynamic lookup

Lexical scoping determines where, but not when to look for values. R looks for values when the function is run, not when the function is created. Together, these two properties tell us that the output of a function can differ depending on the objects outside the function's environment:

```
g12 <- function() x + 1
x <- 15
g12()
#> [1] 16

x <- 20
g12()
#> [1] 21
```

---

[4]g11() returns 1 every time it's called.

This behaviour can be quite annoying. If you make a spelling mistake in your code, you won't get an error message when you create the function. And depending on the variables defined in the global environment, you might not even get an error message when you run the function.

To detect this problem, use codetools::findGlobals(). This function lists all the external dependencies (unbound symbols) within a function:

```
codetools::findGlobals(g12)
#> [1] "+" "x"
```

To solve this problem, you can manually change the function's environment to the emptyenv(), an environment which contains nothing:

```
environment(g12) <- emptyenv()
g12()
#> Error in x + 1: could not find function "+"
```

The problem and its solution reveal why this seemingly undesirable behaviour exists: R relies on lexical scoping to find *everything*, from the obvious, like mean(), to the less obvious, like + or even {. This gives R's scoping rules a rather beautiful simplicity.

6.4.5 Exercises

1. What does the following code return? Why? Describe how each of the three c's is interpreted.

    ```
    c <- 10
    c(c = c)
    ```

2. What are the four principles that govern how R looks for values?

3. What does the following function return? Make a prediction before running the code yourself.

    ```
    f <- function(x) {
      f <- function(x) {
        f <- function() {
          x ^ 2
        }
        f() + 1
    ```

```
  }
    f(x) * 2
  }
  f(10)
```

6.5   Lazy evaluation

In R, function arguments are **lazily evaluated**: they're only evaluated if
accessed. For example, this code doesn't generate an error because x is never
used:

```
h01 <- function(x) {
  10
}
h01(stop("This is an error!"))
#> [1] 10
```

This is an important feature because it allows you to do things like include
potentially expensive computations in function arguments that will only be
evaluated if needed.

6.5.1   Promises

Lazy evaluation is powered by a data structure called a **promise**, or (less
commonly) a thunk. It's one of the features that makes R such an interesting
programming language (we'll return to promises again in Section 20.3).

A promise has three components:

- An expression, like x + y, which gives rise to the delayed computation.

- An environment where the expression should be evaluated, i.e. the environment where the function is called. This makes sure that the following
  function returns 11, not 101:

```
y <- 10
h02 <- function(x) {
  y <- 100
  x + 1
```

```
}

h02(y)
#> [1] 11
```

This also means that when you do assignment inside a call to a function, the variable is bound outside of the function, not inside of it.

```
h02(y <- 1000)
#> [1] 1001
y
#> [1] 1000
```

- A value, which is computed and cached the first time a promise is accessed when the expression is evaluated in the specified environment. This ensures that the promise is evaluated at most once, and is why you only see "Calculating..." printed once in the following example.

```
double <- function(x) {
  message("Calculating...")
  x * 2
}

h03 <- function(x) {
  c(x, x)
}

h03(double(x))
#> Calculating...
#> [1] 40 40
```

You cannot manipulate promises with R code. Promises are like a quantum state: any attempt to inspect them with R code will force an immediate evaluation, making the promise disappear. Later, in Section 20.3, you'll learn about quosures, which convert promises into an R object where you can easily inspect the expression and the environment.

6.5.2 Default arguments

Thanks to lazy evaluation, default values can be defined in terms of other arguments, or even in terms of variables defined later in the function:

```
h04 <- function(x = 1, y = x * 2, z = a + b) {
  a <- 10
  b <- 100

  c(x, y, z)
}

h04()
#> [1]   1   2 110
```

Many base R functions use this technique, but I don't recommend it. It makes the code harder to understand: to predict *what* will be returned, you need to know the exact order in which default arguments are evaluated.

The evaluation environment is slightly different for default and user supplied arguments, as default arguments are evaluated inside the function. This means that seemingly identical calls can yield different results. It's easiest to see this with an extreme example:

```
h05 <- function(x = ls()) {
  a <- 1
  x
}

# ls() evaluated inside h05:
h05()
#> [1] "a" "x"

# ls() evaluated in global environment:
h05(ls())
#> [1] "h05"
```

6.5.3 Missing arguments

To determine if an argument's value comes from the user or from a default, you can use missing():

```
h06 <- function(x = 10) {
  list(missing(x), x)
}
str(h06())
#> List of 2
```

```
#>  $ : logi TRUE
#>  $ : num 10
str(h06(10))
#> List of 2
#>  $ : logi FALSE
#>  $ : num 10
```

`missing()` is best used sparingly, however. Take `sample()`, for example. How many arguments are required?

```
args(sample)
#> function (x, size, replace = FALSE, prob = NULL)
#> NULL
```

It looks like both `x` and `size` are required, but if `size` is not supplied, `sample()` uses `missing()` to provide a default. If I were to rewrite sample, I'd use an explicit `NULL` to indicate that `size` is not required but can be supplied:

```
sample <- function(x, size = NULL, replace = FALSE, prob = NULL) {
  if (is.null(size)) {
    size <- length(x)
  }

  x[sample.int(length(x), size, replace = replace, prob = prob)]
}
```

With the binary pattern created by the `%||%` infix function, which uses the left side if it's not `NULL` and the right side otherwise, we can further simplify `sample()`:

```
`%||%` <- function(lhs, rhs) {
  if (!is.null(lhs)) {
    lhs
  } else {
    rhs
  }
}

sample <- function(x, size = NULL, replace = FALSE, prob = NULL) {
  size <- size %||% length(x)
  x[sample.int(length(x), size, replace = replace, prob = prob)]
}
```

Because of lazy evaluation, you don't need to worry about unnecessary computation: the right side of %||% will only be evaluated if the left side is NULL.

6.5.4 Exercises

1. What important property of && makes x_ok() work?

```
x_ok <- function(x) {
  !is.null(x) && length(x) == 1 && x > 0
}

x_ok(NULL)
#> [1] FALSE
x_ok(1)
#> [1] TRUE
x_ok(1:3)
#> [1] FALSE
```

What is different with this code? Why is this behaviour undesirable here?

```
x_ok <- function(x) {
  !is.null(x) & length(x) == 1 & x > 0
}

x_ok(NULL)
#> logical(0)
x_ok(1)
#> [1] TRUE
x_ok(1:3)
#> [1] FALSE FALSE FALSE
```

2. What does this function return? Why? Which principle does it illustrate?

```
f2 <- function(x = z) {
  z <- 100
  x
}
f2()
```

3. What does this function return? Why? Which principle does it illustrate?

```
y <- 10
f1 <- function(x = {y <- 1; 2}, y = 0) {
  c(x, y)
}
f1()
y
```

4. In hist(), the default value of xlim is range(breaks), the default value for breaks is "Sturges", and

```
range("Sturges")
#> [1] "Sturges" "Sturges"
```

Explain how hist() works to get a correct xlim value.

5. Explain why this function works. Why is it confusing?

```
show_time <- function(x = stop("Error!")) {
  stop <- function(...) Sys.time()
  print(x)
}
show_time()
#> [1] "2019-04-04 11:48:56 CDT"
```

6. How many arguments are required when calling library()?

6.6 ... (dot-dot-dot)

Functions can have a special argument ... (pronounced dot-dot-dot). With it, a function can take any number of additional arguments. In other programming languages, this type of argument is often called *varargs* (short for variable arguments), and a function that uses it is said to be variadic.

You can also use ... to pass those additional arguments on to another function.

```
i01 <- function(y, z) {
  list(y = y, z = z)
}

i02 <- function(x, ...) {
  i01(...)
}

str(i02(x = 1, y = 2, z = 3))
#> List of 2
#>  $ y: num 2
#>  $ z: num 3
```

Using a special form, ..N, it's possible (but rarely useful) to refer to elements of ... by position:

```
i03 <- function(...) {
  list(first = ..1, third = ..3)
}
str(i03(1, 2, 3))
#> List of 2
#>  $ first: num 1
#>  $ third: num 3
```

More useful is list(...), which evaluates the arguments and stores them in a list:

```
i04 <- function(...) {
  list(...)
}
str(i04(a = 1, b = 2))
#> List of 2
#>  $ a: num 1
#>  $ b: num 2
```

(See also rlang::list2() to support splicing and to silently ignore trailing commas, and rlang::enquos() to capture unevaluated arguments, the topic of quasiquotation.)

There are two primary uses of ..., both of which we'll come back to later in the book:

- If your function takes a function as an argument, you want some way to pass additional arguments to that function. In this example, lapply() uses ... to pass na.rm on to mean():

```
x <- list(c(1, 3, NA), c(4, NA, 6))
str(lapply(x, mean, na.rm = TRUE))
#> List of 2
#>  $ : num 2
#>  $ : num 5
```

We'll come back to this technique in Section 9.2.3.

- If your function is an S3 generic, you need some way to allow methods to take arbitrary extra arguments. For example, take the `print()` function. Because there are different options for printing depending on the type of object, there's no way to pre-specify every possible argument and ... allows individual methods to have different arguments:

```
print(factor(letters), max.levels = 4)

print(y ~ x, showEnv = TRUE)
```

We'll come back to this use of ... in Section 13.4.3.

Using ... comes with two downsides:

- When you use it to pass arguments to another function, you have to carefully explain to the user where those arguments go. This makes it hard to understand what you can do with functions like `lapply()` and `plot()`.

- A misspelled argument will not raise an error. This makes it easy for typos to go unnoticed:

```
sum(1, 2, NA, na_rm = TRUE)
#> [1] NA
```

6.6.1 Exercises

1. Explain the following results:

```
sum(1, 2, 3)
#> [1] 6
mean(1, 2, 3)
#> [1] 1

sum(1, 2, 3, na.omit = TRUE)
#> [1] 7
```

```
mean(1, 2, 3, na.omit = TRUE)
#> [1] 1
```

2. In the following call, explain how to find the documentation for the named arguments in the following function call:

```
plot(1:10, col = "red", pch = 20, xlab = "x", col.lab = "blue")
```

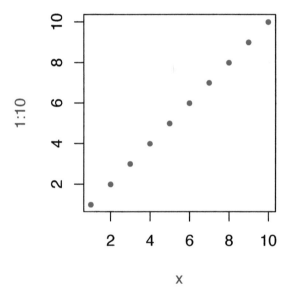

3. Why does plot(1:10, col = "red") only colour the points, not the axes or labels? Read the source code of plot.default() to find out.

6.7   Exiting a function

Most functions exit in one of two ways[5]: they either return a value, indicating success, or they throw an error, indicating failure. This section describes return values (implicit versus explicit; visible versus invisible), briefly discusses errors, and introduces exit handlers, which allow you to run code when a function exits.

---

[5]Functions can exit in other more esoteric ways like signalling a condition that is caught by an exit handler, invoking a restart, or pressing "Q" in an interactive browser.

6.7.1 Implicit versus explicit returns

There are two ways that a function can return a value:

- Implicitly, where the last evaluated expression is the return value:

```
j01 <- function(x) {
  if (x < 10) {
    0
  } else {
    10
  }
}
j01(5)
#> [1] 0
j01(15)
#> [1] 10
```

- Explicitly, by calling `return()`:

```
j02 <- function(x) {
  if (x < 10) {
    return(0)
  } else {
    return(10)
  }
}
```

6.7.2 Invisible values

Most functions return visibly: calling the function in an interactive context prints the result.

```
j03 <- function() 1
j03()
#> [1] 1
```

However, you can prevent automatic printing by applying `invisible()` to the last value:

```
j04 <- function() invisible(1)
j04()
```

To verify that this value does indeed exist, you can explicitly print it or wrap it in parentheses:

```
print(j04())
#> [1] 1
```

```
(j04())
#> [1] 1
```

Alternatively, you can use withVisible() to return the value and a visibility flag:

```
str(withVisible(j04()))
#> List of 2
#>  $ value  : num 1
#>  $ visible: logi FALSE
```

The most common function that returns invisibly is <-:

```
a <- 2
(a <- 2)
#> [1] 2
```

This is what makes it possible to chain assignments:

```
a <- b <- c <- d <- 2
```

In general, any function called primarily for a side effect (like <-, print(), or plot()) should return an invisible value (typically the value of the first argument).

6.7.3 Errors

If a function cannot complete its assigned task, it should throw an error with stop(), which immediately terminates the execution of the function.

```
j05 <- function() {
  stop("I'm an error")
  return(10)
}
j05()
#> Error in j05(): I'm an error
```

An error indicates that something has gone wrong, and forces the user to deal with the problem. Some languages (like C, Go, and Rust) rely on special return values to indicate problems, but in R you should always throw an error. You'll learn more about errors, and how to handle them, in Conditions.

6.7.4 Exit handlers

Sometimes a function needs to make temporary changes to the global state. But having to cleanup those changes can be painful (what happens if there's an error?). To ensure that these changes are undone and that the global state is restored no matter how a function exits, use on.exit() to set up an **exit handler**. The following simple example shows that the exit handler is run regardless of whether the function exits normally or with an error.

```
j06 <- function(x) {
  cat("Hello\n")
  on.exit(cat("Goodbye!\n"), add = TRUE)

  if (x) {
    return(10)
  } else {
    stop("Error")
  }
}

j06(TRUE)
#> Hello
#> Goodbye!
#> [1] 10

j06(FALSE)
#> Hello
#> Error in j06(FALSE): Error
#> Goodbye!
```

Always set add = TRUE when using on.exit(). If you don't, each call to on.exit() will overwrite the previous exit handler. Even when only registering a single handler, it's good practice to set add = TRUE so that you won't get any unpleasant surprises if you later add more exit handlers.

on.exit() is useful because it allows you to place clean-up code directly next to the code that requires clean-up:

```
cleanup <- function(dir, code) {
  old_dir <- setwd(dir)
  on.exit(setwd(old_dir), add = TRUE)

  old_opt <- options(stringsAsFactors = FALSE)
  on.exit(options(old_opt), add = TRUE)
}
```

Coupled with lazy evaluation, this creates a very useful pattern for running a block of code in an altered environment:

```
with_dir <- function(dir, code) {
  old <- setwd(dir)
  on.exit(setwd(old), add = TRUE)

  force(code)
}

getwd()
#> [1] "/Users/hadley/Documents/adv-r/adv-r"
with_dir("~", getwd())
#> [1] "/Users/hadley"
```

The use of force() isn't strictly necessary here as simply referring to code will force its evaluation. However, using force() makes it very clear that we are deliberately forcing the execution. You'll learn other uses of force() in Chapter 10.

The withr package [Hester et al., 2018] provides a collection of other functions for setting up a temporary state.

In R 3.4 and earlier, on.exit() expressions are always run in order of creation:

```
j08 <- function() {
  on.exit(message("a"), add = TRUE)
  on.exit(message("b"), add = TRUE)
}
j08()
#> a
#> b
```

This can make cleanup a little tricky if some actions need to happen in a specific order; typically you want the most recent added expression to be run first. In R 3.5 and later, you can control this by setting after = FALSE:

```
j09 <- function() {
  on.exit(message("a"), add = TRUE, after = FALSE)
  on.exit(message("b"), add = TRUE, after = FALSE)
}
j09()
#> b
#> a
```

6.7.5   Exercises

1. What does `load()` return? Why don't you normally see these values?

2. What does `write.table()` return? What would be more useful?

3. How does the `chdir` parameter of `source()` compare to `in_dir()`? Why might you prefer one to the other?

4. Write a function that opens a graphics device, runs the supplied code, and closes the graphics device (always, regardless of whether or not the plotting code works).

5. We can use `on.exit()` to implement a simple version of capture.output().

```
capture.output2 <- function(code) {
  temp <- tempfile()
  on.exit(file.remove(temp), add = TRUE, after = TRUE)

  sink(temp)
  on.exit(sink(), add = TRUE, after = TRUE)

  force(code)
  readLines(temp)
}
capture.output2(cat("a", "b", "c", sep = "\n"))
#> [1] "a" "b" "c"
```

Compare `capture.output()` to `capture.output2()`. How do the functions differ? What features have I removed to make the key ideas easier to see? How have I rewritten the key ideas so they're easier to understand?

6.8 Function forms

> *To understand computations in R, two slogans are helpful:*
>
> - *Everything that exists is an object.*
> - *Everything that happens is a function call.*
>
> — *John Chambers*

While everything that happens in R is a result of a function call, not all calls look the same. Function calls come in four varieties:

- **prefix**: the function name comes before its arguments, like foofy(a, b, c). These constitute of the majority of function calls in R.

- **infix**: the function name comes in between its arguments, like x + y. Infix forms are used for many mathematical operators, and for user-defined functions that begin and end with %.

- **replacement**: functions that replace values by assignment, like names(df) <- c("a", "b", "c"). They actually look like prefix functions.

- **special**: functions like [[, if, and for. While they don't have a consistent structure, they play important roles in R's syntax.

While there are four forms, you actually only need one because any call can be written in prefix form. I'll demonstrate this property, and then you'll learn about each of the forms in turn.

6.8.1 Rewriting to prefix form

An interesting property of R is that every infix, replacement, or special form can be rewritten in prefix form. Doing so is useful because it helps you better understand the structure of the language, it gives you the real name of every function, and it allows you to modify those functions for fun and profit.

The following example shows three pairs of equivalent calls, rewriting an infix form, replacement form, and a special form into prefix form.

```
x + y
`+`(x, y)

names(df) <- c("x", "y", "z")
`names<-`(df, c("x", "y", "z"))
```

```
for(i in 1:10) print(i)
`for`(i, 1:10, print(i))
```

Suprisingly, in R, for can be called like a regular function! The same is true for basically every operation in R, which means that knowing the function name of a non-prefix function allows you to override its behaviour. For example, if you're ever feeling particularly evil, run the following code while a friend is away from their computer. It will introduce a fun bug: 10% of the time, it will add 1 to any numeric calculation inside the parentheses.

```
`(` <- function(e1) {
  if (is.numeric(e1) && runif(1) < 0.1) {
    e1 + 1
  } else {
    e1
  }
}
replicate(50, (1 + 2))
#>  [1] 3 3 3 3 3 3 3 3 3 3 3 3 4 3 3 3 3 3 3 3 3 3 3 3 3 3 3 3 3 3 3 3
#> [33] 3 3 3 3 3 3 4 3 4 3 3 3 3 4 3 3 3 3
rm("(")
```

Of course, overriding built-in functions like this is a bad idea, but, as you'll learn in Section 21.2.5, it's possible to apply it only to selected code blocks. This provides a clean and elegant approach to writing domain specific languages and translators to other languages.

A more useful application comes up when using functional programming tools. For example, you could use lapply() to add 3 to every element of a list by first defining a function add():

```
add <- function(x, y) x + y
lapply(list(1:3, 4:5), add, 3)
#> [[1]]
#> [1] 4 5 6
#>
#> [[2]]
#> [1] 7 8
```

But we can also get the same result simply by relying on the existing + function:

```
lapply(list(1:3, 4:5), `+`, 3)
#> [[1]]
#> [1] 4 5 6
#>
#> [[2]]
#> [1] 7 8
```

We'll explore this idea in detail in Section 9.

6.8.2   Prefix form

The prefix form is the most common form in R code, and indeed in the majority
of programming languages. Prefix calls in R are a little special because you
can specify arguments in three ways:

- By position, like help(mean).
- Using partial matching, like help(top = mean).
- By name, like help(topic = mean).

As illustrated by the following chunk, arguments are matched by exact name,
then with unique prefixes, and finally by position.

```
k01 <- function(abcdef, bcde1, bcde2) {
  list(a = abcdef, b1 = bcde1, b2 = bcde2)
}
str(k01(1, 2, 3))
#> List of 3
#>  $ a : num 1
#>  $ b1: num 2
#>  $ b2: num 3
str(k01(2, 3, abcdef = 1))
#> List of 3
#>  $ a : num 1
#>  $ b1: num 2
#>  $ b2: num 3

# Can abbreviate long argument names:
str(k01(2, 3, a = 1))
#> List of 3
#>  $ a : num 1
#>  $ b1: num 2
#>  $ b2: num 3

# But this doesn't work because abbreviation is ambiguous
```

```
str(k01(1, 3, b = 1))
#> Error in k01(1, 3, b = 1): argument 3 matches multiple formal
#> arguments
```

In general, use positional matching only for the first one or two arguments; they will be the most commonly used, and most readers will know what they are. Avoid using positional matching for less commonly used arguments, and never use partial matching. Unfortunately you can't disable partial matching, but you can turn it into a warning with the `warnPartialMatchArgs` option:

```
options(warnPartialMatchArgs = TRUE)
x <- k01(a = 1, 2, 3)
#> Warning in k01(a = 1, 2, 3): partial argument match of 'a' to
#> 'abcdef'
```

6.8.3 Infix functions

Infix functions get their name from the fact the function name comes inbetween its arguments, and hence have two arguments. R comes with a number of built-in infix operators: `:`, `::`, `:::`, `$`, `@`, `^`, `*`, `/`, `+`, `-`, `>`, `>=`, `<`, `<=`, `==`, `!=`, `!`, `&`, `&&`, `|`, `||`, `~`, `<-`, and `<<-`. You can also create your own infix functions that start and end with `%`. Base R uses this pattern to define `%%`, `%*%`, `%/%`, `%in%`, `%o%`, and `%x%`.

Defining your own infix function is simple. You create a two argument function and bind it to a name that starts and ends with `%`:

```
`%+%` <- function(a, b) paste0(a, b)
"new " %+% "string"
#> [1] "new string"
```

The names of infix functions are more flexible than regular R functions: they can contain any sequence of characters except for `%`. You will need to escape any special characters in the string used to define the function, but not when you call it:

```
`% %` <- function(a, b) paste(a, b)
`%/\\%` <- function(a, b) paste(a, b)

"a" % % "b"
#> [1] "a b"
```

```
"a" %/\% "b"
#> [1] "a b"
```

R's default precedence rules mean that infix operators are composed left to right:

```
`%-%` <- function(a, b) paste0("(", a, " %-% ", b, ")")
"a" %-% "b" %-% "c"
#> [1] "((a %-% b) %-% c)"
```

There are two special infix functions that can be called with a single argument: + and -.

```
-1
#> [1] -1
+10
#> [1] 10
```

6.8.4 Replacement functions

Replacement functions act like they modify their arguments in place, and have the special name xxx<-. They must have arguments named x and value, and must return the modified object. For example, the following function modifies the second element of a vector:

```
`second<-` <- function(x, value) {
  x[2] <- value
  x
}
```

Replacement functions are used by placing the function call on the left side of <-:

```
x <- 1:10
second(x) <- 5L
x
#> [1]  1  5  3  4  5  6  7  8  9 10
```

I say they act like they modify their arguments in place, because, as explained in Section 2.5, they actually create a modified copy. We can see that by using tracemem():

```
x <- 1:10
tracemem(x)
#> <0x7ffae71bd880>

second(x) <- 6L
#> tracemem[0x7ffae71bd880 -> 0x7ffae61b5480]:
#> tracemem[0x7ffae61b5480 -> 0x7ffae73f0408]: second<-
```

If your replacement function needs additional arguments, place them between x and value, and call the replacement function with additional arguments on the left:

```
`modify<-` <- function(x, position, value) {
  x[position] <- value
  x
}
modify(x, 1) <- 10
x
#> [1] 10  5  3  4  5  6  7  8  9 10
```

When you write `modify(x, 1) <- 10`, behind the scenes R turns it into:

```
x <- `modify<-`(x, 1, 10)
```

Combining replacement with other functions requires more complex translation. For example:

```
x <- c(a = 1, b = 2, c = 3)
names(x)
#> [1] "a" "b" "c"

names(x)[2] <- "two"
names(x)
#> [1] "a"   "two" "c"
```

is translated into:

```
`*tmp*` <- x
x <- `names<-`(`*tmp*`, `[<-`(names(`*tmp*`), 2, "two"))
rm(`*tmp*`)
```

(Yes, it really does create a local variable named *tmp*, which is removed afterwards.)

6.8.5   Special forms

Finally, there are a bunch of language features that are usually written in special ways, but also have prefix forms. These include parentheses:

- `(x)` (`` `(`(x)) ``)
- `{x}` (`` `{`(x)) ``).

The subsetting operators:

- `x[i]` (`` `[`(x, i) ``)
- `x[[i]]` (`` `[[`(x, i) ``)

And the tools of control flow:

- `if (cond) true` (`` `if`(cond, true) ``)
- `if (cond) true else false` (`` `if`(cond, true, false) ``)
- `for(var in seq) action` (`` `for`(var, seq, action) ``)
- `while(cond) action` (`` `while`(cond, action) ``)
- `repeat expr` (`` `repeat`(expr) ``)
- `next` (`` `next`() ``)
- `break` (`` `break`() ``)

Finally, the most complex is the `function` function:

- `function(arg1, arg2) {body}` (`` `function`(alist(arg1, arg2), body, env) ``)

Knowing the name of the function that underlies a special form is useful for getting documentation: `?(` is a syntax error; `` ?`(` `` will give you the documentation for parentheses.

All special forms are implemented as primitive functions (i.e. in C); this means printing these functions is not informative:

```
`for`
#> .Primitive("for")
```

6.8.6   Exercises

1. Rewrite the following code snippets into prefix form:

```
1 + 2 + 3

1 + (2 + 3)

if (length(x) <= 5) x[[5]] else x[[n]]
```

2. Clarify the following list of odd function calls:

```
x <- sample(replace = TRUE, 20, x = c(1:10, NA))
y <- runif(min = 0, max = 1, 20)
cor(m = "k", y = y, u = "p", x = x)
```

3. Explain why the following code fails:

```
modify(get("x"), 1) <- 10
#> Error: target of assignment expands to non-language object
```

4. Create a replacement function that modifies a random location in a vector.

5. Write your own version of + that pastes its inputs together if they are character vectors but behaves as usual otherwise. In other words, make this code work:

```
1 + 2
#> [1] 3

"a" + "b"
#> [1] "ab"
```

6. Create a list of all the replacement functions found in the base package. Which ones are primitive functions? (Hint: use apropos().)

7. What are valid names for user-created infix functions?

8. Create an infix xor() operator.

9. Create infix versions of the set functions intersect(), union(), and setdiff(). You might call them %n%, %u%, and %/% to match conventions from mathematics.

6.9 Quiz answers

1. The three components of a function are its body, arguments, and environment.

2. `f1(1)()` returns 11.

3. You'd normally write it in infix style: `1 + (2 * 3)`.

4. Rewriting the call to `mean(c(1:10, NA), na.rm = TRUE)` is easier to understand.

5. No, it does not throw an error because the second argument is never used so it's never evaluated.

6. See Sections 6.8.3 and 6.8.4.

7. You use `on.exit()`; see Section 6.7.4 for details.

7

Environments

7.1 Introduction

The environment is the data structure that powers scoping. This chapter dives deep into environments, describing their structure in depth, and using them to improve your understanding of the four scoping rules described in Section 6.4. Understanding environments is not necessary for day-to-day use of R. But they are important to understand because they power many important R features like lexical scoping, namespaces, and R6 classes, and interact with evaluation to give you powerful tools for making domain specific languages, like dplyr and ggplot2.

Quiz

If you can answer the following questions correctly, you already know the most important topics in this chapter. You can find the answers at the end of the chapter in Section 7.7.

1. List at least three ways that an environment differs from a list.

2. What is the parent of the global environment? What is the only environment that doesn't have a parent?

3. What is the enclosing environment of a function? Why is it important?

4. How do you determine the environment from which a function was called?

5. How are <- and <<- different?

Outline

- Section 7.2 introduces you to the basic properties of an environment and shows you how to create your own.

- Section 7.3 provides a function template for computing with environments, illustrating the idea with a useful function.

- Section 7.4 describes environments used for special purposes: for packages, within functions, for namespaces, and for function execution.

- Section 7.5 explains the last important environment: the caller environment. This requires you to learn about the call stack, that describes how a function was called. You'll have seen the call stack if you've ever called `traceback()` to aid debugging.

- Section 7.6 briefly discusses three places where environments are useful data structures for solving other problems.

Prerequisites

This chapter will use rlang (`https://rlang.r-lib.org`) functions for working with environments, because it allows us to focus on the essence of environments, rather than the incidental details.

```
library(rlang)
```

The `env_` functions in rlang are designed to work with the pipe: all take an environment as the first argument, and many also return an environment. I won't use the pipe in this chapter in the interest of keeping the code as simple as possible, but you should consider it for your own code.

7.2 Environment basics

Generally, an environment is similar to a named list, with four important exceptions:

- Every name must be unique.

- The names in an environment are not ordered.

- An environment has a parent.

- Environments are not copied when modified.

Let's explore these ideas with code and pictures.

7.2.1 Basics

To create an environment, use `rlang::env()`. It works like `list()`, taking a set of name-value pairs:

```
e1 <- env(
  a = FALSE,
  b = "a",
  c = 2.3,
  d = 1:3,
)
```

In Base R

Use `new.env()` to create a new environment. Ignore the hash and size parameters; they are not needed. You cannot simultaneously create and define values; use $<-, as shown below.

The job of an environment is to associate, or **bind**, a set of names to a set of values. You can think of an environment as a bag of names, with no implied order (i.e. it doesn't make sense to ask which is the first element in an environment). For that reason, we'll draw the environment as so:

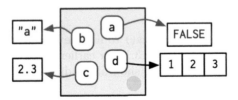

As discussed in Section 2.5.2, environments have reference semantics: unlike most R objects, when you modify them, you modify them in place, and don't create a copy. One important implication is that environments can contain themselves.

```
e1$d <- e1
```

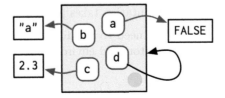

Printing an environment just displays its memory address, which is not terribly useful:

```
e1
#> <environment: 0x7fba6eac4b08>
```

Instead, we'll use env_print() which gives us a little more information:

```
env_print(e1)
#> <environment: 0x7fba6eac4b08>
#> parent: <environment: global>
#> bindings:
#>  * a: <lgl>
#>  * b: <chr>
#>  * c: <dbl>
#>  * d: <env>
```

You can use env_names() to get a character vector giving the current bindings

```
env_names(e1)
#> [1] "a" "b" "c" "d"
```

In Base R

In R 3.2.0 and greater, use names() to list the bindings in an environment. If your code needs to work with R 3.1.0 or earlier, use ls(), but note that you'll need to set all.names = TRUE to show all bindings.

7.2.2   Important environments

We'll talk in detail about special environments in 7.4, but for now we need to mention two. The current environment, or current_env() is the environment in which code is currently executing. When you're experimenting interactively, that's usually the global environment, or global_env(). The global environment is sometimes called your "workspace", as it's where all interactive (i.e. outside of a function) computation takes place.

To compare environments, you need to use identical() and not ==. This is because == is a vectorised operator, and environments are not vectors.

```
identical(global_env(), current_env())
#> [1] TRUE
```

```
global_env() == current_env()
#> Error in global_env() == current_env(): comparison (1) is possible
#> only for atomic and list types
```

In Base R

Access the global environment with `globalenv()` and the current environment
with `environment()`. The global environment is printed as `Rf_GlobalEnv` and
`.GlobalEnv`.

7.2.3 Parents

Every environment has a **parent**, another environment. In diagrams, the parent is shown as a small pale blue circle and arrow that points to another
environment. The parent is what's used to implement lexical scoping: if a
name is not found in an environment, then R will look in its parent (and so
on). You can set the parent environment by supplying an unnamed argument
to `env()`. If you don't supply it, it defaults to the current environment. In the
code below, e2a is the parent of e2b.

```
e2a <- env(d = 4, e = 5)
e2b <- env(e2a, a = 1, b = 2, c = 3)
```

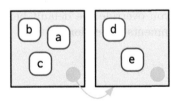

To save space, I typically won't draw all the ancestors; just remember whenever
you see a pale blue circle, there's a parent environment somewhere.

You can find the parent of an environment with `env_parent()`:

```
env_parent(e2b)
#> <environment: 0x7fba74178f48>
env_parent(e2a)
#> <environment: R_GlobalEnv>
```

Only one environment doesn't have a parent: the **empty** environment. I draw the empty environment with a hollow parent environment, and where space allows I'll label it with R_EmptyEnv, the name R uses.

```
e2c <- env(empty_env(), d = 4, e = 5)
e2d <- env(e2c, a = 1, b = 2, c = 3)
```

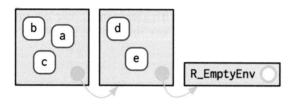

The ancestors of every environment eventually terminate with the empty environment. You can see all ancestors with env_parents():

```
env_parents(e2b)
#> [[1]]   <env: 0x7fba74178f48>
#> [[2]] $ <env: global>
env_parents(e2d)
#> [[1]]   <env: 0x7fba74f70088>
#> [[2]] $ <env: empty>
```

By default, env_parents() stops when it gets to the global environment. This is useful because the ancestors of the global environment include every attached package, which you can see if you override the default behaviour as below. We'll come back to these environments in Section 7.4.1.

```
env_parents(e2b, last = empty_env())
#>  [[1]]   <env: 0x7fba74178f48>
#>  [[2]] $ <env: global>
#>  [[3]] $ <env: package:rlang>
#>  [[4]] $ <env: package:stats>
#>  [[5]] $ <env: package:graphics>
#>  [[6]] $ <env: package:grDevices>
#>  [[7]] $ <env: package:utils>
#>  [[8]] $ <env: package:datasets>
#>  [[9]] $ <env: package:methods>
#> [[10]] $ <env: Autoloads>
#> [[11]] $ <env: package:base>
#> [[12]] $ <env: empty>
```

In Base R

Use `parent.env()` to find the parent of an environment. No base function returns all ancestors.

7.2.4 Super assignment, <<-

The ancestors of an environment have an important relationship to <<-. Regular assignment, <-, always creates a variable in the current environment. Super assignment, <<-, never creates a variable in the current environment, but instead modifies an existing variable found in a parent environment.

```
x <- 0
f <- function() {
  x <<- 1
}
f()
x
#> [1] 1
```

If <<- doesn't find an existing variable, it will create one in the global environment. This is usually undesirable, because global variables introduce non-obvious dependencies between functions. <<- is most often used in conjunction with a function factory, as described in Section 10.2.4.

7.2.5 Getting and setting

You can get and set elements of an environment with $ and [[in the same way as a list:

```
e3 <- env(x = 1, y = 2)
e3$x
#> [1] 1
e3$z <- 3
e3[["z"]]
#> [1] 3
```

But you can't use [[with numeric indices, and you can't use [:

```
e3[[1]]
#> Error in e3[[1]]: wrong arguments for subsetting an environment
```

```
e3[c("x", "y")]
#> Error in e3[c("x", "y")]: object of type 'environment' is not
#> subsettable
```

`$` and `[[` will return `NULL` if the binding doesn't exist. Use `env_get()` if you want an error:

```
e3$xyz
#> NULL

env_get(e3, "xyz")
#> Error in env_get(e3, "xyz"): object 'xyz' not found
```

If you want to use a default value if the binding doesn't exist, you can use the `default` argument.

```
env_get(e3, "xyz", default = NA)
#> [1] NA
```

There are two other ways to add bindings to an environment:

- `env_poke()`[1] takes a name (as string) and a value:

  ```
  env_poke(e3, "a", 100)
  e3$a
  #> [1] 100
  ```

- `env_bind()` allows you to bind multiple values:

  ```
  env_bind(e3, a = 10, b = 20)
  env_names(e3)
  #> [1] "x" "y" "z" "a" "b"
  ```

You can determine if an environment has a binding with `env_has()`:

```
env_has(e3, "a")
#>    a
#> TRUE
```

---

[1]You might wonder why rlang has `env_poke()` instead of `env_set()`. This is for consistency: `_set()` functions return a modified copy; `_poke()` functions modify in place.

Unlike lists, setting an element to NULL does not remove it, because sometimes you want a name that refers to NULL. Instead, use env_unbind():

```
e3$a <- NULL
env_has(e3, "a")
#>    a
#> TRUE

env_unbind(e3, "a")
env_has(e3, "a")
#>     a
#> FALSE
```

Unbinding a name doesn't delete the object. That's the job of the garbage collector, which automatically removes objects with no names binding to them. This process is described in more detail in Section 2.6.

In Base R

See get(), assign(), exists(), and rm(). These are designed interactively for use with the current environment, so working with other environments is a little clunky. Also beware the inherits argument: it defaults to TRUE meaning that the base equivalents will inspect the supplied environment and all its ancestors.

7.2.6 Advanced bindings

There are two more exotic variants of env_bind():

- env_bind_lazy() creates **delayed bindings**, which are evaluated the first time they are accessed. Behind the scenes, delayed bindings create promises, so behave in the same way as function arguments.

```
env_bind_lazy(current_env(), b = {Sys.sleep(1); 1})

system.time(print(b))
#> [1] 1
#>    user  system elapsed
#>   0.001   0.000   1.003
system.time(print(b))
#> [1] 1
#>    user  system elapsed
#>       0       0       0
```

The primary use of delayed bindings is in `autoload()`, which allows R packages to provide datasets that behave like they are loaded in memory, even though they're only loaded from disk when needed.

- `env_bind_active()` creates **active bindings** which are re-computed every time they're accessed:

```
env_bind_active(current_env(), z1 = function(val) runif(1))

z1
#> [1] 0.0808
z1
#> [1] 0.834
```

Active bindings are used to implement R6's active fields, which you'll learn about in Section 14.3.2.

In Base R

See `?delayedAssign()` and `?makeActiveBinding()`.

7.2.7 Exercises

1. List three ways in which an environment differs from a list.

2. Create an environment as illustrated by this picture.

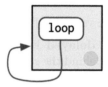

3. Create a pair of environments as illustrated by this picture.

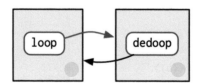

4. Explain why `e[[1]]` and `e[c("a", "b")]` don't make sense when e is an environment.

5. Create a version of env_poke() that will only bind new names, never re-bind old names. Some programming languages only do this, and are known as single assignment languages (http://en.wikipedia.org/wiki/Assignment_(computer_science)#Single_assignment).

6. What does this function do? How does it differ from <<- and why might you prefer it?

```
rebind <- function(name, value, env = caller_env()) {
  if (identical(env, empty_env())) {
    stop("Can't find `", name, "`", call. = FALSE)
  } else if (env_has(env, name)) {
    env_poke(env, name, value)
  } else {
    rebind(name, value, env_parent(env))
  }
}
rebind("a", 10)
#> Error: Can't find `a`
a <- 5
rebind("a", 10)
a
#> [1] 10
```

7.3 Recursing over environments

If you want to operate on every ancestor of an environment, it's often convenient to write a recursive function. This section shows you how, applying your new knowledge of environments to write a function that given a name, finds the environment where() that name is defined, using R's regular scoping rules.

The definition of where() is straightforward. It has two arguments: the name to look for (as a string), and the environment in which to start the search. (We'll learn why caller_env() is a good default in Section 7.5.)

```
where <- function(name, env = caller_env()) {
  if (identical(env, empty_env())) {
    # Base case
    stop("Can't find ", name, call. = FALSE)
```

```
  } else if (env_has(env, name)) {
    # Success case
    env
  } else {
    # Recursive case
    where(name, env_parent(env))
  }
}
```

There are three cases:

- The base case: we've reached the empty environment and haven't found the binding. We can't go any further, so we throw an error.

- The successful case: the name exists in this environment, so we return the environment.

- The recursive case: the name was not found in this environment, so try the parent.

These three cases are illustrated with these three examples:

```
where("yyy")
#> Error: Can't find yyy

x <- 5
where("x")
#> <environment: R_GlobalEnv>

where("mean")
#> <environment: base>
```

It might help to see a picture. Imagine you have two environments, as in the following code and diagram:

```
e4a <- env(empty_env(), a = 1, b = 2)
e4b <- env(e4a, x = 10, a = 11)
```

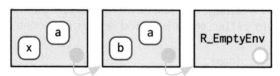

- where("a", e4b) will find a in e4b.

- `where("b", e4b)` doesn't find b in e4b, so it looks in its parent, e4a, and finds it there.

- `where("c", e4b)` looks in e4b, then e4a, then hits the empty environment and throws an error.

It's natural to work with environments recursively, so `where()` provides a useful template. Removing the specifics of `where()` shows the structure more clearly:

```
f <- function(..., env = caller_env()) {
  if (identical(env, empty_env())) {
    # base case
  } else if (success) {
    # success case
  } else {
    # recursive case
    f(..., env = env_parent(env))
  }
}
```

Iteration versus recursion

It's possible to use a loop instead of recursion. I think it's harder to understand than the recursive version, but I include it because you might find it easier to see what's happening if you haven't written many recursive functions.

```
f2 <- function(..., env = caller_env()) {
  while (!identical(env, empty_env())) {
    if (success) {
      # success case
      return()
    }
    # inspect parent
    env <- env_parent(env)
  }

  # base case
}
```

7.3.1 Exercises

1. Modify where() to return *all* environments that contain a binding for name. Carefully think through what type of object the function will need to return.

2. Write a function called fget() that finds only function objects. It should have two arguments, name and env, and should obey the regular scoping rules for functions: if there's an object with a matching name that's not a function, look in the parent. For an added challenge, also add an inherits argument which controls whether the function recurses up the parents or only looks in one environment.

7.4 Special environments

Most environments are not created by you (e.g. with env()) but are instead created by R. In this section, you'll learn about the most important environments, starting with the package environments. You'll then learn about the function environment bound to the function when it is created, and the (usually) ephemeral execution environment created every time the function is called. Finally, you'll see how the package and function environments interact to support namespaces, which ensure that a package always behaves the same way, regardless of what other packages the user has loaded.

7.4.1 Package environments and the search path

Each package attached by library() or require() becomes one of the parents of the global environment. The immediate parent of the global environment is the last package you attached[2], the parent of that package is the second to last package you attached, ...

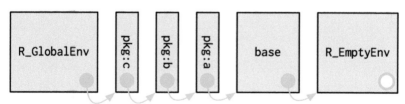

---

[2]Note the difference between attached and loaded. A package is loaded automatically if you access one of its functions using ::; it is only **attached** to the search path by library() or require().

If you follow all the parents back, you see the order in which every package has been attached. This is known as the **search path** because all objects in these environments can be found from the top-level interactive workspace. You can see the names of these environments with `base::search()`, or the environments themselves with `rlang::search_envs()`:

```
search()
#>  [1] ".GlobalEnv"        "package:rlang"     "package:stats"
#>  [4] "package:graphics"  "package:grDevices" "package:utils"
#>  [7] "package:datasets"  "package:methods"   "Autoloads"
#> [10] "package:base"
```

```
search_envs()
#>  [[1]] $ <env: global>
#>  [[2]] $ <env: package:rlang>
#>  [[3]] $ <env: package:stats>
#>  [[4]] $ <env: package:graphics>
#>  [[5]] $ <env: package:grDevices>
#>  [[6]] $ <env: package:utils>
#>  [[7]] $ <env: package:datasets>
#>  [[8]] $ <env: package:methods>
#>  [[9]] $ <env: Autoloads>
#> [[10]] $ <env: package:base>
```

The last two environments on the search path are always the same:

- The `Autoloads` environment uses delayed bindings to save memory by only loading package objects (like big datasets) when needed.

- The base environment, `package:base` or sometimes just `base`, is the environment of the base package. It is special because it has to be able to bootstrap the loading of all other packages. You can access it directly with `base_env()`.

Note that when you attach another package with `library()`, the parent environment of the global environment changes:

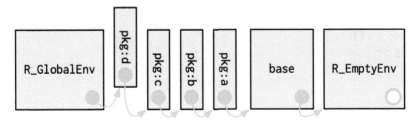

7.4.2   The function environment

A function binds the current environment when it is created. This is called the **function environment**, and is used for lexical scoping. Across computer languages, functions that capture (or enclose) their environments are called **closures**, which is why this term is often used interchangeably with function in R's documentation.

You can get the function environment with `fn_env()`:

```
y <- 1
f <- function(x) x + y
fn_env(f)
#> <environment: R_GlobalEnv>
```

In Base R

Use `environment(f)` to access the environment of function f.

In diagrams, I'll draw a function as a rectangle with a rounded end that binds an environment.

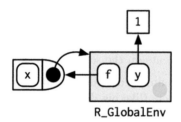

R_GlobalEnv

In this case, `f()` binds the environment that binds the name f to the function. But that's not always the case: in the following example g is bound in a new environment e, but `g()` binds the global environment. The distinction between binding and being bound by is subtle but important; the difference is how we find g versus how g finds its variables.

```
e <- env()
e$g <- function() 1
```

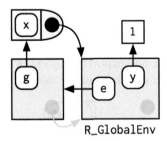

R_GlobalEnv

7.4.3 Namespaces

In the diagram above, you saw that the parent environment of a package varies based on what other packages have been loaded. This seems worrying: doesn't that mean that the package will find different functions if packages are loaded in a different order? The goal of **namespaces** is to make sure that this does not happen, and that every package works the same way regardless of what packages are attached by the user.

For example, take sd():

```
sd
#> function (x, na.rm = FALSE)
#> sqrt(var(if (is.vector(x) || is.factor(x)) x else as.double(x),
#>     na.rm = na.rm))
#> <bytecode: 0x7fba712cc020>
#> <environment: namespace:stats>
```

sd() is defined in terms of var(), so you might worry that the result of sd() would be affected by any function called var() either in the global environment, or in one of the other attached packages. R avoids this problem by taking advantage of the function versus binding environment described above. Every function in a package is associated with a pair of environments: the package environment, which you learned about earlier, and the **namespace** environment.

*   The package environment is the external interface to the package. It's how you, the R user, find a function in an attached package or with ::. Its parent is determined by search path, i.e. the order in which packages have been attached.

*   The namespace environment is the internal interface to the package. The package environment controls how we find the function; the namespace controls how the function finds its variables.

Every binding in the package environment is also found in the namespace environment; this ensures every function can use every other function in the package. But some bindings only occur in the namespace environment. These are known as internal or non-exported objects, which make it possible to hide internal implementation details from the user.

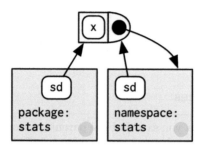

Every namespace environment has the same set of ancestors:

- Each namespace has an **imports** environment that contains bindings to all the functions used by the package. The imports environment is controlled by the package developer with the NAMESPACE file.

- Explicitly importing every base function would be tiresome, so the parent of the imports environment is the base **namespace**. The base namespace contains the same bindings as the base environment, but it has a different parent.

- The parent of the base namespace is the global environment. This means that if a binding isn't defined in the imports environment the package will look for it in the usual way. This is usually a bad idea (because it makes code depend on other loaded packages), so R CMD check automatically warns about such code. It is needed primarily for historical reasons, particularly due to how S3 method dispatch works.

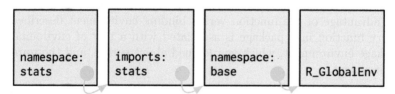

Putting all these diagrams together we get:

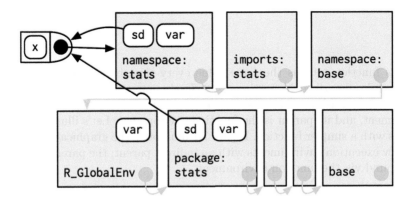

So when sd() looks for the value of var it always finds it in a sequence of environments determined by the package developer, but not by the package user. This ensures that package code always works the same way regardless of what packages have been attached by the user.

There's no direct link between the package and namespace environments; the link is defined by the function environments.

7.4.4 Execution environments

The last important topic we need to cover is the **execution** environment. What will the following function return the first time it's run? What about the second?

```
g <- function(x) {
  if (!env_has(current_env(), "a")) {
    message("Defining a")
    a <- 1
  } else {
    a <- a + 1
  }
  a
}
```

Think about it for a moment before you read on.

```
g(10)
#> Defining a
#> [1] 1
g(10)
```

```
#> Defining a
#> [1] 1
```

This function returns the same value every time because of the fresh start principle, described in Section 6.4.3. Each time a function is called, a new environment is created to host execution. This is called the execution environment, and its parent is the function environment. Let's illustrate that process with a simpler function. Figure 7.1 illustrates the graphical conventions: I draw execution environments with an indirect parent; the parent environment is found via the function environment.

```
h <- function(x) {
  # 1.
  a <- 2 # 2.
  x + a
}
y <- h(1) # 3.
```

An execution environment is usually ephemeral; once the function has completed, the environment will be garbage collected. There are several ways to make it stay around for longer. The first is to explicitly return it:

```
h2 <- function(x) {
  a <- x * 2
  current_env()
}

e <- h2(x = 10)
env_print(e)
#> <environment: 0x7fba76593d20>
#> parent: <environment: global>
#> bindings:
#>  * a: <dbl>
#>  * x: <dbl>
fn_env(h2)
#> <environment: R_GlobalEnv>
```

Another way to capture it is to return an object with a binding to that environment, like a function. The following example illustrates that idea with a function factory, plus(). We use that factory to create a function called plus_one().

There's a lot going on in the diagram because the enclosing environment of plus_one() is the execution environment of plus().

1. Function called with x = 1

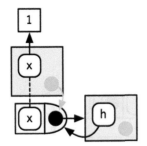

2. a bound to value 2

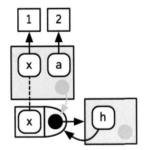

3. Function completes returning value 3.
Execution environment goes away.

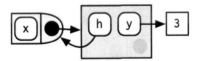

Figure 7.1 The execution environment of a simple function call. Note that the parent of the execution environment is the function environment.

```
plus <- function(x) {
  function(y) x + y
}

plus_one <- plus(1)
plus_one
#> function(y) x + y
#> <environment: 0x7fba76650600>
```

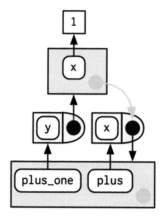

What happens when we call `plus_one()`? Its execution environment will have the captured execution environment of `plus()` as its parent:

```
plus_one(2)
#> [1] 3
```

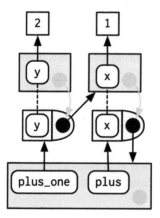

You'll learn more about function factories in Section 10.2.

7.4.5 Exercises

1. How is `search_envs()` different from `env_parents(global_env())`?

2. Draw a diagram that shows the enclosing environments of this function:

```
f1 <- function(x1) {
  f2 <- function(x2) {
    f3 <- function(x3) {
      x1 + x2 + x3
    }
    f3(3)
  }
  f2(2)
}
f1(1)
```

3. Write an enhanced version of str() that provides more information about functions. Show where the function was found and what
   environment it was defined in.

7.5 Call stacks

There is one last environment we need to explain, the **caller** environment,
accessed with rlang::caller_env(). This provides the environment from which
the function was called, and hence varies based on how the function is called,
not how the function was created. As we saw above this is a useful default
whenever you write a function that takes an environment as an argument.

In Base R

parent.frame() is equivalent to caller_env(); just note that it returns an
environment, not a frame.

To fully understand the caller environment we need to discuss two related
concepts: the **call stack**, which is made up of **frames**. Executing a function
creates two types of context. You've learned about one already: the execution
environment is a child of the function environment, which is determined by
where the function was created. There's another type of context created by
where the function was called: this is called the call stack.

7.5.1 Simple call stacks

Let's illustrate this with a simple sequence of calls: f() calls g() calls h().

```
f <- function(x) {
  g(x = 2)
}
g <- function(x) {
  h(x = 3)
}
h <- function(x) {
  stop()
}
```

The way you most commonly see a call stack in R is by looking at the trace-back() after an error has occurred:

```
f(x = 1)
#> Error:
traceback()
#> 4: stop()
#> 3: h(x = 3)
#> 2: g(x = 2)
#> 1: f(x = 1)
```

Instead of stop() + traceback() to understand the call stack, we're going to use lobstr::cst() to print out the call stack tree:

```
h <- function(x) {
  lobstr::cst()
}
f(x = 1)
#> █
#>  └─f(x = 1)
#>     └─g(x = 2)
#>        └─h(x = 3)
#>           └─lobstr::cst()
```

This shows us that cst() was called from h(), which was called from g(), which was called from f(). Note that the order is the opposite from traceback(). As the call stacks get more complicated, I think it's easier to understand the sequence of calls if you start from the beginning, rather than the end (i.e. f() calls g(); rather than g() was called by f()).

7.5.2 Lazy evaluation

The call stack above is simple: while you get a hint that there's some tree-like structure involved, everything happens on a single branch. This is typical of a call stack when all arguments are eagerly evaluated.

Let's create a more complicated example that involves some lazy evaluation. We'll create a sequence of functions, a(), b(), c(), that pass along an argument x.

```
a <- function(x) b(x)
b <- function(x) c(x)
c <- function(x) x

a(f())
#> ▐
#> ├─a(f())
#> │ └─b(x)
#> │   └─c(x)
#> └─f()
#>   └─g(x = 2)
#>     └─h(x = 3)
#>       └─lobstr::cst()
```

x is lazily evaluated so this tree gets two branches. In the first branch a() calls b(), then b() calls c(). The second branch starts when c() evaluates its argument x. This argument is evaluated in a new branch because the environment in which it is evaluated is the global environment, not the environment of c().

7.5.3 Frames

Each element of the call stack is a **frame**[3], also known as an evaluation context. The frame is an extremely important internal data structure, and R code can only access a small part of the data structure because tampering with it will break R. A frame has three key components:

• An expression (labelled with expr) giving the function call. This is what traceback() prints out.

• An environment (labelled with env), which is typically the execution environment of a function. There are two main exceptions: the environment of

---

[3]NB: ?environment uses frame in a different sense: "Environments consist of a *frame*, or collection of named objects, and a pointer to an enclosing environment." We avoid this sense of frame, which comes from S, because it's very specific and not widely used in base R. For example, the frame in parent.frame() is an execution context, not a collection of named objects.

the global frame is the global environment, and calling eval() also generates frames, where the environment can be anything.

- A parent, the previous call in the call stack (shown by a grey arrow).

Figure 7.2 illustrates the stack for the call to f(x = 1) shown in Section 7.5.1.

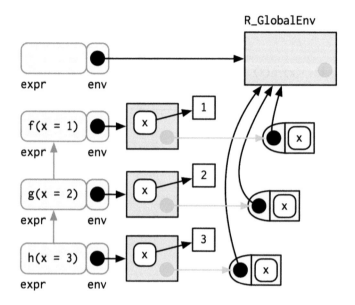

Figure 7.2 The graphical depiction of a simple call stack

(To focus on the calling environments, I have omitted the bindings in the global environment from f, g, and h to the respective function objects.)

The frame also holds exit handlers created with on.exit(), restarts and handlers for the condition system, and which context to return() to when a function completes. These are important internal details that are not accessible with R code.

7.5.4 Dynamic scope

Looking up variables in the calling stack rather than in the enclosing environment is called **dynamic scoping**. Few languages implement dynamic scoping (Emacs Lisp is a notable exception (http://www.gnu.org/software/emacs/emacs-paper.html#SEC15).) This is because dynamic scoping makes it much harder to reason about how a function operates: not only do you need to know how it was defined, you also need to know the context in which it was called. Dynamic scoping is primarily useful for developing functions that aid interactive data analysis, and one of the topics discussed in Chapter 20.

7.5.5 Exercises

1. Write a function that lists all the variables defined in the environment in which it was called. It should return the same results as `ls()`.

7.6 As data structures

As well as powering scoping, environments are also useful data structures in their own right because they have reference semantics. There are three common problems that they can help solve:

- **Avoiding copies of large data**. Since environments have reference semantics, you'll never accidentally create a copy. But bare environments are painful to work with, so instead I recommend using R6 objects, which are built on top of environments. Learn more in Chapter 14.

- **Managing state within a package**. Explicit environments are useful in packages because they allow you to maintain state across function calls. Normally, objects in a package are locked, so you can't modify them directly. Instead, you can do something like this:

```
my_env <- new.env(parent = emptyenv())
my_env$a <- 1

get_a <- function() {
  my_env$a
}
set_a <- function(value) {
  old <- my_env$a
  my_env$a <- value
  invisible(old)
}
```

Returning the old value from setter functions is a good pattern because it makes it easier to reset the previous value in conjunction with `on.exit()` (Section 6.7.4).

- **As a hashmap**. A hashmap is a data structure that takes constant, $O(1)$, time to find an object based on its name. Environments provide this behaviour by default, so can be used to simulate a hashmap. See the hash package [Brown, 2013] for a complete development of this idea.

7.7   Quiz answers

1. There are four ways: every object in an environment must have a name; order doesn't matter; environments have parents; environments have reference semantics.

2. The parent of the global environment is the last package that you loaded. The only environment that doesn't have a parent is the empty environment.

3. The enclosing environment of a function is the environment where it was created. It determines where a function looks for variables.

4. Use `caller_env()` or `parent.frame()`.

5. `<-` always creates a binding in the current environment; `<<-` rebinds an existing name in a parent of the current environment.

8

Conditions

8.1 Introduction

The **condition** system provides a paired set of tools that allow the author of a function to indicate that something unusual is happening, and the user of that function to deal with it. The function author **signals** conditions with functions like `stop()` (for errors), `warning()` (for warnings), and `message()` (for messages), then the function user can handle them with functions like `tryCatch()` and `withCallingHandlers()`. Understanding the condition system is important because you'll often need to play both roles: signalling conditions from the functions you create, and handle conditions signalled by the functions you call.

R offers a very powerful condition system based on ideas from Common Lisp. Like R's approach to object-oriented programming, it is rather different to currently popular programming languages so it is easy to misunderstand, and there has been relatively little written about how to use it effectively. Historically, this has meant that few people (myself included) have taken full advantage of its power. The goal of this chapter is to remedy that situation. Here you will learn about the big ideas of R's condition system, as well as learning a bunch of practical tools that will make your code stronger.

I found two resources particularly useful when writing this chapter. You may also want to read them if you want to learn more about the inspirations and motivations for the system:

- *A prototype of a condition system for R* (http://homepage.stat.uiowa.edu/~luke/R/exceptions/simpcond.html) by Robert Gentleman and Luke Tierney. This describes an early version of R's condition system. While the implementation has changed somewhat since this document was written, it provides a good overview of how the pieces fit together, and some motivation for its design.

- *Beyond exception handling: conditions and restarts* (http://www.gigamonkeys.com/book/beyond-exception-handling-conditions-and-restarts.html) by Peter Seibel. This describes exception handling in Lisp, which happens to be very similar to R's approach. It provides useful motivation and more sophisticated examples. I have provided an R translation of the chapter at `http://adv-r.had.co.nz/beyond-exception-handling.html`.

I also found it helpful to work through the underlying C code that implements these ideas. If you're interested in understanding how it all works, you might find my notes (`https://gist.github.com/hadley/4278d0a6d3a10e42533d59905fbed0ac`) to be useful.

Quiz

Want to skip this chapter? Go for it, if you can answer the questions below. Find the answers at the end of the chapter in Section 8.7.

1. What are the three most important types of condition?
2. What function do you use to ignore errors in block of code?
3. What's the main difference between `tryCatch()` and `withCalling-Handlers()`?
4. Why might you want to create a custom error object?

Outline

- Section 8.2 introduces the basic tools for signalling conditions, and discusses when it is appropriate to use each type.
- Section 8.3 teaches you about the simplest tools for handling conditions: functions like `try()` and `supressMessages()` that swallow conditions and prevent them from getting to the top level.
- Section 8.4 introduces the condition **object**, and the two fundamental tools of condition handling: `tryCatch()` for error conditions, and `withCallingHandlers()` for everything else.
- Section 8.5 shows you how to extend the built-in condition objects to store useful data that condition handlers can use to make more informed decisions.
- Section 8.6 closes out the chapter with a grab bag of practical applications based on the low-level tools found in earlier sections.

8.1.1 Prerequisites

As well as base R functions, this chapter uses condition signalling and handling functions from rlang (`https://rlang.r-lib.org`).

```
library(rlang)
```

8.2 Signalling conditions

There are three conditions that you can signal in code: errors, warnings, and messages.

- Errors are the most severe; they indicate that there is no way for a function to continue and execution must stop.

- Warnings fall somewhat in between errors and message, and typically indicate that something has gone wrong but the function has been able to at least partially recover.

- Messages are the mildest; they are way of informing users that some action has been performed on their behalf.

There is a final condition that can only be generated interactively: an interrupt, which indicates that the user has interrupted execution by pressing Escape, Ctrl + Break, or Ctrl + C (depending on the platform).

Conditions are usually displayed prominently, in a bold font or coloured red, depending on the R interface. You can tell them apart because errors always start with "Error", warnings with "Warning" or "Warning message", and messages with nothing.

```
stop("This is what an error looks like")
#> Error in eval(expr, envir, enclos): This is what an error looks like

warning("This is what a warning looks like")
#> Warning: This is what a warning looks like

message("This is what a message looks like")
#> This is what a message looks like
```

The following three sections describe errors, warnings, and messages in more detail.

8.2.1 Errors

In base R, errors are signalled, or **thrown**, by stop():

```
f <- function() g()
g <- function() h()
h <- function() stop("This is an error!")

f()
#> Error in h(): This is an error!
```

By default, the error message includes the call, but this is typically not useful (and recapitulates information that you can easily get from traceback()), so I think it's good practice to use call. = FALSE[1]:

```
h <- function() stop("This is an error!", call. = FALSE)
f()
#> Error: This is an error!
```

The rlang equivalent to stop(), rlang::abort(), does this automatically. We'll use abort() throughout this chapter, but we won't get to its most compelling feature, the ability to add additional metadata to the condition object, until we're near the end of the chapter.

```
h <- function() abort("This is an error!")
f()
#> Error: This is an error!
```

(NB: stop() pastes together multiple inputs, while abort() does not. To create complex error messages with abort, I recommend using glue::glue(). This allows us to use other arguments to abort() for useful features that you'll learn about in Section 8.5.)

The best error messages tell you what is wrong and point you in the right direction to fix the problem. Writing good error messages is hard because errors usually occur when the user has a flawed mental model of the function. As a developer, it's hard to imagine how the user might be thinking incorrectly about your function, and thus it's hard to write a message that will steer the user in the correct direction. That said, the tidyverse style guide discusses a few general principles that we have found useful: http://style.tidyverse.org/error-messages.html.

---

[1]The trailing . in call. is a peculiarity of stop(); don't read anything into it.

8.2.2 Warnings

Warnings, signalled by `warning()`, are weaker than errors: they signal that something has gone wrong, but the code has been able to recover and continue. Unlike errors, you can have multiple warnings from a single function call:

```
fw <- function() {
  cat("1\n")
  warning("W1")
  cat("2\n")
  warning("W2")
  cat("3\n")
  warning("W3")
}
```

By default, warnings are cached and printed only when control returns to the top level:

```
fw()
#> 1
#> 2
#> 3
#> Warning messages:
#> 1: In f() : W1
#> 2: In f() : W2
#> 3: In f() : W3
```

You can control this behaviour with the `warn` option:

- To make warnings appear immediately, set `options(warn = 1)`.
- To turn warnings into errors, set `options(warn = 2)`. This is usually the easiest way to debug a warning, as once it's an error you can use tools like `traceback()` to find the source.
- Restore the default behaviour with `options(warn = 0)`.

Like `stop()`, `warning()` also has a call argument. It is slightly more useful (since warnings are often more distant from their source), but I still generally suppress it with `call. = FALSE`. Like `rlang::abort()`, the rlang equivalent of `warning()`, `rlang::warn()`, also suppresses the `call.` by default.

Warnings occupy a somewhat challenging place between messages ("you should know about this") and errors ("you must fix this!"), and it's hard to give precise advice on when to use them. Generally, be restrained, as warnings are easy to miss if there's a lot of other output, and you don't want your function to recover too easily from clearly invalid input. In my opinion, base

R tends to overuse warnings, and many warnings in base R would be better off as errors. For example, I think these warnings would be more helpful as errors:

```
formals(1)
#> Warning in formals(fun): argument is not a function
#> NULL

file.remove("this-file-doesn't-exist")
#> Warning in file.remove("this-file-doesn't-exist"): cannot remove file
#> 'this-file-doesn't-exist', reason 'No such file or directory'
#> [1] FALSE

lag(1:3, k = 1.5)
#> Warning in lag.default(1:3, k = 1.5): 'k' is not an integer
#> [1] 1 2 3
#> attr(,"tsp")
#> [1] -1  1  1
```

There are only a couple of cases where using a warning is clearly appropriate:

- When you **deprecate** a function you want to allow older code to continue to work (so ignoring the warning is OK) but you want to encourage the user to switch to a new function.

- When you are reasonably certain you can recover from a problem: If you were 100% certain that you could fix the problem, you wouldn't need any message; if you were more uncertain that you could correctly fix the issue, you'd throw an error.

Otherwise use warnings with restraint, and carefully consider if an error would be more appropriate.

8.2.3 Messages

Messages, signalled by message(), are informational; use them to tell the user that you've done something on their behalf. Good messages are a balancing act: you want to provide just enough information so the user knows what's going on, but not so much that they're overwhelmed.

message()s are displayed immediately and do not have a call. argument:

```
fm <- function() {
  cat("1\n")
  message("M1")
```

```
cat("2\n")
message("M2")
cat("3\n")
message("M3")
}

fm()
#> 1
#> M1
#> 2
#> M2
#> 3
#> M3
```

Good places to use a message are:

- When a default argument requires some non-trivial amount of computation and you want to tell the user what value was used. For example, ggplot2 reports the number of bins used if you don't supply a `binwidth`.

- In functions that are called primarily for their side-effects which would otherwise be silent. For example, when writing files to disk, calling a web API, or writing to a database, it's useful provide regular status messages telling the user what's happening.

- When you're about to start a long running process with no intermediate output. A progress bar (e.g. with progress (https://github.com/r-lib/progress)) is better, but a message is a good place to start.

- When writing a package, you sometimes want to display a message when your package is loaded (i.e. in `.onAttach()`; here you must use `packageStartupMessage()`.

Generally any function that produces a message should have some way to suppress it, like a `quiet = TRUE` argument. It is possible to suppress all messages with `suppressMessages()`, as you'll learn shortly, but it is nice to also give finer grained control.

It's important to compare `message()` to the closely related `cat()`. In terms of usage and result, they appear quite similar[2]:

```
cat("Hi!\n")
#> Hi!
```

---

[2]But note that `cat()` requires an explicit trailing "\n" to print a new line.

```
message("Hi!")
#> Hi!
```

However, the *purposes* of `cat()` and `message()` are different. Use `cat()` when the primary role of the function is to print to the console, like `print()` or `str()` methods. Use `message()` as a side-channel to print to the console when the primary purpose of the function is something else. In other words, `cat()` is for when the user *asks* for something to be printed and `message()` is for when the developer *elects* to print something.

8.2.4 Exercises

1. Write a wrapper around `file.remove()` that throws an error if the file to be deleted does not exist.

2. What does the `appendLF` argument to `message()` do? How is it related to `cat()`?

8.3 Ignoring conditions

The simplest way of handling conditions in R is to simply ignore them:

- Ignore errors with `try()`.
- Ignore warnings with `suppressWarnings()`.
- Ignore messages with `suppressMessages()`.

These functions are heavy handed as you can't use them to suppress a single type of condition that you know about, while allowing everything else to pass through. We'll come back to that challenge later in the chapter.

`try()` allows execution to continue even after an error has occurred. Normally if you run a function that throws an error, it terminates immediately and doesn't return a value:

```
f1 <- function(x) {
  log(x)
  10
}
f1("x")
#> Error in log(x): non-numeric argument to mathematical function
```

However, if you wrap the statement that creates the error in try(), the error message will be displayed[3] but execution will continue:

```
f2 <- function(x) {
  try(log(x))
  10
}
f2("a")
#> Error in log(x) : non-numeric argument to mathematical function
#> [1] 10
```

It is possible, but not recommended, to save the result of try() and perform different actions based on whether or not the code succeeded or failed[4]. Instead, it is better to use tryCatch() or a higher-level helper; you'll learn about those shortly.

A simple, but useful, pattern is to do assignment inside the call: this lets you define a default value to be used if the code does not succeed. This works because the argument is evaluated in the calling environment, not inside the function. (See Section 6.5.1 for more details.)

```
default <- NULL
try(default <- read.csv("possibly-bad-input.csv"), silent = TRUE)
```

suppressWarnings() and suppressMessages() suppress all warnings and messages. Unlike errors, messages and warnings don't terminate execution, so there may be multiple warnings and messages signalled in a single block.

```
suppressWarnings({
  warning("Uhoh!")
  warning("Another warning")
  1
})
#> [1] 1

suppressMessages({
  message("Hello there")
  2
})
#> [1] 2

suppressWarnings({
```

---

[3]You can suppress the message with try(..., silent = TRUE).

[4]You can tell if the expression failed because the result will have class try-error.

```
  message("You can still see me")
  3
})
#> You can still see me
#> [1] 3
```

8.4 Handling conditions

Every condition has default behaviour: errors stop execution and return to the top level, warnings are captured and displayed in aggregate, and messages are immediately displayed. Condition **handlers** allow us to temporarily override or supplement the default behaviour.

Two functions, tryCatch() and withCallingHandlers(), allow us to register handlers, functions that take the signalled condition as their single argument. The registration functions have the same basic form:

```
tryCatch(
  error = function(cnd) {
    # code to run when error is thrown
  },
  code_to_run_while_handlers_are_active
)

withCallingHandlers(
  warning = function(cnd) {
    # code to run when warning is signalled
  },
  message = function(cnd) {
    # code to run when message is signalled
  },
  code_to_run_while_handlers_are_active
)
```

They differ in the type of handlers that they create:

- tryCatch() defines **exiting** handlers; after the condition is handled, control returns to the context where tryCatch() was called. This makes tryCatch() most suitable for working with errors and interrupts, as these have to exit anyway.

- `withCallingHandlers()` defines **calling** handlers; after the condition is captured control returns to the context where the condition was signalled. This makes it most suitable for working with non-error conditions.

But before we can learn about and use these handlers, we need to talk a little bit about condition **objects**. These are created implicitly whenever you signal a condition, but become explicit inside the handler.

8.4.1 Condition objects

So far we've just signalled conditions, and not looked at the objects that are created behind the scenes. The easiest way to see a condition object is to catch one from a signalled condition. That's the job of `rlang::catch_cnd()`:

```
cnd <- catch_cnd(stop("An error"))
str(cnd)
#> List of 2
#>  $ message: chr "An error"
#>  $ call   : language force(expr)
#>  - attr(*, "class")= chr [1:3] "simpleError" "error" "condition"
```

Built-in conditions are lists with two elements:

- `message`, a length-1 character vector containing the text to display to a user. To extract the message, use `conditionMessage(cnd)`.

- `call`, the call which triggered the condition. As described above, we don't use the call, so it will often be `NULL`. To extract it, use `conditionCall(cnd)`.

Custom conditions may contain other components, which we'll discuss in Section 8.5.

Conditions also have a `class` attribute, which makes them S3 objects. We won't discuss S3 until Chapter 13, but fortunately, even if you don't know about S3, condition objects are quite simple. The most important thing to know is that the `class` attribute is a character vector, and it determines which handlers will match the condition.

8.4.2 Exiting handlers

`tryCatch()` registers exiting handlers, and is typically used to handle error conditions. It allows you to override the default error behaviour. For example, the following code will return `NA` instead of throwing an error:

```
f3 <- function(x) {
  tryCatch(
    error = function(cnd) NA,
    log(x)
  )
}

f3("x")
#> [1] NA
```

If no conditions are signalled, or the class of the signalled condition does not match the handler name, the code executes normally:

```
tryCatch(
  error = function(cnd) 10,
  1 + 1
)
#> [1] 2

tryCatch(
  error = function(cnd) 10,
  {
    message("Hi!")
    1 + 1
  }
)
#> Hi!
#> [1] 2
```

The handlers set up by `tryCatch()` are called **exiting** handlers because after the condition is signalled, control passes to the handler and never returns to the original code, effectively meaning that the code exits:

```
tryCatch(
  message = function(cnd) "There",
  {
    message("Here")
    stop("This code is never run!")
  }
)
#> [1] "There"
```

The protected code is evaluated in the environment of `tryCatch()`, but the handler code is not, because the handlers are functions. This is important to remember if you're trying to modify objects in the parent environment.

The handler functions are called with a single argument, the condition object. I call this argument cnd, by convention. This value is only moderately useful for the base conditions because they contain relatively little data. It's more useful when you make your own custom conditions, as you'll see shortly.

```r
tryCatch(
  error = function(cnd) {
    paste0("--", conditionMessage(cnd), "--")
  },
  stop("This is an error")
)
#> [1] "--This is an error--"
```

`tryCatch()` has one other argument: `finally`. It specifies a block of code (not a function) to run regardless of whether the initial expression succeeds or fails. This can be useful for clean up, like deleting files, or closing connections. This is functionally equivalent to using `on.exit()` (and indeed that's how it's implemented) but it can wrap smaller chunks of code than an entire function.

```r
path <- tempfile()
tryCatch(
  {
    writeLines("Hi!", path)
    # ...
  },
  finally = {
    # always run
    unlink(path)
  }
)
```

8.4.3 Calling handlers

The handlers set up by `tryCatch()` are called exiting handlers, because they cause code to exit once the condition has been caught. By contrast, `withCallingHandlers()` sets up **calling** handlers: code execution continues normally once the handler returns. This tends to make `withCallingHandlers()` a more natural pairing with the non-error conditions. Exiting and calling handlers use "handler" in slighty different senses:

- An exiting handler handles a signal like you handle a problem; it makes the problem go away.

- A calling handler handles a signal like you handle a car; the car still exists.

Compare the results of `tryCatch()` and `withCallingHandlers()` in the example below. The messages are not printed in the first case, because the code is terminated once the exiting handler completes. They are printed in the second case, because a calling handler does not exit.

```
tryCatch(
  message = function(cnd) cat("Caught a message!\n"),
  {
    message("Someone there?")
    message("Why, yes!")
  }
)
#> Caught a message!

withCallingHandlers(
  message = function(c) cat("Caught a message!\n"),
  {
    message("Someone there?")
    message("Why, yes!")
  }
)
#> Caught a message!
#> Someone there?
#> Caught a message!
#> Why, yes!
```

Handlers are applied in order, so you don't need to worry getting caught in an infinite loop. In the following example, the `message()` signalled by the handler doesn't also get caught:

```
withCallingHandlers(
  message = function(cnd) message("Second message"),
  message("First message")
)
#> Second message
#> First message
```

(But beware if you have multiple handlers, and some handlers signal conditions that could be captured by another handler: you'll need to think through the order carefully.)

The return value of a calling handler is ignored because the code continues to execute after the handler completes; where would the return value go? That means that calling handlers are only useful for their side-effects.

One important side-effect unique to calling handlers is the ability to **muffle** the signal. By default, a condition will continue to propagate to parent handlers, all the way up to the default handler (or an exiting handler, if provided):

```
# Bubbles all the way up to default handler which generates the message
withCallingHandlers(
  message = function(cnd) cat("Level 2\n"),
  withCallingHandlers(
    message = function(cnd) cat("Level 1\n"),
    message("Hello")
  )
)
#> Level 1
#> Level 2
#> Hello

# Bubbles up to tryCatch
tryCatch(
  message = function(cnd) cat("Level 2\n"),
  withCallingHandlers(
    message = function(cnd) cat("Level 1\n"),
    message("Hello")
  )
)
#> Level 1
#> Level 2
```

If you want to prevent the condition "bubbling up" but still run the rest of the code in the block, you need to explicitly muffle it with `rlang::cnd_muffle()`:

```
# Muffles the default handler which prints the messages
withCallingHandlers(
  message = function(cnd) {
    cat("Level 2\n")
    cnd_muffle(cnd)
  },
  withCallingHandlers(
    message = function(cnd) cat("Level 1\n"),
    message("Hello")
  )
)
```

```
#> Level 1
#> Level 2

# Muffles level 2 handler and the default handler
withCallingHandlers(
  message = function(cnd) cat("Level 2\n"),
  withCallingHandlers(
    message = function(cnd) {
      cat("Level 1\n")
      cnd_muffle(cnd)
    },
    message("Hello")
  )
)
#> Level 1
```

8.4.4   Call stacks

To complete the section, there are some important differences between the call stacks of exiting and calling handlers. These differences are generally not important but I'm including them here because I've occasionally found them useful, and don't want to forget about them!

It's easiest to see the difference by setting up a small example that uses lobstr::cst():

```
f <- function() g()
g <- function() h()
h <- function() message("!")
```

Calling handlers are called in the context of the call that signalled the condition:

```
withCallingHandlers(f(), message = function(cnd) {
  lobstr::cst()
  cnd_muffle(cnd)
})
#>    █
#>  1. ├─base::withCallingHandlers(...)
#>  2. ├─global::f()
#>  3. │ └─global::g()
#>  4. │   └─global::h()
```

```
#>    5. |      └─base::message("!")
#>    6. |        ├─base::withRestarts(...)
#>    7. |        | └─base:::withOneRestart(expr, restarts[[1L]])
#>    8. |        |   └─base:::doWithOneRestart(return(expr), restart)
#>    9. |        └─base::signalCondition(cond)
#>   10. └─(function (cnd) ...
#>   11.   └─lobstr::cst()
```

Whereas exiting handlers are called in the context of the call to `tryCatch()`:

```
tryCatch(f(), message = function(cnd) lobstr::cst())
#>     █
#>   1. └─base::tryCatch(f(), message = function(cnd) lobstr::cst())
#>   2.   └─base:::tryCatchList(expr, classes, parentenv, handlers)
#>   3.     └─base:::tryCatchOne(expr, names, parentenv, handlers[[1L]])
#>   4.       └─value[[3L]](cond)
#>   5.         └─lobstr::cst()
```

8.4.5 Exercises

1. What extra information does the condition generated by `abort()`
   contain compared to the condition generated by `stop()` i.e. what's
   the difference between these two objects? Read the help for `?abort`
   to learn more.

   ```
   catch_cnd(stop("An error"))
   catch_cnd(abort("An error"))
   ```

2. Predict the results of evaluating the following code

   ```
   show_condition <- function(code) {
     tryCatch(
       error = function(cnd) "error",
       warning = function(cnd) "warning",
       message = function(cnd) "message",
       {
         code
         NULL
       }
   ```

```
  )
}

show_condition(stop("!"))
show_condition(10)
show_condition(warning("?!"))
show_condition({
  10
  message("?")
  warning("?!")
})
```

3. Explain the results of running this code:

```
withCallingHandlers(
  message = function(cnd) message("b"),
  withCallingHandlers(
    message = function(cnd) message("a"),
    message("c")
  )
)
#> b
#> a
#> b
#> c
```

4. Read the source code for `catch_cnd()` and explain how it works.

5. How could you rewrite `show_condition()` to use a single handler?

8.5   Custom conditions

One of the challenges of error handling in R is that most functions generate one of the built-in conditions, which contain only a `message` and a `call`. That means that if you want to detect a specific type of error, you can only work with the text of the error message. This is error prone, not only because the message might change over time, but also because messages can be translated into other languages.

Fortunately R has a powerful, but little used feature: the ability to create custom conditions that can contain additional metadata. Creating custom conditions is a little fiddly in base R, but `rlang::abort()` makes it very easy as you can supply a custom `.subclass` and additional metadata.

The following example shows the basic pattern. I recommend using the following call structure for custom conditions. This takes advantage of R's flexible argument matching so that the name of the type of error comes first, followed by the user facing text, followed by custom metadata.

```
abort(
  "error_not_found",
  message = "Path `blah.csv` not found",
  path = "blah.csv"
)
#> Error: Path `blah.csv` not found
```

Custom conditions work just like regular conditions when used interactively, but allow handlers to do much more.

8.5.1 Motivation

To explore these ideas in more depth, let's take `base::log()`. It does the minimum when throwing errors caused by invalid arguments:

```
log(letters)
#> Error in log(letters): non-numeric argument to mathematical function
log(1:10, base = letters)
#> Error in log(1:10, base = letters): non-numeric argument to
#> mathematical function
```

I think we can do better by being explicit about which argument is the problem (i.e. x or base), and saying what the problematic input is (not just what it isn't).

```
my_log <- function(x, base = exp(1)) {
  if (!is.numeric(x)) {
    abort(paste0(
      "`x` must be a numeric vector; not ", typeof(x), "."
    ))
  }
  if (!is.numeric(base)) {
    abort(paste0(
```

```
      "`base` must be a numeric vector; not ", typeof(base), "."
   ))
 }

   base::log(x, base = base)
 }
```

This gives us:

```
my_log(letters)
#> Error: `x` must be a numeric vector; not character.
my_log(1:10, base = letters)
#> Error: `base` must be a numeric vector; not character.
```

This is an improvement for interactive usage as the error messages are more likely to guide the user towards a correct fix. However, they're no better if you want to programmatically handle the errors: all the useful metadata about the error is jammed into a single string.

8.5.2 Signalling

Let's build some infrastructure to improve this situation, We'll start by providing a custom abort() function for bad arguments. This is a little over-generalised for the example at hand, but it reflects common patterns that I've seen across other functions. The pattern is fairly simple. We create a nice error message for the user, using glue::glue(), and store metadata in the condition call for the developer.

```
abort_bad_argument <- function(arg, must, not = NULL) {
  msg <- glue::glue("`{arg}` must {must}")
  if (!is.null(not)) {
    not <- typeof(not)
    msg <- glue::glue("{msg}; not {not}.")
  }

  abort("error_bad_argument",
    message = msg,
    arg = arg,
    must = must,
    not = not
  )
}
```

In Base R

If you want to throw a custom error without adding a dependency on rlang, you can create a condition object "by hand" and then pass it to stop():

```
stop_custom <- function(.subclass, message, call = NULL, ...) {
  err <- structure(
    list(
      message = message,
      call = call,
      ...
    ),
    class = c(.subclass, "error", "condition")
  )
  stop(err)
}

err <- catch_cnd(
  stop_custom("error_new", "This is a custom error", x = 10)
)
class(err)
err$x
```

We can now rewrite my_log() to use this new helper:

```
my_log <- function(x, base = exp(1)) {
  if (!is.numeric(x)) {
    abort_bad_argument("x", must = "be numeric", not = x)
  }
  if (!is.numeric(base)) {
    abort_bad_argument("base", must = "be numeric", not = base)
  }

  base::log(x, base = base)
}
```

my_log() itself is not much shorter, but is a little more meangingful, and it ensures that error messages for bad arguments are consistent across functions. It yields the same interactive error messages as before:

```
my_log(letters)
#> Error: `x` must be numeric; not character.
```

```
my_log(1:10, base = letters)
#> Error: `base` must be numeric; not character.
```

8.5.3   Handling

These structured condition objects are much easier to program with. The first
place you might want to use this capability is when testing your function. Unit
testing is not a subject of this book (see R packages (http://r-pkgs.had.co.
nz/) for details), but the basics are easy to understand. The following code
captures the error, and then asserts it has the structure that we expect.

```
library(testthat)

err <- catch_cnd(my_log("a"))
expect_s3_class(err, "error_bad_argument")
expect_equal(err$arg, "x")
expect_equal(err$not, "character")
```

We can also use the class (error_bad_argument) in tryCatch() to only handle
that specific error:

```
tryCatch(
  error_bad_argument = function(cnd) "bad_argument",
  error = function(cnd) "other error",
  my_log("a")
)
#> [1] "bad_argument"
```

When using tryCatch() with multiple handlers and custom classes, the first
handler to match any class in the signal's class vector is called, not the best
match. For this reason, you need to make sure to put the most specific handlers
first. The following code does not do what you might hope:

```
tryCatch(
  error = function(cnd) "other error",
  error_bad_argument = function(cnd) "bad_argument",
  my_log("a")
)
#> [1] "other error"
```

8.5.4 Exercises

1. Inside a package, it's occasionally useful to check that a package is installed before using it. Write a function that checks if a package is installed (with `requireNamespace("pkg", quietly = FALSE)`) and if not, throws a custom condition that includes the package name in the metadata.

2. Inside a package you often need to stop with an error when something is not right. Other packages that depend on your package might be tempted to check these errors in their unit tests. How could you help these packages to avoid relying on the error message which is part of the user interface rather than the API and might change without notice?

8.6 Applications

Now that you've learned the basic tools of R's condition system, it's time to dive into some applications. The goal of this section is not to show every possible usage of `tryCatch()` and `withCallingHandlers()` but to illustrate some common patterns that frequently crop up. Hopefully these will get your creative juices flowing, so when you encounter a new problem you can come up with a useful solution.

8.6.1 Failure value

There are a few simple, but useful, `tryCatch()` patterns based on returning a value from the error handler. The simplest case is a wrapper to return a default value if an error occurs:

```
fail_with <- function(expr, value = NULL) {
  tryCatch(
    error = function(cnd) value,
    expr
  )
}

fail_with(log(10), NA_real_)
#> [1] 2.3
```

```
fail_with(log("x"), NA_real_)
#> [1] NA
```

A more sophisticated application is base::try(). Below, try2() extracts the
essence of base::try(); the real function is more complicated in order to make
the error message look more like what you'd see if tryCatch() wasn't used.

```
try2 <- function(expr, silent = FALSE) {
  tryCatch(
    error = function(cnd) {
      msg <- conditionMessage(cnd)
      if (!silent) {
        message("Error: ", msg)
      }
      structure(msg, class = "try-error")
    },
    expr
  )
}

try2(1)
#> [1] 1
try2(stop("Hi"))
#> Error: Hi
#> [1] "Hi"
#> attr(,"class")
#> [1] "try-error"
try2(stop("Hi"), silent = TRUE)
#> [1] "Hi"
#> attr(,"class")
#> [1] "try-error"
```

8.6.2   Success and failure values

We can extend this pattern to return one value if the code evaluates successfully (success_val), and another if it fails (error_val). This pattern just
requires one small trick: evaluating the user supplied code, then success_val.
If the code throws an error, we'll never get to success_val and will instead
return error_val.

```
foo <- function(expr) {
  tryCatch(
```

```
    error = function(cnd) error_val,
    {
      expr
      success_val
    }
  )
}
```

We can use this to determine if an expression fails:

```
does_error <- function(expr) {
  tryCatch(
    error = function(cnd) TRUE,
    {
      expr
      FALSE
    }
  )
}
```

Or to capture any condition, like just `rlang::catch_cnd()`:

```
catch_cnd <- function(expr) {
  tryCatch(
    condition = function(cnd) cnd,
    {
      expr
      NULL
    }
  )
}
```

We can also use this pattern to create a `try()` variant. One challenge with `try()` is that it's slightly challenging to determine if the code succeeded or failed. Rather than returning an object with a special class, I think it's slightly nicer to return a list with two components `result` and `error`.

```
safety <- function(expr) {
  tryCatch(
    error = function(cnd) {
      list(result = NULL, error = cnd)
    },
    list(result = expr, error = NULL)
```

```
  )
}

str(safety(1 + 10))
#> List of 2
#>  $ result: num 11
#>  $ error : NULL
str(safety(stop("Error!")))
#> List of 2
#>  $ result: NULL
#>  $ error :List of 2
#>   ..$ message: chr "Error!"
#>   ..$ call   : language doTryCatch(return(expr), name, parentenv, h..
#>   ..- attr(*, "class")= chr [1:3] "simpleError" "error" "condition"
```

(This is closely related to `purrr::safely()`, a function operator, which we'll come back to in Section 11.2.1.)

8.6.3   Resignal

As well as returning default values when a condition is signalled, handlers can be used to make more informative error messages. One simple application is to make a function that works like `options(warn = 2)` for a single block of code. The idea is simple: we handle warnings by throwing an error:

```
warning2error <- function(expr) {
  withCallingHandlers(
    warning = function(cnd) abort(conditionMessage(cnd)),
    expr
  )
}
```

```
warning2error({
  x <- 2 ^ 4
  warn("Hello")
})
#> Error: Hello
```

You could write a similar function if you were trying to find the source of an annoying message. More on this in Section 22.6.

8.6.4 Record

Another common pattern is to record conditions for later investigation. The new challenge here is that calling handlers are called only for their side-effects so we can't return values, but instead need to modify some object in place.

```
catch_cnds <- function(expr) {
  conds <- list()
  add_cond <- function(cnd) {
    conds <<- append(conds, list(cnd))
    cnd_muffle(cnd)
  }

  withCallingHandlers(
    message = add_cond,
    warning = add_cond,
    expr
  )

  conds
}

catch_cnds({
  inform("a")
  warn("b")
  inform("c")
})
#> [[1]]
#> <message: a
#> >
#>
#> [[2]]
#> <warning: b>
#>
#> [[3]]
#> <message: c
#> >
```

What if you also want to capture errors? You'll need to wrap the withCallingHandlers() in a tryCatch(). If an error occurs, it will be the last condition.

```
catch_cnds <- function(expr) {
  conds <- list()
  add_cond <- function(cnd) {
```

```
    conds <<- append(conds, list(cnd))
    cnd_muffle(cnd)
  }

  tryCatch(
    error = function(cnd) {
      conds <<- append(conds, list(cnd))
    },
    withCallingHandlers(
      message = add_cond,
      warning = add_cond,
      expr
    )
  )

  conds
}

catch_cnds({
  inform("a")
  warn("b")
  abort("C")
})
#> [[1]]
#> <message: a
#> >
#>
#> [[2]]
#> <warning: b>
#>
#> [[3]]
#> <error>
#> message: C
#> class:   `rlang_error`
#> backtrace:
#>  1. global::catch_cnds(...)
#>  6. base::withCallingHandlers(...)
#> Call `rlang::last_trace()` to see the full backtrace
```

This is the key idea underlying the evaluate package [Wickham and Xie, 2018]
which powers knitr: it captures every output into a special data structure so
that it can be later replayed. As a whole, the evaluate package is quite a lot
more complicated than the code here because it also needs to handle plots
and text output.

8.6.5 No default behaviour

A final useful pattern is to signal a condition that doesn't inherit from message, warning or error. Because there is no default behaviour, this means the
condition has no effect unless the user specifically requests it. For example,
you could imagine a logging system based on conditions:

```r
log <- function(message, level = c("info", "error", "fatal")) {
  level <- match.arg(level)
  signal(message, "log", level = level)
}
```

When you call log() a condition is signalled, but nothing happens because it
has no default handler:

```r
log("This code was run")
```

To activate logging you need a handler that does something with the log
condition. Below I define a record_log() function that will record all logging
messages to a file:

```r
record_log <- function(expr, path = stdout()) {
  withCallingHandlers(
    log = function(cnd) {
      cat(
        "[", cnd$level, "] ", cnd$message, "\n", sep = "",
        file = path, append = TRUE
      )
    },
    expr
  )
}

record_log(log("Hello"))
#> [info] Hello
```

You could even imagine layering with another function that allows you to
selectively suppress some logging levels.

```r
ignore_log_levels <- function(expr, levels) {
  withCallingHandlers(
    log = function(cnd) {
      if (cnd$level %in% levels) {
```

```
        cnd_muffle(cnd)
    }
  },
  expr
  )
}
```

```
record_log(ignore_log_levels(log("Hello"), "info"))
```

In Base R

If you create a condition object by hand, and signal it with signalCondition(),
cnd_muffle() will not work. Instead you need to call it with a muffle restart
defined, like this:

```
withRestarts(signalCondition(cond), muffle = function() NULL)
```

Restarts are currently beyond the scope of the book, but I suspect will be
included in the third edition.

8.6.6   Exercises

1. Create suppressConditions() that works like suppressMessages()
   and suppressWarnings() but suppresses everything. Think carefully
   about how you should handle errors.

2. Compare the following two implementations of message2error().
   What is the main advantage of withCallingHandlers() in this scenario? (Hint: look carefully at the traceback.)

   ```
   message2error <- function(code) {
     withCallingHandlers(code, message = function(e) stop(e))
   }
   message2error <- function(code) {
     tryCatch(code, message = function(e) stop(e))
   }
   ```

3. How would you modify the catch_cnds() definition if you wanted
   to recreate the original intermingling of warnings and messages?

4. Why is catching interrupts dangerous? Run this code to find out.

```r
bottles_of_beer <- function(i = 99) {
  message(
    "There are ", i, " bottles of beer on the wall, ",
    i, " bottles of beer."
  )
  while(i > 0) {
    tryCatch(
      Sys.sleep(1),
      interrupt = function(err) {
        i <<- i - 1
        if (i > 0) {
          message(
            "Take one down, pass it around, ", i,
            " bottle", if (i > 1) "s", " of beer on the wall."
          )
        }
      }
    )
  }
  message(
    "No more bottles of beer on the wall, ",
    "no more bottles of beer."
  )
}
```

8.7 Quiz answers

1. `error`, `warning`, and `message`.

2. You could use `try()` or `tryCatch()`.

3. `tryCatch()` creates exiting handlers which will terminate the execution of wrapped code; `withCallingHandlers()` creates calling handlers which don't affect the execution of wrapped code.

4. Because you can then capture specific types of error with `tryCatch()`, rather than relying on the comparison of error strings, which is risky, especially when messages are translated.

Part II

Functional programming

Introduction

R, at its heart, is a **functional** language. This means that it has certain technical properties, but more importantly that it lends itself to a style of problem solving centred on functions. Below I'll give a brief overview of the technical definition of a functional *language*, but in this book I will primarily focus on the functional *style* of programming, because I think it is an extremely good fit to the types of problem you commonly encounter when doing data analysis.

Recently, functional techniques have experienced a surge in interest because they can produce efficient and elegant solutions to many modern problems. A functional style tends to create functions that can easily be analysed in isolation (i.e. using only local information), and hence is often much easier to automatically optimise or parallelise. The traditional weaknesses of functional languages, poorer performance and sometimes unpredictable memory usage, have been much reduced in recent years. Functional programming is complementary to object-oriented programming, which has been the dominant programming paradigm for the last several decades.

Functional programming languages

Every programming language has functions, so what makes a programming language functional? There are many definitions for precisely what makes a language functional, but there are two common threads.

Firstly, functional languages have **first-class functions**, functions that behave like any other data structure. In R, this means that you can do many of the things with a function that you can do with a vector: you can assign them to variables, store them in lists, pass them as arguments to other functions, create them inside functions, and even return them as the result of a function.

Secondly, many functional languages require functions to be **pure**. A function is pure if it satisfies two properties:

- The output only depends on the inputs, i.e. if you call it again with the same inputs, you get the same outputs. This excludes functions like `runif()`, `read.csv()`, or `Sys.time()` that can return different values.

- The function has no side-effects, like changing the value of a global variable, writing to disk, or displaying to the screen. This excludes functions like `print()`, `write.csv()` and `<-`.

Pure functions are much easier to reason about, but obviously have significant downsides: imagine doing a data analysis where you couldn't generate random numbers or read files from disk.

Strictly speaking, R isn't a functional programming *language* because it doesn't require that you write pure functions. However, you can certainly adopt a functional style in parts of your code: you don't *have* to write pure functions, but you often *should*. In my experience, partitioning code into functions that are either extremely pure or extremely impure tends to lead to code that is easier to understand and extends to new situations.

Functional style

It's hard to describe exactly what a functional *style* is, but generally I think it means decomposing a big problem into smaller pieces, then solving each piece with a function or combination of functions. When using a functional style, you strive to decompose components of the problem into isolated functions that operate independently. Each function taken by itself is simple and straightforward to understand; complexity is handled by composing functions in various ways.

The following three chapters discuss the three key functional techniques that help you to decompose problems into smaller pieces:

- Chapter 9 shows you how to replace many for loops with **functionals** which are functions (like `lapply()`) that take another function as an argument. Functionals allow you to take a function that solves the problem for a single input and generalise it to handle any number of inputs. Functionals are by far and away the most important technique and you'll use them all the time in data analysis.

- Chapter 10 introduces **function factories**: functions that create functions. Function factories are less commonly used than functionals, but can allow you to elegantly partition work between different parts of your code.

- Chapter 11 shows you how to create **function operators**: functions that take functions as input and produce functions as output. They are like adverbs, because they typically modify the operation of a function.

Collectively, these types of function are called **higher-order functions** and they fill out a two-by-two table:

Out / In	Vector	Function
Vector	Regular function	Function factory
Function	Functional	Function operator

9

Functionals

9.1 Introduction

> *To become significantly more reliable, code must become more*
> *transparent. In particular, nested conditions and loops must be*
> *viewed with great suspicion. Complicated control flows confuse*
> *programmers. Messy code often hides bugs.*
>
> — *Bjarne Stroustrup*

A **functional** is a function that takes a function as an input and returns a vector as output. Here's a simple functional: it calls the function provided as input with 1000 random uniform numbers.

```
randomise <- function(f) f(runif(1e3))
randomise(mean)
#> [1] 0.506
randomise(mean)
#> [1] 0.501
randomise(sum)
#> [1] 489
```

The chances are that you've already used a functional. You might have used for-loop replacements like base R's `lapply()`, `apply()`, and `tapply()`; or purrr's `map()`; or maybe you've used a mathematical functional like `integrate()` or `optim()`.

A common use of functionals is as an alternative to for loops. For loops have a bad rap in R because many people believe they are slow[1], but the real downside of for loops is that they're very flexible: a loop conveys that you're iterating, but not what should be done with the results. Just as it's better to use `while` than `repeat`, and it's better to use `for` than `while` (Section 5.3.2), it's better to use a functional than `for`. Each functional is tailored for a specific

---

[1]Typically it's not the for loop itself that's slow, but what you're doing inside of it. A common culprit of slow loops is modifying a data structure, where each modification generates a copy. See Sections 2.5.1 and 24.6 for more details.

task, so when you recognise the functional you immediately know why it's being used.

If you're an experienced for loop user, switching to functionals is typically a pattern matching exercise. You look at the for loop and find a functional that matches the basic form. If one doesn't exist, don't try and torture an existing functional to fit the form you need. Instead, just leave it as a for loop! (Or once you've repeated the same loop two or more times, maybe think about writing your own functional).

Outline

- Section 9.2 introduces your first functional: `purrr::map()`.

- Section 9.3 demonstrates how you can combine multiple simple functionals to solve a more complex problem and discusses how purrr style differs from other approaches.

- Section 9.4 teaches you about 18 (!!) important variants of `purrr::map()`. Fortunately, their orthogonal design makes them easy to learn, remember, and master.

- Section 9.5 introduces a new style of functional: `purrr::reduce()`. `reduce()` systematically reduces a vector to a single result by applying a function that takes two inputs.

- Section 9.6 teaches you about predicates: functions that return a single TRUE or FALSE, and the family of functionals that use them to solve common problems.

- Section 9.7 reviews some functionals in base R that are not members of the map, reduce, or predicate families.

Prerequisites

This chapter will focus on functionals provided by the purrr package (https://purrr.tidyverse.org) [Henry and Wickham, 2018a]. These functions have a consistent interface that makes it easier to understand the key ideas than their base equivalents, which have grown organically over many years. I'll compare and contrast base R functions as we go, and then wrap up the chapter with a discussion of base functionals that don't have purrr equivalents.

```
library(purrr)
```

9.2 My first functional: `map()`

The most fundamental functional is `purrr::map()`[2]. It takes a vector and a function, calls the function once for each element of the vector, and returns the results in a list. In other words, `map(1:3, f)` is equivalent to `list(f(1)`, `f(2)`, `f(3))`.

```
triple <- function(x) x * 3
map(1:3, triple)
#> [[1]]
#> [1] 3
#>
#> [[2]]
#> [1] 6
#>
#> [[3]]
#> [1] 9
```

Or, graphically:

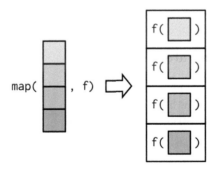

You might wonder why this function is called `map()`. What does it have to do with depicting physical features of land or sea ▮▮? In fact, the meaning comes from mathematics where *map* refers to "an operation that associates each element of a given set with one or more elements of a second set". This makes sense here because `map()` defines a mapping from one vector to another. (*"Map"* also has the nice property of being short, which is useful for such a fundamental building block.)

---

[2]Not to be confused with `base::Map()`, which is considerably more complex. I'll discuss `Map()` in Section 9.4.5.

The implementation of `map()` is quite simple. We allocate a list the same length as the input, and then fill in the list with a for loop. The heart of the implementation is only a handful of lines of code:

```
simple_map <- function(x, f, ...) {
  out <- vector("list", length(x))
  for (i in seq_along(x)) {
    out[[i]] <- f(x[[i]], ...)
  }
  out
}
```

The real `purrr::map()` function has a few differences: it is written in C to eke out every last iota of performance, preserves names, and supports a few shortcuts that you'll learn about in Section 9.2.2.

In Base R

The base equivalent to `map()` is `lapply()`. The only difference is that `lapply()` does not support the helpers that you'll learn about below, so if you're only using `map()` from purrr, you can skip the additional dependency and use `lapply()` directly.

9.2.1 Producing atomic vectors

`map()` returns a list, which makes it the most general of the map family because you can put anything in a list. But it is inconvenient to return a list when a simpler data structure would do, so there are four more specific variants: `map_lgl()`, `map_int()`, `map_dbl()`, and `map_chr()`. Each returns an atomic vector of the specified type:

```
# map_chr() always returns a character vector
map_chr(mtcars, typeof)
#>      mpg      cyl     disp       hp     drat       wt     qsec
#> "double" "double" "double" "double" "double" "double" "double"
#>       vs       am     gear     carb
#> "double" "double" "double" "double"

# map_lgl() always returns a logical vector
map_lgl(mtcars, is.double)
#>  mpg  cyl disp   hp drat   wt qsec   vs   am gear carb
#> TRUE TRUE TRUE TRUE TRUE TRUE TRUE TRUE TRUE TRUE TRUE
```

```
# map_int() always returns a integer vector
n_unique <- function(x) length(unique(x))
map_int(mtcars, n_unique)
#>  mpg cyl disp   hp drat   wt qsec   vs   am gear carb
#>   25    3   27   22   22   29   30    2    2    3    6

# map_dbl() always returns a double vector
map_dbl(mtcars, mean)
#>      mpg      cyl     disp       hp     drat       wt     qsec       vs
#>   20.091    6.188  230.722  146.688    3.597    3.217   17.849    0.438
#>       am     gear     carb
#>    0.406    3.688    2.812
```

purrr uses the convention that suffixes, like _dbl(), refer to the output. All
map_*() functions can take any type of vector as input. These examples rely on
two facts: mtcars is a data frame, and data frames are lists containing vectors
of the same length. This is more obvious if we draw a data frame with the
same orientation as vector:

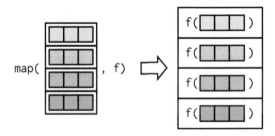

All map functions always return an output vector the same length as the
input, which implies that each call to .f must return a single value. If it does
not, you'll get an error:

```
pair <- function(x) c(x, x)
map_dbl(1:2, pair)
#> Error: Result 1 must be a single double, not an integer vector of
#> length 2
```

This is similar to the error you'll get if .f returns the wrong type of result:

```
map_dbl(1:2, as.character)
#> Error: Can't coerce element 1 from a character to a double
```

In either case, it's often useful to switch back to map(), because map() can
accept any type of output. That allows you to see the problematic output,
and figure out what to do with it.

```
map(1:2, pair)
#> [[1]]
#> [1] 1 1
#>
#> [[2]]
#> [1] 2 2
map(1:2, as.character)
#> [[1]]
#> [1] "1"
#>
#> [[2]]
#> [1] "2"
```

In Base R

Base R has two apply functions that can return atomic vectors: `sapply()` and `vapply()`. I recommend that you avoid `sapply()` because it tries simplify the result, so it can return a list, a vector, or a matrix. This makes it difficult to program with, and it should be avoided in non-interactive settings. `vapply()` is safer because it allows you to provide a template, `FUN.VALUE`, that describes the output shape. If you don't want to use purrr, I recommend you always use `vapply()` in your functions, not `sapply()`. The primary downside of `vapply()` is its verbosity: for example, the equivalent to `map_dbl(x, mean, na.rm = TRUE)` is `vapply(x, mean, na.rm = TRUE, FUN.VALUE = double(1))`.

9.2.2 Anonymous functions and shortcuts

Instead of using `map()` with an existing function, you can create an inline anonymous function (as mentioned in Section 6.2.3):

```
map_dbl(mtcars, function(x) length(unique(x)))
#>  mpg  cyl disp   hp drat   wt qsec   vs   am gear carb
#>   25    3   27   22   22   29   30    2    2    3    6
```

Anonymous functions are very useful, but the syntax is verbose. So purrr supports a special shortcut:

```
map_dbl(mtcars, ~ length(unique(.x)))
#>  mpg  cyl disp   hp drat   wt qsec   vs   am gear carb
#>   25    3   27   22   22   29   30    2    2    3    6
```

This works because all purrr functions translate formulas, created by ~ (pronounced "twiddle"), into functions. You can see what's happening behind the scenes by calling `as_mapper()`:

```
as_mapper(~ length(unique(.x)))
#> <lambda>
#> function (..., .x = ..1, .y = ..2, . = ..1)
#> length(unique(.x))
#> attr(,"class")
#> [1] "rlang_lambda_function"
```

The function arguments look a little quirky but allow you to refer to . for one argument functions, .x and .y. for two argument functions, and ..1, ..2, ..3, etc, for functions with an arbitrary number of arguments. . remains for backward compatibility but I don't recommend using it because it's easily confused with the . use by magrittr's pipe.

This shortcut is particularly useful for generating random data:

```
x <- map(1:3, ~ runif(2))
str(x)
#> List of 3
#>  $ : num [1:2] 0.281 0.53
#>  $ : num [1:2] 0.433 0.917
#>  $ : num [1:2] 0.0275 0.8249
```

Reserve this syntax for short and simple functions. A good rule of thumb is that if your function spans lines or uses {}, it's time to give it a name.

The map functions also have shortcuts for extracting elements from a vector, powered by `purrr::pluck()`. You can use a character vector to select elements by name, an integer vector to select by position, or a list to select by both name and position. These are very useful for working with deeply nested lists, which often arise when working with JSON.

```
x <- list(
  list(-1, x = 1, y = c(2), z = "a"),
  list(-2, x = 4, y = c(5, 6), z = "b"),
  list(-3, x = 8, y = c(9, 10, 11))
)

# Select by name
map_dbl(x, "x")
#> [1] 1 4 8
```

```
# Or by position
map_dbl(x, 1)
#> [1] -1 -2 -3

# Or by both
map_dbl(x, list("y", 1))
#> [1] 2 5 9

# You'll get an error if a component doesn't exist:
map_chr(x, "z")
#> Error: Result 3 must be a single string, not NULL of length 0

# Unless you supply a .default value
map_chr(x, "z", .default = NA)
#> [1] "a" "b" NA
```

In Base R

In base R functions, like `lapply()`, you can provide the name of the function as a string. This isn't tremendously useful as `lapply(x, "f")` is almost always equivalent to `lapply(x, f)` and is more typing.

9.2.3   Passing arguments with ...

It's often convenient to pass along additional arguments to the function that you're calling. For example, you might want to pass `na.rm = TRUE` along to `mean()`. One way to do that is with an anonymous function:

```
x <- list(1:5, c(1:10, NA))
map_dbl(x, ~ mean(.x, na.rm = TRUE))
#> [1] 3.0 5.5
```

But because the map functions pass ... along, there's a simpler form available:

```
map_dbl(x, mean, na.rm = TRUE)
#> [1] 3.0 5.5
```

This is easiest to understand with a picture: any arguments that come after `f` in the call to `map()` are inserted *after* the data in individual calls to `f()`:

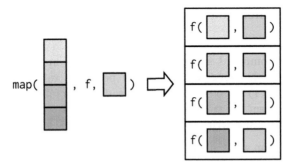

It's important to note that these arguments are not decomposed; or said another way, `map()` is only vectorised over its first argument. If an argument after `f` is a vector, it will be passed along as is:

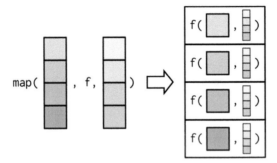

(You'll learn about map variants that *are* vectorised over multiple arguments in Sections 9.4.2 and 9.4.5.)

Note there's a subtle difference between placing extra arguments inside an anonymous function compared with passing them to `map()`. Putting them in an anonymous function means that they will be evaluated every time `f()` is executed, not just once when you call `map()`. This is easiest to see if we make the additional argument random:

```
plus <- function(x, y) x + y

x <- c(0, 0, 0, 0)
map_dbl(x, plus, runif(1))
#> [1] 0.0625 0.0625 0.0625 0.0625
map_dbl(x, ~ plus(.x, runif(1)))
#> [1] 0.903 0.132 0.629 0.945
```

9.2.4   Argument names

In the diagrams, I've omitted argument names to focus on the overall structure. But I recommend writing out the full names in your code, as it makes it easier to read. map(x, mean, 0.1) is perfectly valid code, but will call mean(x[[1]], 0.1) so it relies on the reader remembering that the second argument to mean() is trim. To avoid unnecessary burden on the brain of the reader[3], be kind and write map(x, mean, trim = 0.1).

This is the reason why the arguments to map() are a little odd: instead of being x and f, they are .x and .f. It's easiest to see the problem that leads to these names using simple_map() defined above. simple_map() has arguments x and f so you'll have problems whenever the function you are calling has arguments x or f:

```
boostrap_summary <- function(x, f) {
  f(sample(x, replace = TRUE))
}

simple_map(mtcars, boostrap_summary, f = mean)
#> Error in mean.default(x[[i]], ...): 'trim' must be numeric of length
#> one
```

The error is a little bewildering until you remember that the call to simple_map() is equivalent to simple_map(x = mtcars, f = mean, boostrap_summary) because named matching beats positional matching.

purrr functions reduce the likelihood of such a clash by using .f and .x instead of the more common f and x. Of course this technique isn't perfect (because the function you are calling might still use .f and .x), but it avoids 99% of issues. The remaining 1% of the time, use an anonymous function.

In Base R

Base functions that pass along ... use a variety of naming conventions to prevent undesired argument matching:

- The apply family mostly uses capital letters (e.g. X and FUN).
- transform() uses the more exotic prefix _: this makes the name non-syntactic so it must always be surrounded in `, as described in Section 2.2.1. This makes undesired matches extremely unlikely.
- Other functionals like uniroot() and optim() make no effort to avoid clashes but they tend to be used with specially created functions so clashes are less likely.

---

[3]Who is highly likely to be future you!

9.2.5 Varying another argument

So far the first argument to `map()` has always become the first argument to the function. But what happens if the first argument should be constant, and you want to vary a different argument? How do you get the result in this picture?

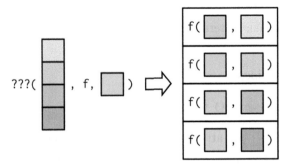

It turns out that there's no way to do it directly, but there are two tricks you can use instead. To illustrate them, imagine I have a vector that contains a few unusual values, and I want to explore the effect of different amounts of trimming when computing the mean. In this case, the first argument to `mean()` will be constant, and I want to vary the second argument, `trim`.

```
trims <- c(0, 0.1, 0.2, 0.5)
x <- rcauchy(1000)
```

- The simplest technique is to use an anonymous function to rearrange the argument order:

```
map_dbl(trims, ~ mean(x, trim = .x))
#> [1] -0.3500  0.0434  0.0354  0.0502
```

This is still a little confusing because I'm using both x and .x. You can make it a little clearer by abandoning the ~ helper:

```
map_dbl(trims, function(trim) mean(x, trim = trim))
#> [1] -0.3500  0.0434  0.0354  0.0502
```

- Sometimes, if you want to be (too) clever, you can take advantage of R's flexible argument matching rules (as described in Section 6.8.2). For example, in this example you can rewrite `mean(x, trim = 0.1)` as `mean(0.1, x = x)`, so you could write the call to `map_dbl()` as:

```
map_dbl(trims, mean, x = x)
#> [1] -0.3500  0.0434  0.0354  0.0502
```

I don't recommend this technique as it relies on the reader's familiarity with both the argument order to .f, and R's argument matching rules.

You'll see one more alternative in Section 9.4.5.

9.2.6 Exercises

1. Use `as_mapper()` to explore how purrr generates anonymous functions for the integer, character, and list helpers. What helper allows you to extract attributes? Read the documentation to find out.

2. `map(1:3, ~ runif(2))` is a useful pattern for generating random numbers, but `map(1:3, runif(2))` is not. Why not? Can you explain why it returns the result that it does?

3. Use the appropriate `map()` function to:

   a) Compute the standard deviation of every column in a numeric data frame.

   b) Compute the standard deviation of every numeric column in a mixed data frame. (Hint: you'll need to do it in two steps.)

   c) Compute the number of levels for every factor in a data frame.

4. The following code simulates the performance of a t-test for non-normal data. Extract the p-value from each test, then visualise.

```
trials <- map(1:100, ~ t.test(rpois(10, 10), rpois(7, 10)))
```

5. The following code uses a map nested inside another map to apply a function to every element of a nested list. Why does it fail, and what do you need to do to make it work?

```
x <- list(
  list(1, c(3, 9)),
  list(c(3, 6), 7, c(4, 7, 6))
)

triple <- function(x) x * 3
map(x, map, .f = triple)
#> Error in .f(.x[[i]], ...): unused argument (map)
```

6. Use `map()` to fit linear models to the `mtcars` dataset using the formulas stored in this list:

```
formulas <- list(
  mpg ~ disp,
  mpg ~ I(1 / disp),
  mpg ~ disp + wt,
  mpg ~ I(1 / disp) + wt
)
```

7. Fit the model `mpg ~ disp` to each of the bootstrap replicates of `mtcars` in the list below, then extract the R^2 of the model fit (Hint: you can compute the R^2 with `summary()`.)

```
bootstrap <- function(df) {
  df[sample(nrow(df), replace = TRUE), , drop = FALSE]
}

bootstraps <- map(1:10, ~ bootstrap(mtcars))
```

9.3 Purrr style

Before we go on to explore more map variants, let's take a quick look at how you tend to use multiple purrr functions to solve a moderately realistic problem: fitting a model to each subgroup and extracting a coefficient of the model. For this toy example, I'm going to break the `mtcars` data set down into groups defined by the number of cylinders, using the base `split` function:

```
by_cyl <- split(mtcars, mtcars$cyl)
```

This creates a list of three data frames: the cars with 4, 6, and 8 cylinders respectively.

Now imagine we want to fit a linear model, then extract the second coefficient (i.e. the slope). The following code shows how you might do that with purrr:

```
by_cyl %>%
  map(~ lm(mpg ~ wt, data = .x)) %>%
```

```
  map(coef) %>%
  map_dbl(2)
#>    4     6     8
#> -5.65 -2.78 -2.19
```

(If you haven't seen %>%, the pipe, before, it's described in Section 6.3.)

I think this code is easy to read because each line encapsulates a single step, you can easily distinguish the functional from what it does, and the purrr helpers allow us to very concisely describe what to do in each step.

How would you attack this problem with base R? You certainly *could* replace each purrr function with the equivalent base function:

```
by_cyl %>%
  lapply(function(data) lm(mpg ~ wt, data = data)) %>%
  lapply(coef) %>%
  vapply(function(x) x[[2]], double(1))
#>    4     6     8
#> -5.65 -2.78 -2.19
```

But this isn't really base R since we're using the pipe. To tackle purely in base I think you'd use an intermediate variable, and do more in each step:

```
models <- lapply(by_cyl, function(data) lm(mpg ~ wt, data = data))
vapply(models, function(x) coef(x)[[2]], double(1))
#>    4     6     8
#> -5.65 -2.78 -2.19
```

Or, of course, you could use a for loop:

```
intercepts <- double(length(by_cyl))
for (i in seq_along(by_cyl)) {
  model <- lm(mpg ~ wt, data = by_cyl[[i]])
  intercepts[[i]] <- coef(model)[[2]]
}
intercepts
#> [1] -5.65 -2.78 -2.19
```

It's interesting to note that as you move from purrr to base apply functions to for loops you tend to do more and more in each iteration. In purrr we iterate 3 times (map(), map(), map_dbl()), with apply functions we iterate twice (lapply(), vapply()), and with a for loop we iterate once. I prefer more, but simpler, steps because I think it makes the code easier to understand and later modify.

9.4 Map variants

There are 23 primary variants of map(). So far, you've learned about five (map(), map_lgl(), map_int(), map_dbl() and map_chr()). That means that you've got 18 (!!) more to learn. That sounds like a lot, but fortunately the design of purrr means that you only need to learn five new ideas:

- Output same type as input with modify()
- Iterate over two inputs with map2().
- Iterate with an index using imap()
- Return nothing with walk().
- Iterate over any number of inputs with pmap().

The map family of functions has orthogonal input and outputs, meaning that we can organise all the family into a matrix, with inputs in the rows and outputs in the columns. Once you've mastered the idea in a row, you can combine it with any column; once you've mastered the idea in a column, you can combine it with any row. That relationship is summarised in the following table:

	List	Atomic	Same type	Nothing
One argument	map()	map_lgl(), ...	modify()	walk()
Two arguments	map2()	map2_lgl(), ...	modify2()	walk2()
One argument + index	imap()	imap_lgl(), ...	imodify()	iwalk()
N arguments	pmap()	pmap_lgl(), ...	—	pwalk()

9.4.1 Same type of output as input: modify()

Imagine you wanted to double every column in a data frame. You might first try using map(), but map() always returns a list:

```
df <- data.frame(
  x = 1:3,
  y = 6:4
)

map(df, ~ .x * 2)
#> $x
#> [1] 2 4 6
#>
```

```
#> $y
#> [1] 12 10  8
```

If you want to keep the output as a data frame, you can use `modify()`, which
always returns the same type of output as the input:

```
modify(df, ~ .x * 2)
#>    x  y
#> 1  2 12
#> 2  4 10
#> 3  6  8
```

Despite the name, `modify()` doesn't modify in place, it returns a modified
copy, so if you wanted to permanently modify `df`, you'd need to assign it:

```
df <- modify(df, ~ .x * 2)
```

As usual, the basic implementation of `modify()` is simple, and in fact it's even
simpler than `map()` because we don't need to create a new output vector; we
can just progressively replace the input. (The real code is a little complex to
handle edge cases more gracefully.)

```
simple_modify <- function(x, f, ...) {
  for (i in seq_along(x)) {
    x[[i]] <- f(x[[i]], ...)
  }
  x
}
```

In Section 9.6.2 you'll learn about a very useful variant of `modify()`, called
`modify_if()`. This allows you to (e.g.) only double *numeric* columns of a data
frame with `modify_if(df, is.numeric, ~ .x * 2)`.

9.4.2 Two inputs: `map2()` and friends

`map()` is vectorised over a single argument, `.x`. This means it only varies `.x`
when calling `.f`, and all other arguments are passed along unchanged, thus
making it poorly suited for some problems. For example, how would you find
a weighted mean when you have a list of observations and a list of weights?
Imagine we have the following data:

```
xs <- map(1:8, ~ runif(10))
xs[[1]][[1]] <- NA
ws <- map(1:8, ~ rpois(10, 5) + 1)
```

You can use `map_dbl()` to compute the unweighted means:

```
map_dbl(xs, mean)
#> [1]    NA 0.463 0.551 0.453 0.564 0.501 0.371 0.443
```

But passing `ws` as an additional argument doesn't work because arguments after `.f` are not transformed:

```
map_dbl(xs, weighted.mean, w = ws)
#> Error in weighted.mean.default(.x[[i]], ...): 'x' and 'w' must have
#> the same length
```

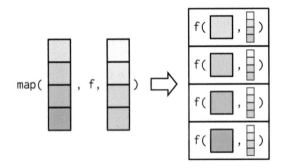

We need a new tool: a `map2()`, which is vectorised over two arguments. This means both `.x` and `.y` are varied in each call to `.f`:

```
map2_dbl(xs, ws, weighted.mean)
#> [1]    NA 0.451 0.603 0.452 0.563 0.510 0.342 0.464
```

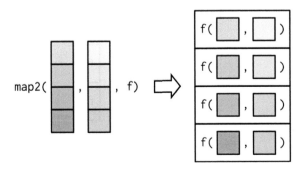

The arguments to `map2()` are slightly different to the arguments to `map()` as two vectors come before the function, rather than one. Additional arguments still go afterwards:

```
map2_dbl(xs, ws, weighted.mean, na.rm = TRUE)
#> [1] 0.504 0.451 0.603 0.452 0.563 0.510 0.342 0.464
```

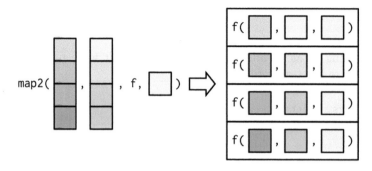

The basic implementation of `map2()` is simple, and quite similar to that of `map()`. Instead of iterating over one vector, we iterate over two in parallel:

```
simple_map2 <- function(x, y, f, ...) {
  out <- vector("list", length(xs))
  for (i in seq_along(x)) {
    out[[i]] <- f(x[[i]], y[[i]], ...)
  }
  out
}
```

One of the big differences between `map2()` and the simple function above is that `map2()` recycles its inputs to make sure that they're the same length:

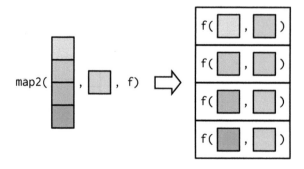

In other words, map2(x, y, f) will automatically behave like map(x, f, y) when needed. This is helpful when writing functions; in scripts you'd generally just use the simpler form directly.

In Base R

The closest base equivalent to map2() is Map(), which is discussed in Section 9.4.5.

9.4.3 No outputs: walk() and friends

Most functions are called for the value that they return, so it makes sense to capture and store the value with a map() function. But some functions are called primarily for their side-effects (e.g. cat(), write.csv(), or ggsave()) and it doesn't make sense to capture their results. Take this simple example that displays a welcome message using cat(). cat() returns NULL, so while map() works (in the sense that it generates the desired welcomes), it also returns list(NULL, NULL).

```r
welcome <- function(x) {
  cat("Welcome ", x, "!\n", sep = "")
}
names <- c("Hadley", "Jenny")

# As well as generate the welcomes, it also shows
# the return value of cat()
map(names, welcome)
#> Welcome Hadley!
#> Welcome Jenny!
#> [[1]]
#> NULL
#>
#> [[2]]
#> NULL
```

You could avoid this problem by assigning the results of map() to a variable that you never use, but that would muddy the intent of the code. Instead, purrr provides the walk family of functions that ignore the return values of the .f and instead return .x invisibly[4].

---

[4]In brief, invisible values are only printed if you explicitly request it. This makes them well suited for functions called primarily for their side-effects, as it allows their output to be ignored by default, while still giving an option to capture it. See Section 6.7.2 for more details.

```
walk(names, welcome)
#> Welcome Hadley!
#> Welcome Jenny!
```

My visual depiction of walk attempts to capture the important difference from map(): the outputs are ephemeral, and the input is returned invisibly.

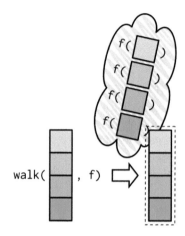

One of the most useful walk() variants is walk2() because a very common side-effect is saving something to disk, and when saving something to disk you always have a pair of values: the object and the path that you want to save it to.

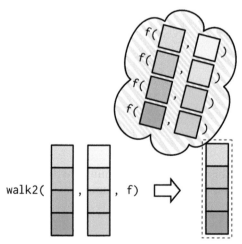

For example, imagine you have a list of data frames (which I've created here using split()), and you'd like to save each one to a separate CSV file. That's easy with walk2():

```
temp <- tempfile()
dir.create(temp)

cyls <- split(mtcars, mtcars$cyl)
paths <- file.path(temp, paste0("cyl-", names(cyls), ".csv"))
walk2(cyls, paths, write.csv)

dir(temp)
#> [1] "cyl-4.csv" "cyl-6.csv" "cyl-8.csv"
```

Here the walk2() is equivalent to write.csv(cyls[[1]], paths[[1]]),
write.csv(cyls[[2]], paths[[2]]), write.csv(cyls[[3]], paths[[3]]).

In Base R

There is no base equivalent to walk(); either wrap the result of lapply() in
invisible() or save it to a variable that is never used.

9.4.4 Iterating over values and indices

There are three basic ways to loop over a vector with a for loop:

- Loop over the elements: for (x in xs)
- Loop over the numeric indices: for (i in seq_along(xs))
- Loop over the names: for (nm in names(xs))

The first form is analogous to the map() family. The second and third forms
are equivalent to the imap() family which allows you to iterate over the values
and the indices of a vector in parallel.

imap() is like map2() in the sense that your .f gets called with two arguments,
but here both are derived from the vector. imap(x, f) is equivalent to map2(x,
names(x), f) if x has names, and map2(x, seq_along(x), f) if it does not.

imap() is often useful for constructing labels:

```
imap_chr(iris, ~ paste0("The first value of ", .y, " is ", .x[[1]]))
#>                       Sepal.Length
#> "The first value of Sepal.Length is 5.1"
#>                       Sepal.Width
#>   "The first value of Sepal.Width is 3.5"
#>                       Petal.Length
#> "The first value of Petal.Length is 1.4"
#>                       Petal.Width
```

```
#>   "The first value of Petal.Width is 0.2"
#>                                       Species
#>   "The first value of Species is setosa"
```

If the vector is unnamed, the second argument will be the index:

```
x <- map(1:6, ~ sample(1000, 10))
imap_chr(x, ~ paste0("The highest value of ", .y, " is ", max(.x)))
#> [1] "The highest value of 1 is 885" "The highest value of 2 is 808"
#> [3] "The highest value of 3 is 942" "The highest value of 4 is 966"
#> [5] "The highest value of 5 is 857" "The highest value of 6 is 671"
```

imap() is a useful helper if you want to work with the values in a vector along with their positions.

9.4.5   Any number of inputs: pmap() and friends

Since we have map() and map2(), you might expect map3(), map4(), map5(), ... But where would you stop? Instead of generalising map2() to an arbitrary number of arguments, purrr takes a slightly different tack with pmap(): you supply it a single list, which contains any number of arguments. In most cases, that will be a list of equal-length vectors, i.e. something very similar to a data frame. In diagrams, I'll emphasise that relationship by drawing the input similar to a data frame.

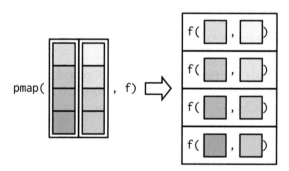

There's a simple equivalence between map2() and pmap(): map2(x, y, f) is the same as pmap(list(x, y), f). The pmap() equivalent to the map2_dbl(xs, ws, weighted.mean) used above is:

```
pmap_dbl(list(xs, ws), weighted.mean)
#> [1]    NA 0.451 0.603 0.452 0.563 0.510 0.342 0.464
```

As before, the varying arguments come before `.f` (although now they must be wrapped in a list), and the constant arguments come afterwards.

```
pmap_dbl(list(xs, ws), weighted.mean, na.rm = TRUE)
#> [1] 0.504 0.451 0.603 0.452 0.563 0.510 0.342 0.464
```

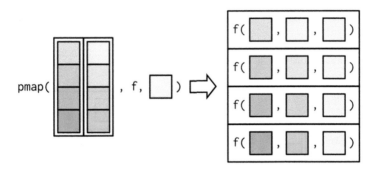

A big difference between `pmap()` and the other map functions is that `pmap()` gives you much finer control over argument matching because you can name the components of the list. Returning to our example from Section 9.2.5, where we wanted to vary the `trim` argument to x, we could instead use `pmap()`:

```
trims <- c(0, 0.1, 0.2, 0.5)
x <- rcauchy(1000)

pmap_dbl(list(trim = trims), mean, x = x)
#> [1] -6.6754  0.0192  0.0228  0.0151
```

I think it's good practice to name the components of the list to make it very clear how the function will be called.

It's often convenient to call `pmap()` with a data frame. A handy way to create that data frame is with `tibble::tribble()`, which allows you to describe a data frame row-by-row (rather than column-by-column, as usual): thinking about the parameters to a function as a data frame is a very powerful pattern. The following example shows how you might draw random uniform numbers with varying parameters:

```
params <- tibble::tribble(
  ~ n, ~ min, ~ max,
   1L,    0,     1,
   2L,   10,   100,
   3L,  100,  1000
)
```

```
pmap(params, runif)
#> [[1]]
#> [1] 0.718
#>
#> [[2]]
#> [1] 19.5 39.9
#>
#> [[3]]
#> [1] 535 476 231
```

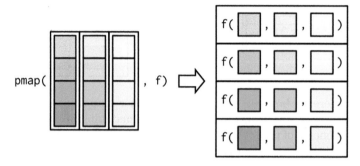

Here, the column names are critical: I've carefully chosen to match them to the arguments to runif(), so the pmap(params, runif) is equivalent to runif(n = 1L, min = 0, max = 1), runif(n = 2, min = 10, max = 100), runif(n = 3L, min = 100, max = 1000). (If you have a data frame in hand, and the names don't match, use dplyr::rename() or similar.)

In Base R

There are two base equivalents to the pmap() family: Map() and mapply(). Both have significant drawbacks:

- Map() vectorises over all arguments so you cannot supply arguments that do not vary.
- mapply() is the multidimensional version of sapply(); conceptually it takes the output of Map() and simplifies it if possible. This gives it similar issues to sapply(). There is no multi-input equivalent of vapply().

9.4.6   Exercises

1. Explain the results of modify(mtcars, 1).

2. Rewrite the following code to use `iwalk()` instead of `walk2()`. What are the advantages and disadvantages?

```
cyls <- split(mtcars, mtcars$cyl)
paths <- file.path(temp, paste0("cyl-", names(cyls), ".csv"))
walk2(cyls, paths, write.csv)
```

3. Explain how the following code transforms a data frame using functions stored in a list.

```
trans <- list(
  disp = function(x) x * 0.0163871,
  am = function(x) factor(x, labels = c("auto", "manual"))
)

nm <- names(trans)
mtcars[nm] <- map2(trans, mtcars[nm], function(f, var) f(var))
```

Compare and contrast the `map2()` approach to this `map()` approach:

```
mtcars[vars] <- map(vars, ~ trans[[.x]](mtcars[[.x]]))
```

4. What does `write.csv()` return? i.e. what happens if you use it with `map2()` instead of `walk2()`?

9.5 Reduce family

After the map family, the next most important family of functions is the reduce family. This family is much smaller, with only two main variants, and is used less commonly, but it's a powerful idea, gives us the opportunity to discuss some useful algebra, and powers the map-reduce framework frequently used for processing very large datasets.

9.5.1   Basics

reduce() takes a vector of length n and produces a vector of length 1 by calling a function with a pair of values at a time: reduce(1:4, f) is equivalent to f(f(f(1, 2), 3), 4).

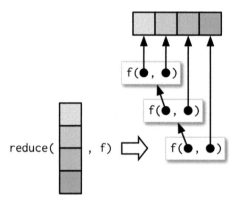

reduce() is a useful way to generalise a function that works with two inputs (a **binary** function) to work with any number of inputs. Imagine you have a list of numeric vectors, and you want to find the values that occur in every element. First we generate some sample data:

```
l <- map(1:4, ~ sample(1:10, 15, replace = T))
str(l)
#> List of 4
#>  $ : int [1:15] 7 5 9 7 9 9 5 10 5 5 ...
#>  $ : int [1:15] 6 3 6 10 3 4 4 2 9 9 ...
#>  $ : int [1:15] 5 3 4 6 1 1 9 9 6 8 ...
#>  $ : int [1:15] 4 2 6 6 8 5 10 6 7 1 ...
```

To solve this challenge we need to use intersect() repeatedly:

```
out <- l[[1]]
out <- intersect(out, l[[2]])
out <- intersect(out, l[[3]])
out <- intersect(out, l[[4]])
out
#> [1] 5 1
```

reduce() automates this solution for us, so we can write:

```
reduce(l, intersect)
#> [1] 5 1
```

We could apply the same idea if we wanted to list all the elements that appear in at least one entry. All we have to do is switch from `intersect()` to `union()`:

```
reduce(l, union)
#>  [1]  7  5  9 10  1  6  3  4  2  8
```

Like the map family, you can also pass additional arguments. `intersect()` and `union()` don't take extra arguments so I can't demonstrate them here, but the principle is straightforward and I drew you a picture.

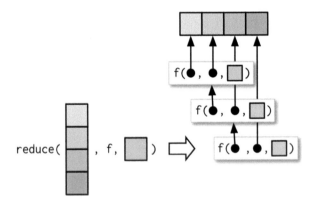

As usual, the essence of `reduce()` can be reduced to a simple wrapper around a for loop:

```
simple_reduce <- function(x, f) {
  out <- x[[1]]
  for (i in seq(2, length(x))) {
    out <- f(out, x[[i]])
  }
  out
}
```

In Base R

The base equivalent is `Reduce()`. Note that the argument order is different: the function comes first, followed by the vector, and there is no way to supply additional arguments.

9.5.2 Accumulate

The first `reduce()` variant, `accumulate()`, is useful for understanding how reduce works, because instead of returning just the final result, it returns all the intermediate results as well:

```
accumulate(l, intersect)
#> [[1]]
#>  [1]  7  5  9  7  9  9  5 10  5  5  5 10  9  9  1
#>
#> [[2]]
#> [1]  5  9 10  1
#>
#> [[3]]
#> [1] 5 9 1
#>
#> [[4]]
#> [1] 5 1
```

Another useful way to understand reduce is to think about `sum()`: `sum(x)` is equivalent to `x[[1]] + x[[2]] + x[[3]] + ...`, i.e. `reduce(x, `+`)`. Then `accumulate(x, `+`)` is the cumulative sum:

```
x <- c(4, 3, 10)
reduce(x, `+`)
#> [1] 17

accumulate(x, `+`)
#> [1]  4  7 17
```

9.5.3 Output types

In the above example using +, what should `reduce()` return when x is short, i.e. length 1 or 0? Without additional arguments, `reduce()` just returns the input when x is length 1:

```
reduce(1, `+`)
#> [1] 1
```

This means that `reduce()` has no way to check that the input is valid:

```
reduce("a", `+`)
#> [1] "a"
```

What if it's length 0? We get an error that suggests we need to use the `.init`
argument:

```
reduce(integer(), `+`)
#> Error: `.x` is empty, and no `.init` supplied
```

What should `.init` be here? To figure that out, we need to see what happens
when `.init` is supplied:

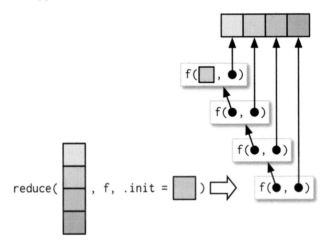

So if we call `reduce(1, `+`, init)` the result will be 1 + init. Now we know
that the result should be just 1, so that suggests that `.init` should be 0:

```
reduce(integer(), `+`, .init = 0)
#> [1] 0
```

This also ensures that `reduce()` checks that length 1 inputs are valid for the
function that you're calling:

```
reduce("a", `+`, .init = 0)
#> Error in .x + .y: non-numeric argument to binary operator
```

If you want to get algebraic about it, 0 is called the **identity** of the real
numbers under the operation of addition: if you add a 0 to any number, you
get the same number back. R applies the same principle to determine what a
summary function with a zero length input should return:

```
sum(integer())  # x + 0 = x
#> [1] 0
prod(integer()) # x * 1 = x
#> [1] 1
min(integer())  # min(x, Inf) = x
#> [1] Inf
max(integer())  # max(x, -Inf) = x
#> [1] -Inf
```

If you're using reduce() in a function, you should always supply .init. Think carefully about what your function should return when you pass a vector of length 0 or 1, and make sure to test your implementation.

9.5.4 Multiple inputs

Very occasionally you need to pass two arguments to the function that you're reducing. For example, you might have a list of data frames that you want to join together, and the variables you use to join will vary from element to element. This is a very specialised scenario, so I don't want to spend much time on it, but I do want you to know that reduce2() exists.

The length of the second argument varies based on whether or not .init is supplied: if you have four elements of x, f will only be called three times. If you supply init, f will be called four times.

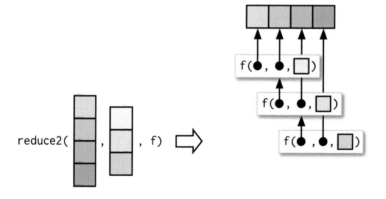

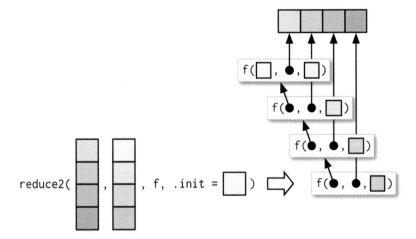

9.5.5 Map-reduce

You might have heard of map-reduce, the idea that powers technology like Hadoop. Now you can see how simple and powerful the underlying idea is: map-reduce is a map combined with a reduce. The difference for large data is that the data is spread over multiple computers. Each computer performs the map on the data that it has, then it sends the result to back to a coordinator which *reduces* the individual results back to a single result.

As a simple example, imagine computing the mean of a very large vector, so large that it has to be split over multiple computers. You could ask each computer to calculate the sum and the length, and then return those to the coordinator which computes the overall mean by dividing the total sum by the total length.

9.6 Predicate functionals

A **predicate** is a function that returns a single TRUE or FALSE, like is.character(), is.null(), or all(), and we say a predicate **matches** a vector if it returns TRUE.

9.6.1   Basics

A **predicate functional** applies a predicate to each element of a vector. purrr provides six useful functions which come in three pairs:

- `some(.x, .p)` returns `TRUE` if *any* element matches; `every(.x, .p)` returns `TRUE` if *all* elements match.

  These are similar to `any(map_lgl(.x, .p))` and `all(map_lgl(.x, .p))` but they terminate early: `some()` returns `TRUE` when it sees the first `TRUE`, and `every()` returns `FALSE` when it sees the first `FALSE`.

- `detect(.x, .p)` returns the *value* of the first match; `detect_index(.x, .p)` returns the *location* of the first match.

- `keep(.x, .p)` *keeps* all matching elements; `discard(.x, .p)` *drops* all matching elements.

The following example shows how you might use these functionals with a data frame:

```
df <- data.frame(x = 1:3, y = c("a", "b", "c"))
detect(df, is.factor)
#> [1] a b c
#> Levels: a b c
detect_index(df, is.factor)
#> [1] 2

str(keep(df, is.factor))
#> 'data.frame':    3 obs. of  1 variable:
#>  $ y: Factor w/ 3 levels "a","b","c": 1 2 3
str(discard(df, is.factor))
#> 'data.frame':    3 obs. of  1 variable:
#>  $ x: int  1 2 3
```

9.6.2   Map variants

`map()` and `modify()` come in variants that also take predicate functions, transforming only the elements of `.x` where `.p` is `TRUE`.

```
df <- data.frame(
  num1 = c(0, 10, 20),
  num2 = c(5, 6, 7),
  chr1 = c("a", "b", "c"),
  stringsAsFactors = FALSE
```

```
)

str(map_if(df, is.numeric, mean))
#> List of 3
#>  $ num1: num 10
#>  $ num2: num 6
#>  $ chr1: chr [1:3] "a" "b" "c"
str(modify_if(df, is.numeric, mean))
#> 'data.frame':    3 obs. of  3 variables:
#>  $ num1: num  10 10 10
#>  $ num2: num  6 6 6
#>  $ chr1: chr  "a" "b" "c"
str(map(keep(df, is.numeric), mean))
#> List of 2
#>  $ num1: num 10
#>  $ num2: num 6
```

9.6.3 Exercises

1. Why isn't is.na() a predicate function? What base R function is closest to being a predicate version of is.na()?

2. simple_reduce() has a problem when x is length 0 or length 1. Describe the source of the problem and how you might go about fixing it.

```
simple_reduce <- function(x, f) {
  out <- x[[1]]
  for (i in seq(2, length(x))) {
    out <- f(out, x[[i]])
  }
  out
}
```

3. Implement the span() function from Haskell: given a list x and a predicate function f, span(x, f) returns the location of the longest sequential run of elements where the predicate is true. (Hint: you might find rle() helpful.)

4. Implement arg_max(). It should take a function and a vector of inputs, and return the elements of the input where the function returns the highest value. For example, arg_max(-10:5, function(x)

x ^ 2) should return -10. arg_max(-5:5, function(x) x ^ 2) should
return c(-5, 5). Also implement the matching arg_min() function.

5. The function below scales a vector so it falls in the range $[0, 1]$. How
would you apply it to every column of a data frame? How would
you apply it to every numeric column in a data frame?

```r
scale01 <- function(x) {
  rng <- range(x, na.rm = TRUE)
  (x - rng[1]) / (rng[2] - rng[1])
}
```

9.7   Base functionals

To finish up the chapter, here I provide a survey of important base functionals
that are not members of the map, reduce, or predicate families, and hence have
no equivalent in purrr. This is not to say that they're not important, but they
have more of a mathematical or statistical flavour, and they are generally less
useful in data analysis.

9.7.1   Matrices and arrays

map() and friends are specialised to work with one-dimensional vectors.
base::apply() is specialised to work with two-dimensional and higher vectors, i.e. matrices and arrays. You can think of apply() as an operation that
summarises a matrix or array by collapsing each row or column to a single
value. It has four arguments:

- X, the matrix or array to summarise.

- MARGIN, an integer vector giving the dimensions to summarise over, 1 =
  rows, 2 = columns, etc. (The argument name comes from thinking about
  the margins of a joint distribution.)

- FUN, a summary function.

- ... other arguments passed on to FUN.

A typical example of apply() looks like this

```
a2d <- matrix(1:20, nrow = 5)
apply(a2d, 1, mean)
#> [1]  8.5  9.5 10.5 11.5 12.5
apply(a2d, 2, mean)
#> [1]  3  8 13 18
```

You can specify multiple dimensions to MARGIN, which is useful for high-dimensional arrays:

```
a3d <- array(1:24, c(2, 3, 4))
apply(a3d, 1, mean)
#> [1] 12 13
apply(a3d, c(1, 2), mean)
#>      [,1] [,2] [,3]
#> [1,]   10   12   14
#> [2,]   11   13   15
```

There are two caveats to using apply():

- Like base::sapply(), you have no control over the output type; it will automatically be simplified to a list, matrix, or vector. However, you usually use apply() with numeric arrays and a numeric summary function so you are less likely to encounter a problem than with sapply().

- apply() is also not idempotent in the sense that if the summary function is the identity operator, the output is not always the same as the input.

```
a1 <- apply(a2d, 1, identity)
identical(a2d, a1)
#> [1] FALSE

a2 <- apply(a2d, 2, identity)
identical(a2d, a2)
#> [1] TRUE
```

- Never use apply() with a data frame. It always coerces it to a matrix, which will lead to undesirable results if your data frame contains anything other than numbers.

```
df <- data.frame(x = 1:3, y = c("a", "b", "c"))
apply(df, 2, mean)
#> Warning in mean.default(newX[, i], ...): argument is not numeric or
#> logical: returning NA
```

```
#> Warning in mean.default(newX[, i], ...): argument is not numeric or
#> logical: returning NA
#>   x  y
#> NA NA
```

9.7.2 Mathematical concerns

Functionals are very common in mathematics. The limit, the maximum, the roots (the set of points where f(x) = 0), and the definite integral are all functionals: given a function, they return a single number (or vector of numbers). At first glance, these functions don't seem to fit in with the theme of eliminating loops, but if you dig deeper you'll find out that they are all implemented using an algorithm that involves iteration.

Base R provides a useful set:

- integrate() finds the area under the curve defined by f()
- uniroot() finds where f() hits zero
- optimise() finds the location of the lowest (or highest) value of f()

The following example shows how functionals might be used with a simple function, sin():

```
integrate(sin, 0, pi)
#> 2 with absolute error < 2.2e-14
str(uniroot(sin, pi * c(1 / 2, 3 / 2)))
#> List of 5
#>  $ root     : num 3.14
#>  $ f.root   : num 1.22e-16
#>  $ iter     : int 2
#>  $ init.it  : int NA
#>  $ estim.prec: num 6.1e-05
str(optimise(sin, c(0, 2 * pi)))
#> List of 2
#>  $ minimum  : num 4.71
#>  $ objective: num -1
str(optimise(sin, c(0, pi), maximum = TRUE))
#> List of 2
#>  $ maximum  : num 1.57
#>  $ objective: num 1
```

9.7.3 Exercises

1. How does `apply()` arrange the output? Read the documentation and perform some experiments.

2. What do `eapply()` and `rapply()` do? Does purrr have equivalents?

3. Challenge: read about the fixed point algorithm (`https://mitpress.mit.edu/sites/default/files/sicp/full-text/book/book-Z-H-12.html#%25_idx_1096`). Complete the exercises using R.

10

Function factories

10.1 Introduction

A **function factory** is a function that makes functions. Here's a very simple example: we use a function factory (power1()) to make two child functions (square() and cube()):

```r
power1 <- function(exp) {
  function(x) {
    x ^ exp
  }
}

square <- power1(2)
cube <- power1(3)
```

Don't worry if this doesn't make sense yet, it should by the end of the chapter!

I'll call square() and cube() **manufactured functions**, but this is just a term to ease communication with other humans: from R's perspective they are no different to functions created any other way.

```r
square(3)
#> [1] 9
cube(3)
#> [1] 27
```

You have already learned about the individual components that make function factories possible:

- In Section 6.2.3, you learned about R's first-class functions. In R, you bind a function to a name in the same way as you bind any object to a name: with <-.

- In Section 7.4.2, you learned that a function captures (encloses) the environment in which it is created.

- In Section 7.4.4, you learned that a function creates a new execution environment every time it is run. This environment is usually ephemeral, but here it becomes the enclosing environment of the manufactured function.

In this chapter, you'll learn how the non-obvious combination of these three features leads to the function factory. You'll also see examples of their usage in visualisation and statistics.

Of the three main functional programming tools (functionals, function factories, and function operators), function factories are the least used. Generally, they don't tend to reduce overall code complexity but instead partition complexity into more easily digested chunks. Function factories are also an important building block for the very useful function operators, which you'll learn about in Chapter 11.

Outline

- Section 10.2 begins the chapter with an explanation of how function factories work, pulling together ideas from scoping and environments. You'll also see how function factories can be used to implement a memory for functions, allowing data to persist across calls.

- Section 10.3 illustrates the use of function factories with examples from ggplot2. You'll see two examples of how ggplot2 works with user supplied function factories, and one example of where ggplot2 uses a function factory internally.

- Section 10.4 uses function factories to tackle three challenges from statistics: understanding the Box-Cox transform, solving maximum likelihood problems, and drawing bootstrap resamples.

- Section 10.5 shows how you can combine function factories and functionals to rapidly generate a family of functions from data.

Prerequisites

Make sure you're familiar with the contents of Sections 6.2.3 (first-class functions), 7.4.2 (the function environment), and 7.4.4 (execution environments) mentioned above.

Function factories only need base R. We'll use a little rlang (https://rlang.r-lib.org) to peek inside of them more easily, and we'll use ggplot2 (https://ggplot2.tidyverse.org) and scales (https://scales.r-lib.org) to explore the use of function factories in visualisation.

```
library(rlang)
library(ggplot2)
library(scales)
```

10.2 Factory fundamentals

The key idea that makes function factories work can be expressed very concisely:

> *The enclosing environment of the manufactured function is*
> *an execution environment of the function factory.*

It only takes few words to express these big ideas, but it takes a lot more work to really understand what this means. This section will help you put the pieces together with interactive exploration and some diagrams.

10.2.1 Environments

Let's start by taking a look at square() and cube():

```
square
#> function(x) {
#>     x ^ exp
#>   }
#> <environment: 0x7fa43cc11b00>

cube
#> function(x) {
#>     x ^ exp
#>   }
#> <bytecode: 0x7fa43d0e1740>
#> <environment: 0x7fa439297d50>
```

It's obvious where x comes from, but how does R find the value associated with exp? Simply printing the manufactured functions is not revealing because the bodies are identical; the contents of the enclosing environment are the important factors. We can get a little more insight by using rlang::env_print(). That shows us that we have two different environments (each of which was originally an execution environment of power1()). The environments have the same parent, which is the enclosing environment of power1(), the global environment.

```
env_print(square)
#> <environment: 0x7fa43cc11b00>
#> parent: <environment: global>
#> bindings:
#>  * exp: <dbl>
```

```
env_print(cube)
#> <environment: 0x7fa439297d50>
#> parent: <environment: global>
#> bindings:
#>  * exp: <dbl>
```

env_print() shows us that both environments have a binding to exp, but we want to see its value[1]. We can do that by first getting the environment of the function, and then extracting the values:

```
fn_env(square)$exp
#> [1] 2
```

```
fn_env(cube)$exp
#> [1] 3
```

This is what makes manufactured functions behave differently from one another: names in the enclosing environment are bound to different values.

10.2.2 Diagram conventions

We can also show these relationships in a diagram:

---

[1]A future version of env_print() is likely to do better at summarising the contents so you don't need this step.

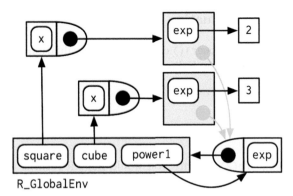

There's a lot going on this diagram and some of the details aren't that important. We can simplify considerably by using two conventions:

- Any free floating symbol lives in the global environment.

- Any environment without an explicit parent inherits from the global environment.

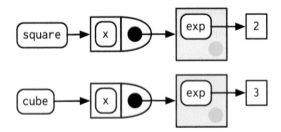

This view, which focuses on the environments, doesn't show any direct link between cube() and square(). That's because the link is the through the body of the function, which is identical for both, but is not shown in this diagram.

To finish up, let's look at the execution environment of square(10). When square() executes x ^ exp it finds x in the execution environment and exp in its enclosing environment.

```
square(10)
#> [1] 100
```

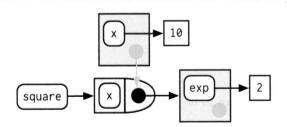

10.2.3   Forcing evaluation

There's a subtle bug in power1() caused by lazy evaluation. To see the problem
we need to introduce some indirection:

```
x <- 2
square <- power1(x)
x <- 3
```

What should square(2) return? You would hope it returns 4:

```
square(2)
#> [1] 8
```

Unfortunately it doesn't because x is only evaluated lazily when square() is
run, not when power1() is run. In general, this problem will arise whenever
a binding changes in between calling the factory function and calling the
manufactured function. This is likely to only happen rarely, but when it does,
it will lead to a real head-scratcher of a bug.

We can fix this problem by **forcing** evaluation with force():

```
power2 <- function(exp) {
  force(exp)
  function(x) {
    x ^ exp
  }
}

x <- 2
square <- power2(x)
x <- 3
square(2)
#> [1] 4
```

Whenever you create a function factory, make sure every argument is evaluated, using force() as necessary if the argument is only used by the manufactured function.

10.2.4   Stateful functions

Function factories also allow you to maintain state across function invocations,
which is generally hard to do because of the fresh start principle described in
Section 6.4.3.

There are two things that make this possible:

- The enclosing environment of the manufactured function is unique and constant.

- R has a special assignment operator, <<-, which modifies bindings in the enclosing environment.

The usual assignment operator, <-, always creates a binding in the current environment. The **super assignment operator**, <<- rebinds an existing name found in a parent environment.

The following example shows how we can combine these ideas to create a function that records how many times it has been called:

```
new_counter <- function() {
  i <- 0

  function() {
    i <<- i + 1
    i
  }
}

counter_one <- new_counter()
counter_two <- new_counter()
```

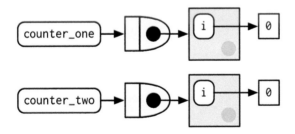

When the manufactured function is run i <<- i + 1 will modify i in its enclosing environment. Because manufactured functions have independent enclosing environments, they have independent counts:

```
counter_one()
#> [1] 1
counter_one()
#> [1] 2
counter_two()
#> [1] 1
```

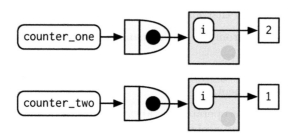

Stateful functions are best used in moderation. As soon as your function starts managing the state of multiple variables, it's better to switch to R6, the topic of Chapter 14.

10.2.5   Garbage collection

With most functions, you can rely on the garbage collector to clean up any large temporary objects created inside a function. However, manufactured functions hold on to the execution environment, so you'll need to explicitly unbind any large temporary objects with rm(). Compare the sizes of g1() and g2() in the example below:

```
f1 <- function(n) {
  x <- runif(n)
  m <- mean(x)
  function() m
}

g1 <- f1(1e6)
lobstr::obj_size(g1)
#> 8,013,120 B

f2 <- function(n) {
  x <- runif(n)
  m <- mean(x)
  rm(x)
  function() m
}

g2 <- f2(1e6)
lobstr::obj_size(g2)
#> 12,960 B
```

10.2.6   Exercises

1. The definition of `force()` is simple:

    ```
    force
    #> function (x)
    #> x
    #> <bytecode: 0x7fa4381626a0>
    #> <environment: namespace:base>
    ```

    Why is it better to `force(x)` instead of just x?

2. Base R contains two function factories, `approxfun()` and `ecdf()`.
    Read their documentation and experiment to figure out what the
    functions do and what they return.

3. Create a function `pick()` that takes an index, i, as an argument
    and returns a function with an argument x that subsets x with i.

    ```
    pick(1)(x)
    # should be equivalent to
    x[[1]]

    lapply(mtcars, pick(5))
    # should be equivalent to
    lapply(mtcars, function(x) x[[5]])
    ```

4. Create a function that creates functions that compute the ith cen-
    tral moment (http://en.wikipedia.org/wiki/Central_moment) of a
    numeric vector. You can test it by running the following code:

    ```
    m1 <- moment(1)
    m2 <- moment(2)

    x <- runif(100)
    stopifnot(all.equal(m1(x), 0))
    stopifnot(all.equal(m2(x), var(x) * 99 / 100))
    ```

5. What happens if you don't use a closure? Make predictions, then
    verify with the code below.

```
i <- 0
new_counter2 <- function() {
  i <<- i + 1
  i
}
```

6. What happens if you use <- instead of <<-? Make predictions, then verify with the code below.

```
new_counter3 <- function() {
  i <- 0
  function() {
    i <- i + 1
    i
  }
}
```

10.3 Graphical factories

We'll begin our exploration of useful function factories with a few examples from ggplot2.

10.3.1 Labelling

One of the goals of the scales (http://scales.r-lib.org) package is to make it easy to customise the labels on ggplot2. It provides many functions to control the fine details of axes and legends. The formatter functions[2] are a useful class of functions which make it easier to control the appearance of axis breaks. The design of these functions might initially seem a little odd: they all return a function, which you have to call in order to format a number.

```
y <- c(12345, 123456, 1234567)
comma_format()(y)
```

---

[2]It's an unfortunate accident of history that scales uses function suffixes instead of function prefixes. That's because it was written before I understood the autocomplete advantages to using common prefixes instead of common suffixes.

```
#> [1] "12,345"    "123,456"    "1,234,567"

number_format(scale = 1e-3, suffix = " K")(y)
#> [1] "12 K"     "123 K"    "1 235 K"
```

In other words, the primary interface is a function factory. At first glance, this
seems to add extra complexity for little gain. But it enables a nice interaction
with ggplot2's scales, because they accept functions in the label argument:

```
df <- data.frame(x = 1, y = y)
core <- ggplot(df, aes(x, y)) +
  geom_point() +
  scale_x_continuous(breaks = 1, labels = NULL) +
  labs(x = NULL, y = NULL)

core
core + scale_y_continuous(
  labels = comma_format()
)
core + scale_y_continuous(
  labels = number_format(scale = 1e-3, suffix = " K")
)
core + scale_y_continuous(
  labels = scientific_format()
)
```

10.3.2 Histogram bins

A little known feature of geom_histogram() is that the binwidth argument
can be a function. This is particularly useful because the function is executed
once for each group, which means you can have different binwidths in different
facets, which is otherwise not possible.

To illustrate this idea, and see where variable binwidth might be useful, I'm
going to construct an example where a fixed binwidth isn't great.

```
# construct some sample data with very different numbers in each cell
sd <- c(1, 5, 15)
n <- 100

df <- data.frame(x = rnorm(3 * n, sd = sd), sd = rep(sd, n))

ggplot(df, aes(x)) +
  geom_histogram(binwidth = 2) +
  facet_wrap(~ sd, scales = "free_x") +
  labs(x = NULL)
```

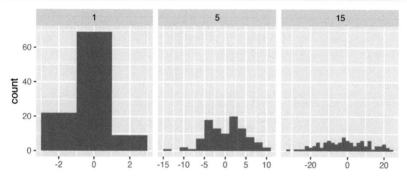

Here each facet has the same number of observations, but the variability is very different. It would be nice if we could request that the binwidths vary so we get approximately the same number of observations in each bin. One way to do that is with a function factory that inputs the desired number of bins (n), and outputs a function that takes a numeric vector and returns a binwidth:

```
binwidth_bins <- function(n) {
  force(n)

  function(x) {
    (max(x) - min(x)) / n
  }
}

ggplot(df, aes(x)) +
  geom_histogram(binwidth = binwidth_bins(20)) +
  facet_wrap(~ sd, scales = "free_x") +
  labs(x = NULL)
```

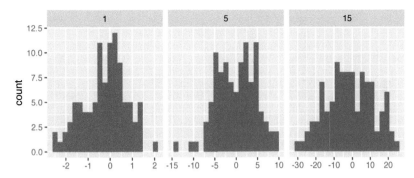

We could use this same pattern to wrap around the base R functions that automatically find the so-called optimal[3] binwidth, `nclass.Sturges()`, `nclass.scott()`, and `nclass.FD()`:

```
base_bins <- function(type) {
  fun <- switch(type,
    Sturges = nclass.Sturges,
    scott = nclass.scott,
    FD = nclass.FD,
    stop("Unknown type", call. = FALSE)
  )

  function(x) {
    (max(x) - min(x)) / fun(x)
  }
}

ggplot(df, aes(x)) +
  geom_histogram(binwidth = base_bins("FD")) +
  facet_wrap(~ sd, scales = "free_x") +
  labs(x = NULL)
```

---

[3]ggplot2 doesn't expose these functions directly because I don't think the definition of optimality needed to make the problem mathematically tractable is a good match to the actual needs of data exploration.

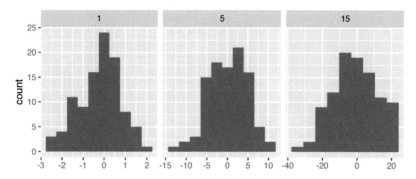

10.3.3 `ggsave()`

Finally, I want to show a function factory used internally by ggplot2. gg-plot2:::plot_dev() is used by `ggsave()` to go from a file extension (e.g. png, jpeg etc) to a graphics device function (e.g. `png()`, `jpeg()`). The challenge here arises because the base graphics devices have some minor inconsistencies which we need to paper over:

- Most have `filename` as first argument but some have `file`.

- The `width` and `height` of raster graphic devices use pixels units by default, but the vector graphics use inches.

A mildly simplified version of `plot_dev()` is shown below:

```
plot_dev <- function(ext, dpi = 96) {
  force(dpi)

  switch(ext,
    eps =  ,
    ps  =  function(path, ...) {
      grDevices::postscript(
        file = filename, ..., onefile = FALSE,
        horizontal = FALSE, paper = "special"
      )
    },
    pdf = function(filename, ...) grDevices::pdf(file = filename, ...),
    svg = function(filename, ...) svglite::svglite(file = filename, ...),
    emf =  ,
    wmf = function(...) grDevices::win.metafile(...),
    png = function(...) grDevices::png(..., res = dpi, units = "in"),
    jpg =  ,
    jpeg = function(...) grDevices::jpeg(..., res = dpi, units = "in"),
    bmp = function(...) grDevices::bmp(..., res = dpi, units = "in"),
```

```
    tiff = function(...) grDevices::tiff(..., res = dpi, units = "in"),
    stop("Unknown graphics extension: ", ext, call. = FALSE)
  )
}

plot_dev("pdf")
#> function(filename, ...) grDevices::pdf(file = filename, ...)
#> <bytecode: 0x7fa781166af8>
#> <environment: 0x7fa78049e698>
plot_dev("png")
#> function(...) grDevices::png(..., res = dpi, units = "in")
#> <bytecode: 0x7fa7812d2708>
#> <environment: 0x7fa780a65a40>
```

10.3.4 Exercises

1. Compare and contrast `ggplot2::label_bquote()` with `scales::number_format()`

10.4 Statistical factories

More motivating examples for function factories come from statistics:

- The Box-Cox transformation.
- Bootstrap resampling.
- Maximum likelihood estimation.

All of these examples can be tackled without function factories, but I think function factories are a good fit for these problems and provide elegant solutions. These examples expect some statistical background, so feel free to skip if they don't make much sense to you.

10.4.1 Box-Cox transformation

The Box-Cox transformation (a type of power transformation (https://en.wikipedia.org/wiki/Power_transform)) is a flexible transformation often used to transform data towards normality. It has a single parameter, λ, which controls the strength of the transformation. We could express the transformation as a simple two argument function:

```
boxcox1 <- function(x, lambda) {
  stopifnot(length(lambda) == 1)

  if (lambda == 0) {
    log(x)
  } else {
    (x ^ lambda - 1) / lambda
  }
}
```

But re-formulating as a function factory makes it easy to explore its behaviour with stat_function():

```
boxcox2 <- function(lambda) {
  if (lambda == 0) {
    function(x) log(x)
  } else {
    function(x) (x ^ lambda - 1) / lambda
  }
}
```

```
stat_boxcox <- function(lambda) {
  stat_function(aes(colour = lambda), fun = boxcox2(lambda), size = 1)
}
```

```
ggplot(data.frame(x = c(0, 5)), aes(x)) +
  lapply(c(0.5, 1, 1.5), stat_boxcox) +
  scale_colour_viridis_c(limits = c(0, 1.5))

# visually, log() does seem to make sense as the transformation
# for lambda = 0; as values get smaller and smaller, the function
# gets close and closer to a log transformation
ggplot(data.frame(x = c(0.01, 1)), aes(x)) +
  lapply(c(0.5, 0.25, 0.1, 0), stat_boxcox) +
  scale_colour_viridis_c(limits = c(0, 1.5))
```

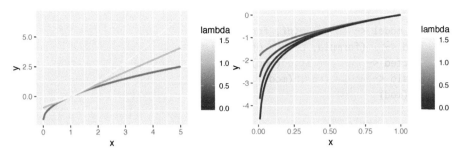

In general, this allows you to use a Box-Cox transformation with any function that accepts a unary transformation function: you don't have to worry about that function providing ... to pass along additional arguments. I also think that the partitioning of lambda and x into two different function arguments is natural since lambda plays quite a different role than x.

10.4.2 Bootstrap generators

Function factories are a useful approach for bootstrapping. Instead of thinking about a single bootstrap (you always need more than one!), you can think about a bootstrap **generator**, a function that yields a fresh bootstrap every time it is called:

```
boot_permute <- function(df, var) {
  n <- nrow(df)
  force(var)

  function() {
    col <- df[[var]]
    col[sample(n, replace = TRUE)]
  }
}

boot_mtcars1 <- boot_permute(mtcars, "mpg")
head(boot_mtcars1())
#> [1] 18.1 22.8 21.5 14.7 21.4 17.3
head(boot_mtcars1())
#> [1] 19.2 19.2 14.3 21.0 13.3 21.4
```

The advantage of a function factory is more clear with a parametric bootstrap where we have to first fit a model. We can do this setup step once, when the factory is called, rather than once every time we generate the bootstrap:

```
boot_model <- function(df, formula) {
  mod <- lm(formula, data = df)
  fitted <- unname(fitted(mod))
  resid <- unname(resid(mod))
  rm(mod)

  function() {
    fitted + sample(resid)
  }
}

boot_mtcars2 <- boot_model(mtcars, mpg ~ wt)
head(boot_mtcars2())
#> [1] 23.1 24.3 23.0 19.1 19.1 16.2
head(boot_mtcars2())
#> [1] 30.2 17.4 31.3 26.1 17.8 16.7
```

I use rm(mod) because linear model objects are quite large (they include complete copies of the model matrix and input data) and I want to keep the manufactured function as small as possible.

10.4.3 Maximum likelihood estimation

The goal of maximum likelihood estimation (MLE) is to find the parameter values for a distribution that make the observed data most likely. To do MLE, you start with a probability function. For example, take the Poisson distribution. If we know λ, we can compute the probability of getting a vector $\mathbf{x}$ of values $(x_1, x_2, ..., x_n)$ by multiplying the Poisson probability function as follows:

$$P(\lambda, \mathbf{x}) = \prod_{i=1}^{n} \frac{\lambda^{x_i} e^{-\lambda}}{x_i!}$$

In statistics, we almost always work with the log of this function. The log is a monotonic transformation which preserves important properties (i.e. the extrema occur in the same place), but has specific advantages:

- The log turns a product into a sum, which is easier to work with.

- Multiplying small numbers yields even smaller numbers, which makes the floating point approximation used by a computer less accurate.

Let's apply a log transformation to this probability function and simplify it as much as possible:

$$\log(P(\lambda, \mathbf{x})) = \sum_{i=1}^{n} \log(\frac{\lambda^{x_i} e^{-\lambda}}{x_i!})$$

$$\log(P(\lambda, \mathbf{x})) = \sum_{i=1}^{n} (x_i \log(\lambda) - \lambda - \log(x_i!))$$

$$\log(P(\lambda, \mathbf{x})) = \sum_{i=1}^{n} x_i \log(\lambda) - \sum_{i=1}^{n} \lambda - \sum_{i=1}^{n} \log(x_i!)$$

$$\log(P(\lambda, \mathbf{x})) = \log(\lambda) \sum_{i=1}^{n} x_i - n\lambda - \sum_{i=1}^{n} \log(x_i!)$$

We can now turn this function into an R function. The R function is quite elegant because R is vectorised and, because it's a statistical programming language, R comes with built-in functions like the log-factorial (`lfactorial()`).

```
lprob_poisson <- function(lambda, x) {
  n <- length(x)
  (log(lambda) * sum(x)) - (n * lambda) - sum(lfactorial(x))
}
```

Consider this vector of observations:

```
x1 <- c(41, 30, 31, 38, 29, 24, 30, 29, 31, 38)
```

We can use `lprob_poisson()` to compute the (logged) probability of x1 for different values of `lambda`.

```
lprob_poisson(10, x1)
#> [1] -184
lprob_poisson(20, x1)
#> [1] -61.1
lprob_poisson(30, x1)
#> [1] -31
```

So far we've been thinking of `lambda` as fixed and known and the function told us the probability of getting different values of x. But in real-life, we observe x and it is `lambda` that is unknown. The likelihood is the probability function seen through this lens: we want to find the `lambda` that makes the observed x the most likely. That is, given x, what value of `lambda` gives us the highest value of `lprob_poisson()`?

In statistics, we highlight this change in perspective by writing $f_{\mathbf{x}}(\lambda)$ instead of $f(\lambda, \mathbf{x})$. In R, we can use a function factory. We provide x and generate a function with a single parameter, `lambda`:

```
ll_poisson1 <- function(x) {
  n <- length(x)

  function(lambda) {
    log(lambda) * sum(x) - n * lambda - sum(lfactorial(x))
  }
}
```

(We don't need force() because length() implicitly forces evaluation of x.)

One nice thing about this approach is that we can do some precomputation: any term that only involves x can be computed once in the factory. This is useful because we're going to need to call this function many times to find the best lambda.

```
ll_poisson2 <- function(x) {
  n <- length(x)
  sum_x <- sum(x)
  c <- sum(lfactorial(x))

  function(lambda) {
    log(lambda) * sum_x - n * lambda - c
  }
}
```

Now we can use this function to find the value of lambda that maximizes the (log) likelihood:

```
lll <- ll_poisson2(x1)

lll(10)
#> [1] -184
lll(20)
#> [1] -61.1
lll(30)
#> [1] -31
```

Rather than trial and error, we can automate the process of finding the best value with optimise(). It will evaluate lll() many times, using mathematical tricks to narrow in on the largest value as quickly as possible. The results tell us that the highest value is -30.27 which occurs when lambda = 32.1:

```
optimise(ll1, c(0, 100), maximum = TRUE)
#> $maximum
#> [1] 32.1
#>
#> $objective
#> [1] -30.3
```

Now, we could have solved this problem without using a function factory because `optimise()` passes ... on to the function being optimised. That means we could use the log-probability function directly:

```
optimise(lprob_poisson, c(0, 100), x = x1, maximum = TRUE)
#> $maximum
#> [1] 32.1
#>
#> $objective
#> [1] -30.3
```

The advantage of using a function factory here is fairly small, but there are two niceties:

- We can precompute some values in the factory, saving computation time in each iteration.

- The two-level design better reflects the mathematical structure of the underlying problem.

These advantages get bigger in more complex MLE problems, where you have multiple parameters and multiple data vectors.

10.4.4 Exercises

1. In `boot_model()`, why don't I need to force the evaluation of `df` or `model`?

2. Why might you formulate the Box-Cox transformation like this?

```
boxcox3 <- function(x) {
  function(lambda) {
    if (lambda == 0) {
      log(x)
    } else {
      (x ^ lambda - 1) / lambda
    }
```

```
    }
}
```

3. Why don't you need to worry that boot_permute() stores a copy of the data inside the function that it generates?

4. How much time does ll_poisson2() save compared to ll_poisson1()? Use bench::mark() to see how much faster the optimisation occurs. How does changing the length of x change the results?

10.5   Function factories + functionals

To finish off the chapter, I'll show how you might combine functionals and function factories to turn data into many functions. The following code creates many specially named power functions by iterating over a list of arguments:

```r
names <- list(
  square = 2,
  cube = 3,
  root = 1/2,
  cuberoot = 1/3,
  reciprocal = -1
)
funs <- purrr::map(names, power1)

funs$root(64)
#> [1] 8
funs$root
#> function(x) {
#>     x ^ exp
#>   }
#> <bytecode: 0x7fa43d0e1740>
#> <environment: 0x7fa43cb6bf90>
```

This idea extends in a straightforward way if your function factory takes two (replace map() with map2()) or more (replace with pmap()) arguments.

One downside of the current construction is that you have to prefix every function call with funs$. There are three ways to eliminate this additional syntax:

- For a very temporary effect, you can use `with()`:

```
with(funs, root(100))
#> [1] 10
```

I recommend this because it makes it very clear when code is being executed in a special context and what that context is.

- For a longer effect, you can `attach()` the functions to the search path, then `detach()` when you're done:

```
attach(funs)
#> The following objects are masked _by_ .GlobalEnv:
#>
#>     cube, square
root(100)
#> [1] 10
detach(funs)
```

You've probably been told to avoid using `attach()`, and that's generally good advice. However, the situation is a little different to the usual because we're attaching a list of functions, not a data frame. It's less likely that you'll modify a function than a column in a data frame, so the some of the worst problems with `attach()` don't apply.

- Finally, you could copy the functions to the global environment with `env_bind()` (you'll learn about `!!!` in Section 19.6). This is mostly permanent:

```
rlang::env_bind(globalenv(), !!!funs)
root(100)
#> [1] 10
```

You can later unbind those same names, but there's no guarantee that they haven't been rebound in the meantime, and you might be deleting an object that someone else created.

```
rlang::env_unbind(globalenv(), names(funs))
```

You'll learn an alternative approach to the same problem in Section 19.7.4. Instead of using a function factory, you could construct the function with quasiquotation. This requires additional knowledge, but generates functions with readable bodies, and avoids accidentally capturing large objects in the enclosing scope. We use that idea in Section 21.2.4 when we work on tools for generating HTML from R.

10.5.1 Exercises

1. Which of the following commands is equivalent to `with(x, f(z))`?

   (a) `x$f(x$z)`.
   (b) `f(x$z)`.
   (c) `x$f(z)`.
   (d) `f(z)`.
   (e) It depends.

2. Compare and contrast the effects of `env_bind()` vs. `attach()` for the following code.

```
funs <- list(
  mean = function(x) mean(x, na.rm = TRUE),
  sum = function(x) sum(x, na.rm = TRUE)
)

attach(funs)
#> The following objects are masked from package:base:
#>
#>     mean, sum
mean <- function(x) stop("Hi!")
detach(funs)

env_bind(globalenv(), !!!funs)
mean <- function(x) stop("Hi!")
env_unbind(globalenv(), names(funs))
```

11

Function operators

11.1 Introduction

In this chapter, you'll learn about function operators. A **function operator** is a function that takes one (or more) functions as input and returns a function as output. The following code shows a simple function operator, chatty(). It wraps a function, making a new function that prints out its first argument. You might create a function like this because it gives you a window to see how functionals, like map_int(), work.

```
chatty <- function(f) {
  force(f)

  function(x, ...) {
    res <- f(x, ...)
    cat("Processing ", x, "\n", sep = "")
    res
  }
}
f <- function(x) x ^ 2
s <- c(3, 2, 1)

purrr::map_dbl(s, chatty(f))
#> Processing 3
#> Processing 2
#> Processing 1
#> [1] 9 4 1
```

Function operators are closely related to function factories; indeed they're just a function factory that takes a function as input. Like factories, there's nothing you can't do without them, but they often allow you to factor out complexity in order to make your code more readable and reusable.

Function operators are typically paired with functionals. If you're using a for-loop, there's rarely a reason to use a function operator, as it will make your code more complex for little gain.

If you're familiar with Python, decorators is just another name for function operators.

Outline

- Section 11.2 introduces you to two extremely useful existing function operators, and shows you how to use them to solve real problems.
- Section 11.3 works through a problem amenable to solution with function operators: downloading many web pages.

Prerequisites

Function operators are a type of function factory, so make sure you're familiar with at least Section 6.2 before you go on.

We'll use purrr (`https://purrr.tidyverse.org`) for a couple of functionals that you learned about in Chapter 9, and some function operators that you'll learn about below. We'll also use the memoise package (`https://memoise.r-lib.org`) [Wickham et al., 2018] for the `memoise()` operator.

```
library(purrr)
library(memoise)
```

11.2   Existing function operators

There are two very useful function operators that will both help you solve common recurring problems, and give you a sense for what function operators can do: `purrr::safely()` and `memoise::memoise()`.

11.2.1   Capturing errors with `purrr::safely()`

One advantage of for-loops is that if one of the iterations fails, you can still access all the results up to the failure:

```
x <- list(
  c(0.512, 0.165, 0.717),
  c(0.064, 0.781, 0.427),
  c(0.890, 0.785, 0.495),
  "oops"
)
```

```
out <- rep(NA_real_, length(x))
for (i in seq_along(x)) {
  out[[i]] <- sum(x[[i]])
}
#> Error in sum(x[[i]]): invalid 'type' (character) of argument
out
#> [1] 1.39 1.27 2.17   NA
```

If you do the same thing with a functional, you get no output, making it hard to figure out where the problem lies:

```
map_dbl(x, sum)
#> Error in .Primitive("sum")(..., na.rm = na.rm): invalid 'type'
#> (character) of argument
```

`purrr::safely()` provides a tool to help with this problem. `safely()` is a function operator that transforms a function to turn errors into data. (You can learn the basic idea that makes it work in Section 8.6.2.) Let's start by taking a look at it outside of `map_dbl()`:

```
safe_sum <- safely(sum)
safe_sum
#> function (...)
#> capture_error(.f(...), otherwise, quiet)
#> <bytecode: 0x7fe46b20f348>
#> <environment: 0x7fe46b20eeb0>
```

Like all function operators, `safely()` takes a function and returns a wrapped function which we can call as usual:

```
str(safe_sum(x[[1]]))
#> List of 2
#>  $ result: num 1.39
#>  $ error : NULL
str(safe_sum(x[[4]]))
#> List of 2
```

```
#>  $ result: NULL
#>  $ error :List of 2
#>   ..$ message: chr "invalid 'type' (character) of argument"
#>   ..$ call   : language .Primitive("sum")(..., na.rm = na.rm)
#>   ..- attr(*, "class")= chr [1:3] "simpleError" "error" "condition"
```

You can see that a function transformed by safely() always returns a list with two elements, result and error. If the function runs successfully, error is NULL and result contains the result; if the function fails, result is NULL and error contains the error.

Now lets use safely() with a functional:

```
out <- map(x, safely(sum))
str(out)
#> List of 4
#>  $ :List of 2
#>   ..$ result: num 1.39
#>   ..$ error : NULL
#>  $ :List of 2
#>   ..$ result: num 1.27
#>   ..$ error : NULL
#>  $ :List of 2
#>   ..$ result: num 2.17
#>   ..$ error : NULL
#>  $ :List of 2
#>   ..$ result: NULL
#>   ..$ error :List of 2
#>   .. ..$ message: chr "invalid 'type' (character) of argument"
#>   .. ..$ call   : language .Primitive("sum")(..., na.rm = na.rm)
#>   .. ..- attr(*, "class")= chr [1:3] "simpleError" "error" "condit"..
```

The output is in a slightly inconvenient form, since we have four lists, each of which is a list containing the result and the error. We can make the output easier to use by turning it "inside-out" with purrr::transpose(), so that we get a list of results and a list of errors:

```
out <- transpose(map(x, safely(sum)))
str(out)
#> List of 2
#>  $ result:List of 4
#>   ..$ : num 1.39
#>   ..$ : num 1.27
#>   ..$ : num 2.17
```

```
#>    ..$ : NULL
#>    $ error :List of 4
#>    ..$ : NULL
#>    ..$ : NULL
#>    ..$ : NULL
#>    ..$ :List of 2
#>    .. ..$ message: chr "invalid 'type' (character) of argument"
#>    .. ..$ call    : language .Primitive("sum")(..., na.rm = na.rm)
#>    .. ..- attr(*, "class")= chr [1:3] "simpleError" "error" "condit"..
```

Now we can easily find the results that worked, or the inputs that failed:

```
ok <- map_lgl(out$error, is.null)
ok
#> [1] TRUE  TRUE  TRUE FALSE

x[!ok]
#> [[1]]
#> [1] "oops"

out$result[ok]
#> [[1]]
#> [1] 1.39
#>
#> [[2]]
#> [1] 1.27
#>
#> [[3]]
#> [1] 2.17
```

You can use this same technique in many different situations. For example, imagine you're fitting a generalised linear model (GLM) to a list of data frames. GLMs can sometimes fail because of optimisation problems, but you still want to be able to try to fit all the models, and later look back at those that failed:

```
fit_model <- function(df) {
  glm(y ~ x1 + x2 * x3, data = df)
}


models <- transpose(map(datasets, safely(fit_model)))
ok <- map_lgl(models$error, is.null)
```

```
# which data failed to converge?
datasets[!ok]

# which models were successful?
models[ok]
```

I think this is a great example of the power of combining functionals and function operators: `safely()` lets you succinctly express what you need to solve a common data analysis problem.

purrr comes with three other function operators in a similar vein:

- `possibly()`: returns a default value when there's an error. It provides no way to tell if an error occured or not, so it's best reserved for cases when there's some obvious sentinel value (like `NA`).

- `quietly()`: turns output, messages, and warning side-effects into `output`, `message`, and `warning` components of the output.

- `auto_browser()`: automatically executes `browser()` inside the function when there's an error.

See their documentation for more details.

11.2.2 Caching computations with `memoise::memoise()`

Another handy function operator is `memoise::memoise()`. It **memoises** a function, meaning that the function will remember previous inputs and return cached results. Memoisation is an example of the classic computer science tradeoff of memory versus speed. A memoised function can run much faster, but because it stores all of the previous inputs and outputs, it uses more memory.

Let's explore this idea with a toy function that simulates an expensive operation:

```
slow_function <- function(x) {
  Sys.sleep(1)
  x * 10 * runif(1)
}
system.time(print(slow_function(1)))
#> [1] 0.808
#>    user  system elapsed
#>    0.00    0.00    1.01

system.time(print(slow_function(1)))
```

```
#> [1] 8.34
#>    user  system elapsed
#>   0.002   0.000   1.006
```

When we memoise this function, it's slow when we call it with new arguments.
But when we call it with arguments that it's seen before it's instantaneous: it
retrieves the previous value of the computation.

```
fast_function <- memoise::memoise(slow_function)
system.time(print(fast_function(1)))
#> [1] 6.01
#>    user  system elapsed
#>       0       0       1

system.time(print(fast_function(1)))
#> [1] 6.01
#>    user  system elapsed
#>   0.019   0.000   0.019
```

A relatively realistic use of memoisation is computing the Fibonacci series.
The Fibonacci series is defined recursively: the first two values are defined by
convention, $f(0) = 0$, $f(n) = 1$, and then $f(n) = f(n-1) + f(n-2)$ (for
any positive integer). A naive version is slow because, for example, fib(10)
computes fib(9) and fib(8), and fib(9) computes fib(8) and fib(7), and so
on.

```
fib <- function(n) {
  if (n < 2) return(1)
  fib(n - 2) + fib(n - 1)
}
system.time(fib(23))
#>    user  system elapsed
#>   0.040   0.002   0.043
system.time(fib(24))
#>    user  system elapsed
#>   0.064   0.003   0.068
```

Memoising fib() makes the implementation much faster because each value
is computed only once:

```
fib2 <- memoise::memoise(function(n) {
  if (n < 2) return(1)
  fib2(n - 2) + fib2(n - 1)
```

```
})
system.time(fib2(23))
#>    user  system elapsed
#>    0.01    0.00    0.01
```

And future calls can rely on previous computations:

```
system.time(fib2(24))
#>    user  system elapsed
#>   0.001   0.000   0.000
```

This is an example of **dynamic programming**, where a complex problem can be broken down into many overlapping subproblems, and remembering the results of a subproblem considerably improves performance.

Think carefully before memoising a function. If the function is not **pure**, i.e. the output does not depend only on the input, you will get misleading and confusing results. I created a subtle bug in devtools because I memoised the results of available.packages(), which is rather slow because it has to download a large file from CRAN. The available packages don't change that frequently, but if you have an R process that's been running for a few days, the changes can become important, and because the problem only arose in long-running R processes, the bug was very painful to find.

11.2.3 Exercises

1. Base R provides a function operator in the form of Vectorize(). What does it do? When might you use it?

2. Read the source code for possibly(). How does it work?

3. Read the source code for safely(). How does it work?

11.3 Case study: Creating your own function operators

meomoise() and safely() are very useful but also quite complex. In this case study you'll learn how to create your own simpler function operators. Imagine you have a named vector of URLs and you'd like to download each one to disk. That's pretty simple with walk2() and file.download():

```
urls <- c(
  "adv-r" = "https://adv-r.hadley.nz",
  "r4ds" = "http://r4ds.had.co.nz/"
  # and many many more
)
path <- paste(tempdir(), names(urls), ".html")

walk2(urls, path, download.file, quiet = TRUE)
```

This approach is fine for a handful of URLs, but as the vector gets longer, you might want to add a couple more features:

- Add a small delay between each request to avoid hammering the server.

- Display a . every few URLs so that we know that the function is still working.

It's relatively easy to add these extra features if we're using a for loop:

```
for(i in seq_along(urls)) {
  Sys.sleep(0.1)
  if (i %% 10 == 0) cat(".")
  download.file(urls[[i]], paths[[i]])
}
```

I think this for loop is suboptimal because it interleaves different concerns: pausing, showing progress, and downloading. This makes the code harder to read, and it makes it harder to reuse the components in new situations. Instead, let's see if we can use function operators to extract out pausing and showing progress and make them reusable.

First, let's write an function operator that adds a small delay. I'm going to call it `delay_by()` for reasons that will be more clear shortly, and it has two arguments: the function to wrap, and the amount of delay to add. The actual implementation is quite simple. The main trick is forcing evaluation of all arguments as described in Section 10.2.5, because function operators are a special type of function factory:

```
delay_by <- function(f, amount) {
  force(f)
  force(amount)

  function(...) {
    Sys.sleep(amount)
    f(...)
  }
}
```

```
}
system.time(runif(100))
#>    user  system elapsed
#>       0       0       0
system.time(delay_by(runif, 0.1)(100))
#>    user  system elapsed
#>   0.000   0.000   0.103
```

And we can use it with the original walk2():

```
walk2(urls, path, delay_by(download.file, 0.1), quiet = TRUE)
```

Creating a function to display the occasional dot is a little harder, because we can no longer rely on the index from the loop. We could pass the index along as another argument, but that breaks encapsulation: a concern of the progress function now becomes a problem that the higher level wrapper needs to handle. Instead, we'll use another function factory trick (from Section 10.2.4), so that the progress wrapper can manage its own internal counter:

```
dot_every <- function(f, n) {
  force(f)
  force(n)

  i <- 0
  function(...) {
    i <<- i + 1
    if (i %% n == 0) cat(".")
    f(...)
  }
}
walk(1:100, runif)
walk(1:100, dot_every(runif, 10))
#> ..........
```

Now we can express our original for loop as:

```
walk2(
  urls, path,
  dot_every(delay_by(download.file, 0.1), 10),
  quiet = TRUE
)
```

This is starting to get a little hard to read because we are composing many function calls, and the arguments are getting spread out. One way to resolve that is to use the pipe:

```
walk2(
  urls, path,
  download.file %>% dot_every(10) %>% delay_by(0.1),
  quiet = TRUE
)
```

The pipe works well here because I've carefully chosen the function names to yield an (almost) readable sentence: take `download.file` then (add) a dot every 10 iterations, then delay by 0.1s. The more clearly you can express the intent of your code through function names, the more easily others (including future you!) can read and understand the code.

11.3.1 Exercises

1. Weigh the pros and cons of `download.file %>% dot_every(10) %>% delay_by(0.1)` versus `download.file %>% delay_by(0.1) %>% dot_every(10)`.

2. Should you memoise `file.download()`? Why or why not?

3. Create a function operator that reports whenever a file is created or deleted in the working directory, using `dir()` and `setdiff()`. What other global function effects might you want to track?

4. Write a function operator that logs a timestamp and message to a file every time a function is run.

5. Modify `delay_by()` so that instead of delaying by a fixed amount of time, it ensures that a certain amount of time has elapsed since the function was last called. That is, if you called `g <- delay_by(1, f); g(); Sys.sleep(2); g()` there shouldn't be an extra delay.

Part III

Object-oriented programming

Introduction

In the following five chapters you'll learn about **object-oriented programming** (OOP). OOP is a little more challenging in R than in other languages because:

- There are multiple OOP systems to choose from. In this book, I'll focus on the three that I believe are most important: **S3**, **R6**, and **S4**. S3 and S4 are provided by base R. R6 is provided by the R6 package, and is similar to the Reference Classes, or **RC** for short, from base R.

- There is disagreement about the relative importance of the OOP systems. I think S3 is most important, followed by R6, then S4. Others believe that S4 is most important, followed by RC, and that S3 should be avoided. This means that different R communities use different systems.

- S3 and S4 use generic function OOP which is rather different from the encapsulated OOP used by most languages popular today[1]. We'll come back to precisely what those terms mean shortly, but basically, while the underlying ideas of OOP are the same across languages, their expressions are rather different. This means that you can't immediately transfer your existing OOP skills to R.

Generally in R, functional programming is much more important than object-oriented programming, because you typically solve complex problems by decomposing them into simple functions, not simple objects. Nevertheless, there are important reasons to learn each of the three systems:

- S3 allows your functions to return rich results with user-friendly display and programmer-friendly internals. S3 is used throughout base R, so it's important to master if you want to extend base R functions to work with new types of input.

- R6 provides a standardised way to escape R's copy-on-modify semantics. This is particularly important if you want to model objects that exist independently of R. Today, a common need for R6 is to model data that comes from a web API, and where changes come from inside or outside of R.

- S4 is a rigorous system that forces you to think carefully about program design. It's particularly well-suited for building large systems that evolve

---

[1]The exception is Julia, which also uses generic function OOP. Compared to R, Julia's implementation is fully developed and extremely performant.

over time and will receive contributions from many programmers. This is why it is used by the Bioconductor project, so another reason to learn S4 is to equip you to contribute to that project.

The goal of this brief introductory chapter is to give you some important vocabulary and some tools to identify OOP systems in the wild. The following chapters then dive into the details of R's OOP systems:

1. Chapter 12 details the base types which form the foundation underlying all other OO system.

2. Chapter 13 introduces S3, the simplest and most commonly used OO system.

3. Chapter 14 discusses R6, a encapsulated OO system built on top of environments.

4. Chapter 15 introduces S4, which is similar to S3 but more formal and more strict.

5. Chapter 16 compares these three main OO systems. By understanding the trade-offs of each system you can appreciate when to use one or the other.

This book focusses on the mechanics of OOP, not its effective use, and it may be challenging to fully understand if you have not done object-oriented programming before. You might wonder why I chose not to provide more immediately useful coverage. I have focussed on mechanics here because they need to be well described somewhere (writing these chapters required a considerable amount of reading, exploration, and synthesis on my behalf), and using OOP effectively is sufficiently complex to require book-length treatment; there's simply not enough room in Advanced R to cover it in the depth required.

OOP systems

Different people use OOP terms in different ways, so this section provides a quick overview of important vocabulary. The explanations are necessarily compressed, but we will come back to these ideas multiple times.

The main reason to use OOP is **polymorphism** (literally: many shapes). Polymorphism means that a developer can consider a function's interface separately from its implementation, making it possible to use the same function form for different types of input. This is closely related to the idea of **encapsulation**: the user doesn't need to worry about details of an object because they are encapsulated behind a standard interface.

To be concrete, polymorphism is what allows `summary()` to produce different outputs for numeric and factor variables:

```
diamonds <- ggplot2::diamonds

summary(diamonds$carat)
#>    Min. 1st Qu.  Median    Mean 3rd Qu.    Max.
#>    0.20    0.40    0.70    0.80    1.04    5.01

summary(diamonds$cut)
#>      Fair      Good Very Good   Premium     Ideal
#>      1610      4906     12082     13791     21551
```

You could imagine `summary()` containing a series of if-else statements, but that would mean only the original author could add new implementations. An OOP system makes it possible for any developer to extend the interface with implementations for new types of input.

To be more precise, OO systems call the type of an object its **class**, and an implementation for a specific class is called a **method**. Roughly speaking, a class defines what an object *is* and methods describe what that object can *do*. The class defines the **fields**, the data possessed by every instance of that class. Classes are organised in a hierarchy so that if a method does not exist for one class, its parent's method is used, and the child is said to **inherit** behaviour. For example, in R, an ordered factor inherits from a regular factor, and a generalised linear model inherits from a linear model. The process of finding the correct method given a class is called **method dispatch**.

There are two main paradigms of object-oriented programming which differ in how methods and classes are related. In this book, we'll borrow the terminology of *Extending R* [Chambers, 2016] and call these paradigms encapsulated and functional:

- In **encapsulated** OOP, methods belong to objects or classes, and method calls typically look like `object.method(arg1, arg2)`. This is called encapsulated because the object encapsulates both data (with fields) and behaviour (with methods), and is the paradigm found in most popular languages.

- In **functional** OOP, methods belong to **generic** functions, and method calls look like ordinary function calls: `generic(object, arg2, arg3)`. This is called functional because from the outside it looks like a regular function call, and internally the components are also functions.

With this terminology in hand, we can now talk precisely about the different OO systems available in R.

_____

OOP in R

Base R provides three OOP systems: S3, S4, and reference classes (RC):

- **S3** is R's first OOP system, and is described in *Statistical Models in S* [Chambers and Hastie, 1992]. S3 is an informal implementation of functional OOP and relies on common conventions rather than ironclad guarantees. This makes it easy to get started with, providing a low cost way of solving many simple problems.

- **S4** is a formal and rigorous rewrite of S3, and was introduced in *Programming with Data* [Chambers, 1998]. It requires more upfront work than S3, but in return provides more guarantees and greater encapsulation. S4 is implemented in the base **methods** package, which is always installed with R.

  (You might wonder if S1 and S2 exist. They don't: S3 and S4 were named according to the versions of S that they accompanied. The first two versions of S didn't have any OOP framework.)

- **RC** implements encapsulated OO. RC objects are a special type of S4 objects that are also **mutable**, i.e., instead of using R's usual copy-on-modify semantics, they can be modified in place. This makes them harder to reason about, but allows them to solve problems that are difficult to solve in the functional OOP style of S3 and S4.

A number of other OOP systems are provided by CRAN packages:

- **R6** [Chang, 2017] implements encapsulated OOP like RC, but resolves some important issues. In this book, you'll learn about R6 instead of RC, for reasons described in Section 14.5.

- **R.oo** [Bengtsson, 2003] provides some formalism on top of S3, and makes it possible to have mutable S3 objects.

- **proto** [Grothendieck et al., 2016] implements another style of OOP based on the idea of **prototypes**, which blur the distinctions between classes and instances of classes (objects). I was briefly enamoured with prototype based programming [Wickham, 2011] and used it in ggplot2, but now think it's better to stick with the standard forms.

Apart from R6, which is widely used, these systems are primarily of theoretical interest. They do have their strengths, but few R users know and understand them, so it is hard for others to read and contribute to your code.

sloop

Before we go on I want to introduce the sloop package:

```
library(sloop)
```

The sloop package (think "sail the seas of OOP") provides a number of helpers that fill in missing pieces in base R. The first of these is `sloop::otype()`. It makes it easy to figure out the OOP system used by a wild-caught object:

```
otype(1:10)
#> [1] "base"

otype(mtcars)
#> [1] "S3"

mle_obj <- stats4::mle(function(x = 1) (x - 2) ^ 2)
otype(mle_obj)
#> [1] "S4"
```

Use this function to figure out which chapter to read to understand how to work with an existing object.

12

Base types

12.1 Introduction

To talk about objects and OOP in R we first need to clear up a fundamental confusion about two uses of the word "object". So far in this book, we've used the word in the general sense captured by John Chambers' pithy quote: "Everything that exists in R is an object". However, while everything *is* an object, not everything is object-oriented. This confusion arises because the base objects come from S, and were developed before anyone thought that S might need an OOP system. The tools and nomenclature evolved organically over many years without a single guiding principle.

Most of the time, the distinction between objects and object-oriented objects is not important. But here we need to get into the nitty gritty details so we'll use the terms **base objects** and **OO objects** to distinguish them.

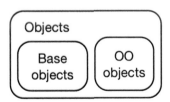

Outline

- Section 12.2 shows you how to identify base and OO objects.
- Section 12.3 gives a complete set of the base types used to build all objects.

12.2   Base versus OO objects

To tell the difference between a base and OO object, use `is.object()` or
`sloop::otype()`:

```r
# A base object:
is.object(1:10)
#> [1] FALSE
sloop::otype(1:10)
#> [1] "base"

# An OO object
is.object(mtcars)
#> [1] TRUE
sloop::otype(mtcars)
#> [1] "S3"
```

Technically, the difference between base and OO objects is that OO objects
have a "class" attribute:

```r
attr(1:10, "class")
#> NULL

attr(mtcars, "class")
#> [1] "data.frame"
```

You may already be familiar with the `class()` function. This function is safe
to apply to S3 and S4 objects, but it returns misleading results when applied
to base objects. It's safer to use `sloop::s3_class()`, which returns the implicit
class that the S3 and S4 systems will use to pick methods. You'll learn more
about `s3_class()` in Section 13.7.1.

```r
x <- matrix(1:4, nrow = 2)
class(x)
#> [1] "matrix"
sloop::s3_class(x)
#> [1] "matrix"  "integer" "numeric"
```

12.3 Base types

While only OO objects have a class attribute, every object has a **base type**:

```
typeof(1:10)
#> [1] "integer"

typeof(mtcars)
#> [1] "list"
```

Base types do not form an OOP system because functions that behave differently for different base types are primarily written in C code that uses switch statements. This means that only R-core can create new types, and creating a new type is a lot of work because every switch statement needs to be modified to handle a new case. As a consequence, new base types are rarely added. The most recent change, in 2011, added two exotic types that you never see in R itself, but are needed for diagnosing memory problems. Prior to that, the last type added was a special base type for S4 objects added in 2005.

In total, there are 25 different base types. They are listed below, loosely grouped according to where they're discussed in this book. These types are most important in C code, so you'll often see them called by their C type names. I've included those in parentheses.

- Vectors, Chapter 3, include types NULL (NULLSXP), logical (LGLSXP), integer (INTSXP), double (REALSXP), complex (CPLSXP), character (STRSXP), list (VECSXP), and raw (RAWSXP).

```
typeof(NULL)
#> [1] "NULL"
typeof(1L)
#> [1] "integer"
typeof(1i)
#> [1] "complex"
```

- Functions, Chapter 6, include types closure (regular R functions, CLOSXP), special (internal functions, SPECIALSXP), and builtin (primitive functions, BUILTINSXP).

```
typeof(mean)
#> [1] "closure"
typeof(`[`)
```

```
#> [1] "special"
typeof(sum)
#> [1] "builtin"
```

Internal and primitive functions are described in Section 6.2.2.

- Environments, Chapter 7, have type environment (ENVSXP).

```
typeof(globalenv())
#> [1] "environment"
```

- The S4 type (S4SXP), Chapter 15, is used for S4 classes that don't inherit from an existing base type.

```
mle_obj <- stats4::mle(function(x = 1) (x - 2) ^ 2)
typeof(mle_obj)
#> [1] "S4"
```

- Language components, Chapter 18, include symbol (aka name, SYMSXP), language (usually called calls, LANGSXP), and pairlist (used for function arguments, LISTSXP) types.

```
typeof(quote(a))
#> [1] "symbol"
typeof(quote(a + 1))
#> [1] "language"
typeof(formals(mean))
#> [1] "pairlist"
```

   expression (EXPRSXP) is a special purpose type that's only returned by parse() and expression(). Expressions are generally not needed in user code.

- The remaining types are esoteric and rarely seen in R. They are important primarily for C code: externalptr (EXTPTRSXP), weakref (WEAKREFSXP), byte-code (BCODESXP), promise (PROMSXP), ... (DOTSXP), and any (ANYSXP).

You may have heard of mode() and storage.mode(). Do not use these functions: they exist only to provide type names that are compatible with S.

12.3.1 Numeric type

Be careful when talking about the numeric type, because R uses "numeric" to mean three slightly different things:

1. In some places numeric is used as an alias for the double type. For example as.numeric() is identical to as.double(), and numeric() is identical to double().

    (R also occasionally uses real instead of double; NA_real_ is the one place that you're likely to encounter this in practice.)

2. In the S3 and S4 systems, numeric is used as a shorthand for either integer or double type, and is used when picking methods:

```
sloop::s3_class(1)
#> [1] "double"  "numeric"
sloop::s3_class(1L)
#> [1] "integer" "numeric"
```

3. is.numeric() tests for objects that *behave* like numbers. For example, factors have type "integer" but don't behave like numbers (i.e. it doesn't make sense to take the mean of factor).

```
typeof(factor("x"))
#> [1] "integer"
is.numeric(factor("x"))
#> [1] FALSE
```

In this book, I consistently use numeric to mean an object of type integer or double.

13

S3

13.1 Introduction

S3 is R's first and simplest OO system. S3 is informal and ad hoc, but there is a certain elegance in its minimalism: you can't take away any part of it and still have a useful OO system. For these reasons, you should use it, unless you have a compelling reason to do otherwise. S3 is the only OO system used in the base and stats packages, and it's the most commonly used system in CRAN packages.

S3 is very flexible, which means it allows you to do things that are quite ill-advised. If you're coming from a strict environment like Java this will seem pretty frightening, but it gives R programmers a tremendous amount of freedom. It may be very difficult to prevent people from doing something you don't want them to do, but your users will never be held back because there is something you haven't implemented yet. Since S3 has few built-in constraints, the key to its successful use is applying the constraints yourself. This chapter will therefore teach you the conventions you should (almost) always follow.

The goal of this chapter is to show you how the S3 system works, not how to use it effectively to create new classes and generics. I'd recommend coupling the theoretical knowledge from this chapter with the practical knowledge encoded in the vctrs package (`https://vctrs.r-lib.org`).

Outline

- Section 13.2 gives a rapid overview of all the main components of S3: classes, generics, and methods. You'll also learn about `sloop::s3_dispatch()`, which we'll use throughout the chapter to explore how S3 works.

- Section 13.3 goes into the details of creating a new S3 class, including the three functions that should accompany most classes: a constructor, a helper, and a validator.

- Section 13.4 describes how S3 generics and methods work, including the basics of method dispatch.

- Section 13.5 discusses the four main styles of S3 objects: vector, record, data frame, and scalar.

- Section 13.6 demonstrates how inheritance works in S3, and shows you what you need to make a class "subclassable".

- Section 13.7 concludes the chapter with a discussion of the finer details of method dispatch including base types, internal generics, group generics, and double dispatch.

Prerequisites

S3 classes are implemented using attributes, so make sure you're familiar with the details described in Section 3.3. We'll use existing base S3 vectors for examples and exploration, so make sure that you're familiar with the factor, Date, difftime, POSIXct, and POSIXlt classes described in Section 3.4.

We'll use the sloop (`https://sloop.r-lib.org`) package for its interactive helpers.

```
library(sloop)
```

13.2   Basics

An S3 object is a base type with at least a `class` attribute (other attributes may be used to store other data). For example, take the factor. Its base type is the integer vector, it has a `class` attribute of "factor", and a `levels` attribute that stores the possible levels:

```
f <- factor(c("a", "b", "c"))

typeof(f)
#> [1] "integer"
attributes(f)
#> $levels
#> [1] "a" "b" "c"
#>
#> $class
#> [1] "factor"
```

You can get the underlying base type by `unclass()`ing it, which strips the class attribute, causing it to lose its special behaviour:

```
unclass(f)
#> [1] 1 2 3
#> attr(,"levels")
#> [1] "a" "b" "c"
```

An S3 object behaves differently from its underlying base type whenever it's passed to a **generic** (short for generic function). The easiest way to tell if a function is a generic is to use `sloop::ftype()` and look for "generic" in the output:

```
ftype(print)
#> [1] "S3"       "generic"
ftype(str)
#> [1] "S3"       "generic"
ftype(unclass)
#> [1] "primitive"
```

A generic function defines an interface, which uses a different implementation depending on the class of an argument (almost always the first argument). Many base R functions are generic, including the important `print()`:

```
print(f)
#> [1] a b c
#> Levels: a b c

# stripping class reverts to integer behaviour
print(unclass(f))
#> [1] 1 2 3
#> attr(,"levels")
#> [1] "a" "b" "c"
```

Beware that `str()` is generic, and some S3 classes use that generic to hide the internal details. For example, the `POSIXlt` class used to represent date-time data is actually built on top of a list, a fact which is hidden by its `str()` method:

```
time <- strptime(c("2017-01-01", "2020-05-04 03:21"), "%Y-%m-%d")
str(time)
#>  POSIXlt[1:2], format: "2017-01-01" "2020-05-04"
```

```
str(unclass(time))
#> List of 11
#>  $ sec   : num [1:2] 0 0
#>  $ min   : int [1:2] 0 0
#>  $ hour  : int [1:2] 0 0
#>  $ mday  : int [1:2] 1 4
#>  $ mon   : int [1:2] 0 4
#>  $ year  : int [1:2] 117 120
#>  $ wday  : int [1:2] 0 1
#>  $ yday  : int [1:2] 0 124
#>  $ isdst : int [1:2] 0 1
#>  $ zone  : chr [1:2] "CST" "CDT"
#>  $ gmtoff: int [1:2] NA NA
#>  - attr(*, "tzone")= chr [1:3] "America/Chicago" "CST" "CDT"
```

The generic is a middleman: its job is to define the interface (i.e. the arguments) then find the right implementation for the job. The implementation for a specific class is called a **method**, and the generic finds that method by performing **method dispatch**.

You can use `sloop::s3_dispatch()` to see the process of method dispatch:

```
s3_dispatch(print(f))
#> => print.factor
#>  * print.default
```

We'll come back to the details of dispatch in Section 13.4.1, for now note that S3 methods are functions with a special naming scheme, generic.class(). For example, the factor method for the print() generic is called print.factor(). You should never call the method directly, but instead rely on the generic to find it for you.

Generally, you can identify a method by the presence of . in the function name, but there are a number of important functions in base R that were written before S3, and hence use . to join words. If you're unsure, check with sloop::ftype():

```
ftype(t.test)
#> [1] "S3"       "generic"
ftype(t.data.frame)
#> [1] "S3"       "method"
```

Unlike most functions, you can't see the source code for most S3 methods[1] just by typing their names. That's because S3 methods are not usually exported: they live only inside the package, and are not available from the global environment. Instead, you can use `sloop::s3_get_method()`, which will work regardless of where the method lives:

```
weighted.mean.Date
#> Error in eval(expr, envir, enclos): object 'weighted.mean.Date' not
#> found

s3_get_method(weighted.mean.Date)
#> function (x, w, ...)
#> structure(weighted.mean(unclass(x), w, ...), class = "Date")
#> <bytecode: 0x7fed4af8d778>
#> <environment: namespace:stats>
```

13.2.1 Exercises

1. Describe the difference between `t.test()` and `t.data.frame()`. When is each function called?

2. Make a list of commonly used base R functions that contain . in their name but are not S3 methods.

3. What does the `as.data.frame.data.frame()` method do? Why is it confusing? How could you avoid this confusion in your own code?

4. Describe the difference in behaviour in these two calls.

   ```
   set.seed(1014)
   some_days <- as.Date("2017-01-31") + sample(10, 5)

   mean(some_days)
   #> [1] "2017-02-05"
   mean(unclass(some_days))
   #> [1] 17202
   ```

5. What class of object does the following code return? What base type is it built on? What attributes does it use?

---

[1]The exceptions are methods found in the base package, like `t.data.frame`, and methods that you've created.

```
x <- ecdf(rpois(100, 10))
x
#> Empirical CDF
#> Call: ecdf(rpois(100, 10))
#>  x[1:18] =   2,   3,   4,  ...,  2e+01, 2e+01
```

6. What class of object does the following code return? What base
   type is it built on? What attributes does it use?

```
x <- table(rpois(100, 5))
x
#>
#>  1  2  3  4  5  6  7  8  9 10
#>  8  5 18 14 12 19 12  3  5  4
```

13.3 Classes

If you have done object-oriented programming in other languages, you may
be surprised to learn that S3 has no formal definition of a class: to make an
object an instance of a class, you simply set the **class attribute**. You can do
that during creation with structure(), or after the fact with class<-():

```
# Create and assign class in one step
x <- structure(list(), class = "my_class")

# Create, then set class
x <- list()
class(x) <- "my_class"
```

You can determine the class of an S3 object with class(x), and see if an object
is an instance of a class using inherits(x, "classname").

```
class(x)
#> [1] "my_class"
inherits(x, "my_class")
#> [1] TRUE
```

```
inherits(x, "your_class")
#> [1] FALSE
```

The class name can be any string, but I recommend using only letters and _.
Avoid . because (as mentioned earlier) it can be confused with the . separator
between a generic name and a class name. When using a class in a package, I
recommend including the package name in the class name. That ensures you
won't accidentally clash with a class defined by another package.

S3 has no checks for correctness which means you can change the class of
existing objects:

```
# Create a linear model
mod <- lm(log(mpg) ~ log(disp), data = mtcars)
class(mod)
#> [1] "lm"
print(mod)
#>
#> Call:
#> lm(formula = log(mpg) ~ log(disp), data = mtcars)
#>
#> Coefficients:
#> (Intercept)    log(disp)
#>       5.381       -0.459

# Turn it into a date (?!)
class(mod) <- "Date"

# Unsurprisingly this doesn't work very well
print(mod)
#> Error in as.POSIXlt.Date(x): (list) object cannot be coerced to type
#> 'double'
```

If you've used other OO languages, this might make you feel queasy, but in
practice this flexibility causes few problems. R doesn't stop you from shooting
yourself in the foot, but as long as you don't aim the gun at your toes and
pull the trigger, you won't have a problem.

To avoid foot-bullet intersections when creating your own class, I recommend
that you usually provide three functions:

- A low-level **constructor**, new_myclass(), that efficiently creates new objects
  with the correct structure.

- A **validator**, validate_myclass(), that performs more computationally expensive checks to ensure that the object has correct values.

- A user-friendly **helper**, `myclass()`, that provides a convenient way for others to create objects of your class.

You don't need a validator for very simple classes, and you can skip the helper if the class is for internal use only, but you should always provide a constructor.

13.3.1 Constructors

S3 doesn't provide a formal definition of a class, so it has no built-in way to ensure that all objects of a given class have the same structure (i.e. the same base type and the same attributes with the same types). Instead, you must enforce a consistent structure by using a **constructor**.

The constructor should follow three principles:

- Be called `new_myclass()`.

- Have one argument for the base object, and one for each attribute.

- Check the type of the base object and the types of each attribute.

I'll illustrate these ideas by creating constructors for base classes[2] that you're already familiar with. To start, lets make a constructor for the simplest S3 class: `Date`. A `Date` is just a double with a single attribute: its class is "Date". This makes for a very simple constructor:

```
new_Date <- function(x = double()) {
  stopifnot(is.double(x))
  structure(x, class = "Date")
}

new_Date(c(-1, 0, 1))
#> [1] "1969-12-31" "1970-01-01" "1970-01-02"
```

The purpose of constructors is to help you, the developer. That means you can keep them simple, and you don't need to optimise error messages for public consumption. If you expect users to also create objects, you should create a friendly helper function, called `class_name()`, which I'll describe shortly.

A slightly more complicated constructor is that for `difftime`, which is used to represent time differences. It is again built on a double, but has a `units` attribute that must take one of a small set of values:

---

[2]Recent versions of R have `.Date()`, `.difftime()`, `.POSIXct()`, and `.POSIXlt()` constructors but they are internal, not well documented, and do not follow the principles that I recommend.

```
new_difftime <- function(x = double(), units = "secs") {
  stopifnot(is.double(x))
  units <- match.arg(units, c("secs", "mins", "hours", "days", "weeks"))

  structure(x,
    class = "difftime",
    units = units
  )
}

new_difftime(c(1, 10, 3600), "secs")
#> Time differences in secs
#> [1]    1   10 3600
new_difftime(52, "weeks")
#> Time difference of 52 weeks
```

The constructor is a developer function: it will be called in many places, by an experienced user. That means it's OK to trade a little safety in return for performance, and you should avoid potentially time-consuming checks in the constructor.

13.3.2 Validators

More complicated classes require more complicated checks for validity. Take factors, for example. A constructor only checks that types are correct, making it possible to create malformed factors:

```
new_factor <- function(x = integer(), levels = character()) {
  stopifnot(is.integer(x))
  stopifnot(is.character(levels))

  structure(
    x,
    levels = levels,
    class = "factor"
  )
}

new_factor(1:5, "a")
#> Error in as.character.factor(x): malformed factor
new_factor(0:1, "a")
#> Error in as.character.factor(x): malformed factor
```

Rather than encumbering the constructor with complicated checks, it's better to put them in a separate function. Doing so allows you to cheaply create new objects when you know that the values are correct, and easily re-use the checks in other places.

```r
validate_factor <- function(x) {
  values <- unclass(x)
  levels <- attr(x, "levels")

  if (!all(!is.na(values) & values > 0)) {
    stop(
      "All `x` values must be non-missing and greater than zero",
      call. = FALSE
    )
  }

  if (length(levels) < max(values)) {
    stop(
      "There must at least as many `levels` as possible values in `x`",
      call. = FALSE
    )
  }

  x
}

validate_factor(new_factor(1:5, "a"))
#> Error: There must at least as many `levels` as possible values in
#> `x`
validate_factor(new_factor(0:1, "a"))
#> Error: All `x` values must be non-missing and greater than zero
```

This validator function is called primarily for its side-effects (throwing an error if the object is invalid) so you'd expect it to invisibly return its primary input (as described in Section 6.7.2). However, it's useful for validation methods to return visibly, as we'll see next.

13.3.3 Helpers

If you want users to construct objects from your class, you should also provide a helper method that makes their life as easy as possible. A helper should always:

- Have the same name as the class, e.g. myclass().

- Finish by calling the constructor, and the validator, if it exists.

- Create carefully crafted error messages tailored towards an end-user.

- Have a thoughtfully crafted user interface with carefully chosen default values and useful conversions.

The last bullet is the trickiest, and it's hard to give general advice. However, there are three common patterns:

- Sometimes all the helper needs to do is coerce its inputs to the desired type. For example, `new_difftime()` is very strict, and violates the usual convention that you can use an integer vector wherever you can use a double vector:

```
new_difftime(1:10)
#> Error in new_difftime(1:10): is.double(x) is not TRUE
```

It's not the job of the constructor to be flexible, so here we create a helper that just coerces the input to a double.

```
difftime <- function(x = double(), units = "secs") {
  x <- as.double(x)
  new_difftime(x, units = units)
}

difftime(1:10)
#> Time differences in secs
#>  [1]  1  2  3  4  5  6  7  8  9 10
```

- Often, the most natural representation of a complex object is a string. For example, it's very convenient to specify factors with a character vector. The code below shows a simple version of `factor()`: it takes a character vector, and guesses that the levels should be the unique values. This is not always correct (since some levels might not be seen in the data), but it's a useful default.

```
factor <- function(x = character(), levels = unique(x)) {
  ind <- match(x, levels)
  validate_factor(new_factor(ind, levels))
}

factor(c("a", "a", "b"))
#> [1] a a b
#> Levels: a b
```

- Some complex objects are most naturally specified by multiple simple components. For example, I think it's natural to construct a date-time by

supplying the individual components (year, month, day etc). That leads me
to this POSIXct() helper that resembles the existing ISODatetime() function[3]:

```
POSIXct <- function(year = integer(),
                    month = integer(),
                    day = integer(),
                    hour = 0L,
                    minute = 0L,
                    sec = 0,
                    tzone = "") {
  ISOdatetime(year, month, day, hour, minute, sec, tz = tzone)
}

POSIXct(2020, 1, 1, tzone = "America/New_York")
#> [1] "2020-01-01 EST"
```

For more complicated classes, you should feel free to go beyond these patterns
to make life as easy as possible for your users.

13.3.4 Exercises

1. Write a constructor for data.frame objects. What base type is a data
   frame built on? What attributes does it use? What are the restrictions placed on the individual elements? What about the names?

2. Enhance my factor() helper to have better behaviour when one or
   more values is not found in levels. What does base::factor() do
   in this situation?

3. Carefully read the source code of factor(). What does it do that
   my constructor does not?

4. Factors have an optional "contrasts" attribute. Read the help for
   C(), and briefly describe the purpose of the attribute. What type
   should it have? Rewrite the new_factor() constructor to include this
   attribute.

5. Read the documentation for utils::as.roman(). How would you
   write a constructor for this class? Does it need a validator? What
   might a helper do?

---

[3]This helper is not efficient: behind the scenes ISODatetime() works by pasting the components into a string and then using strptime(). A more efficient equivalent is available in
lubridate::make_datetime().

13.4 Generics and methods

The job of an S3 generic is to perform method dispatch, i.e. find the specific implementation for a class. Method dispatch is performed by `UseMethod()`, which every generic calls[4]. `UseMethod()` takes two arguments: the name of the generic function (required), and the argument to use for method dispatch (optional). If you omit the second argument, it will dispatch based on the first argument, which is almost always what is desired.

Most generics are very simple, and consist of only a call to `UseMethod()`. Take `mean()` for example:

```
mean
#> function (x, ...)
#> UseMethod("mean")
#> <bytecode: 0x7fed45725430>
#> <environment: namespace:base>
```

Creating your own generic is similarly simple:

```
my_new_generic <- function(x) {
  UseMethod("my_new_generic")
}
```

(If you wonder why we have to repeat `my_new_generic` twice, think back to Section 6.2.3.)

You don't pass any of the arguments of the generic to `UseMethod()`; it uses deep magic to pass to the method automatically. The precise process is complicated and frequently surprising, so you should avoid doing any computation in a generic. To learn the full details, carefully read the Technical Details section in `?UseMethod`.

13.4.1 Method dispatch

How does `UseMethod()` work? It basically creates a vector of method names, `paste0("generic", ".", c(class(x), "default"))`, and then looks for each potential method in turn. We can see this in action with `sloop::s3_dispatch()`. You give it a call to an S3 generic, and it lists all the possible methods. For example, what method is called when you print a `Date` object?

---

[4]The exception is internal generics, which are implemented in C, and are the topic of Section 13.7.2.

```
x <- Sys.Date()
s3_dispatch(print(x))
#> => print.Date
#>  * print.default
```

The output here is simple:

- => indicates the method that is called, here print.Date()
- * indicates a method that is defined, but not called, here print.default().

The "default" class is a special **pseudo-class**. This is not a real class, but is included to make it possible to define a standard fallback that is found whenever a class-specific method is not available.

The essence of method dispatch is quite simple, but as the chapter proceeds you'll see it get progressively more complicated to encompass inheritance, base types, internal generics, and group generics. The code below shows a couple of more complicated cases which we'll come back to in Sections 14.2.4 and 13.7.

```
x <- matrix(1:10, nrow = 2)
s3_dispatch(mean(x))
#>     mean.matrix
#>     mean.integer
#>     mean.numeric
#> => mean.default

s3_dispatch(sum(Sys.time()))
#>     sum.POSIXct
#>     sum.POSIXt
#>     sum.default
#> => Summary.POSIXct
#>     Summary.POSIXt
#>     Summary.default
#> -> sum (internal)
```

13.4.2 Finding methods

sloop::s3_dispatch() lets you find the specific method used for a single call. What if you want to find all methods defined for a generic or associated with a class? That's the job of sloop::s3_methods_generic() and sloop::s3_methods_class():

```
s3_methods_generic("mean")
#> # A tibble: 6 x 4
#>   generic class    visible source
#>   <chr>   <chr>    <lgl>   <chr>
#> 1 mean    Date     TRUE    base
#> 2 mean    default  TRUE    base
#> 3 mean    difftime TRUE    base
#> 4 mean    POSIXct  TRUE    base
#> 5 mean    POSIXlt  TRUE    base
#> 6 mean    quosure  FALSE   registered S3method

s3_methods_class("ordered")
#> # A tibble: 4 x 4
#>   generic       class   visible source
#>   <chr>         <chr>   <lgl>   <chr>
#> 1 as.data.frame ordered TRUE    base
#> 2 Ops           ordered TRUE    base
#> 3 relevel       ordered FALSE   registered S3method
#> 4 Summary       ordered TRUE    base
```

13.4.3 Creating methods

There are two wrinkles to be aware of when you create a new method:

- First, you should only ever write a method if you own the generic or the class. R will allow you to define a method even if you don't, but it is exceedingly bad manners. Instead, work with the author of either the generic or the class to add the method in their code.

- A method must have the same arguments as its generic. This is enforced in packages by R CMD check, but it's good practice even if you're not creating a package.

  There is one exception to this rule: if the generic has ..., the method can contain a superset of the arguments. This allows methods to take arbitrary additional arguments. The downside of using ..., however, is that any misspelled arguments will be silently swallowed[5], as mentioned in Section 6.6.

---

[5]See https://github.com/hadley/ellipsis for an experimental way of warning when methods fail to use all the arguments in ..., providing a potential resolution of this issue.

13.4.4 Exercises

1. Read the source code for t() and t.test() and confirm that
   t.test() is an S3 generic and not an S3 method. What happens
   if you create an object with class test and call t() with it? Why?

   ```
   x <- structure(1:10, class = "test")
   t(x)
   ```

2. What generics does the table class have methods for?

3. What generics does the ecdf class have methods for?

4. Which base generic has the greatest number of defined methods?

5. Carefully read the documentation for UseMethod() and explain why
   the following code returns the results that it does. What two usual
   rules of function evaluation does UseMethod() violate?

   ```
   g <- function(x) {
     x <- 10
     y <- 10
     UseMethod("g")
   }
   g.default <- function(x) c(x = x, y = y)

   x <- 1
   y <- 1
   g(x)
   #>  x  y
   #>  1 10
   ```

6. What are the arguments to [? Why is this a hard question to answer?

13.5 Object styles

So far I've focussed on vector style classes like Date and factor. These have
the key property that length(x) represents the number of observations in the
vector. There are three variants that do not have this property:

- Record style objects use a list of equal-length vectors to represent individual components of the object. The best example of this is POSIXlt, which underneath the hood is a list of 11 date-time components like year, month, and day. Record style classes override length() and subsetting methods to conceal this implementation detail.

```
x <- as.POSIXlt(ISOdatetime(2020, 1, 1, 0, 0, 1:3))
x
#> [1] "2020-01-01 00:00:01 CST" "2020-01-01 00:00:02 CST"
#> [3] "2020-01-01 00:00:03 CST"

length(x)
#> [1] 3
length(unclass(x))
#> [1] 11

x[[1]] # the first date time
#> [1] "2020-01-01 00:00:01 CST"
unclass(x)[[1]] # the first component, the number of seconds
#> [1] 1 2 3
```

- Data frames are similar to record style objects in that both use lists of equal length vectors. However, data frames are conceptually two dimensional, and the individual components are readily exposed to the user. The number of observations is the number of rows, not the length:

```
x <- data.frame(x = 1:100, y = 1:100)
length(x)
#> [1] 2
nrow(x)
#> [1] 100
```

- Scalar objects typically use a list to represent a single thing. For example, an lm object is a list of length 12 but it represents one model.

```
mod <- lm(mpg ~ wt, data = mtcars)
length(mod)
#> [1] 12
```

Scalar objects can also be built on top of functions, calls, and environments[6]. This is less generally useful, but you can see applications in stats::ecdf(), R6 (Chapter 14), and rlang::quo() (Chapter 19).

---

[6]You can also build an object on top of a pairlist, but I have yet to find a good reason to do so.

Unfortunately, describing the appropriate use of each of these object styles is beyond the scope of this book. However, you can learn more from the documentation of the vctrs package (`https://vctrs.r-lib.org`); the package also provides constructors and helpers that make implementation of the different styles easier.

13.5.1 Exercises

1. Categorise the objects returned by `lm()`, `factor()`, `table()`, `as.Date()`, `as.POSIXct()` `ecdf()`, `ordered()`, `I()` into the styles described above.

2. What would a constructor function for `lm` objects, `new_lm()`, look like? Use `?lm` and experimentation to figure out the required fields and their types.

13.6 Inheritance

S3 classes can share behaviour through a mechanism called **inheritance**. Inheritance is powered by three ideas:

- The class can be a character *vector*. For example, the ordered and POSIXct classes have two components in their class:

```
class(ordered("x"))
#> [1] "ordered" "factor"
class(Sys.time())
#> [1] "POSIXct" "POSIXt"
```

- If a method is not found for the class in the first element of the vector, R looks for a method for the second class (and so on):

```
s3_dispatch(print(ordered("x")))
#>     print.ordered
#> => print.factor
#>   * print.default
s3_dispatch(print(Sys.time()))
#> => print.POSIXct
#>     print.POSIXt
#>   * print.default
```

- A method can delegate work by calling `NextMethod()`. We'll come back to that very shortly; for now, note that `s3_dispatch()` reports delegation with `->`.

```
s3_dispatch(ordered("x")[1])
#>    [.ordered
#> => [.factor
#>    [.default
#> -> [ (internal)
s3_dispatch(Sys.time()[1])
#> => [.POSIXct
#>    [.POSIXt
#>    [.default
#> -> [ (internal)
```

Before we continue we need a bit of vocabulary to describe the relationship between the classes that appear together in a class vector. We'll say that `ordered` is a **subclass** of `factor` because it always appears before it in the class vector, and, conversely, we'll say `factor` is a **superclass** of `ordered`.

S3 imposes no restrictions on the relationship between sub- and superclasses but your life will be easier if you impose some. I recommend that you adhere to two simple principles when creating a subclass:

- The base type of the subclass should be that same as the superclass.

- The attributes of the subclass should be a superset of the attributes of the superclass.

`POSIXt` does not adhere to these principles because `POSIXct` has type double, and `POSIXlt` has type list. This means that `POSIXt` is not a superclass, and illustrates that it's quite possible to use the S3 inheritance system to implement other styles of code sharing (here `POSIXt` plays a role more like an interface), but you'll need to figure out safe conventions yourself.

13.6.1 `NextMethod()`

`NextMethod()` is the hardest part of inheritance to understand, so we'll start with a concrete example for the most common use case: `[`. We'll start by creating a simple toy class: a `secret` class that hides its output when printed:

```
new_secret <- function(x = double()) {
  stopifnot(is.double(x))
  structure(x, class = "secret")
}
```

```
print.secret <- function(x, ...) {
  print(strrep("x", nchar(x)))
  invisible(x)
}

x <- new_secret(c(15, 1, 456))
x
#> [1] "xx"  "x"   "xxx"
```

This works, but the default [method doesn't preserve the class:

```
s3_dispatch(x[1])
#>     [.secret
#>     [.default
#> => [ (internal)
x[1]
#> [1] 15
```

To fix this, we need to provide a [.secret method. How could we implement this method? The naive approach won't work because we'll get stuck in an infinite loop:

```
`[.secret` <- function(x, i) {
  new_secret(x[i])
}
```

Instead, we need some way to call the underlying [code, i.e. the implementation that would get called if we didn't have a [.secret method. One approach would be to unclass() the object:

```
`[.secret` <- function(x, i) {
  x <- unclass(x)
  new_secret(x[i])
}
x[1]
#> [1] "xx"
```

This works, but is inefficient because it creates a copy of x. A better approach is to use NextMethod(), which concisely solves the problem delegating to the method that would've have been called if [.secret didn't exist:

```
`[.secret` <- function(x, i) {
  new_secret(NextMethod())
}
x[1]
#> [1] "xx"
```

We can see what's going on with `sloop::s3_dispatch()`:

```
s3_dispatch(x[1])
#> => [.secret
#>    [.default
#> -> [ (internal)
```

The `=>` indicates that `[.secret` is called, but that `NextMethod()` delegates work to the underlying internal `[` method, as shown by the `->`.

As with `UseMethod()`, the precise semantics of `NextMethod()` are complex. In particular, it tracks the list of potential next methods with a special variable, which means that modifying the object that's being dispatched upon will have no impact on which method gets called next.

13.6.2 Allowing subclassing

When you create a class, you need to decide if you want to allow subclasses, because it requires some changes to the constructor and careful thought in your methods.

To allow subclasses, the parent constructor needs to have ... and `class` arguments:

```
new_secret <- function(x, ..., class = character()) {
  stopifnot(is.double(x))

  structure(
    x,
    ...,
    class = c(class, "secret")
  )
}
```

Then the subclass constructor can just call to the parent class constructor with additional arguments as needed. For example, imagine we want to create a supersecret class which also hides the number of characters:

```
new_supersecret <- function(x) {
  new_secret(x, class = "supersecret")
}

print.supersecret <- function(x, ...) {
  print(rep("xxxxx", length(x)))
  invisible(x)
}

x2 <- new_supersecret(c(15, 1, 456))
x2
#> [1] "xxxxx" "xxxxx" "xxxxx"
```

To allow inheritance, you also need to think carefully about your methods, as you can no longer use the constructor. If you do, the method will always return the same class, regardless of the input. This forces whoever makes a subclass to do a lot of extra work.

Concretely, this means we need to revise the [.secret method. Currently it always returns a secret(), even when given a supersecret:

```
`[.secret` <- function(x, ...) {
  new_secret(NextMethod())
}

x2[1:3]
#> [1] "xx"  "x"   "xxx"
```

We want to make sure that [.secret returns the same class as x even if it's a subclass. As far as I can tell, there is no way to solve this problem using base R alone. Instead, you'll need to use the vctrs package, which provides a solution in the form of the vctrs::vec_restore() generic. This generic takes two inputs: an object which has lost subclass information, and a template object to use for restoration.

Typically vec_restore() methods are quite simple: you just call the constructor with appropriate arguments:

```
vec_restore.secret <- function(x, to, ...) new_secret(x)
vec_restore.supersecret <- function(x, to, ...) new_supersecret(x)
```

(If your class has attributes, you'll need to pass them from to into the constructor.)

Now we can use vec_restore() in the [.secret method:

```
`[.secret` <- function(x, ...) {
  vctrs::vec_restore(NextMethod(), x)
}
x2[1:3]
#> [1] "xxxxx" "xxxxx" "xxxxx"
```

(I only fully understood this issue quite recently, so at time of writing it is not used in the tidyverse. Hopefully by the time you're reading this, it will have rolled out, making it much easier to (e.g.) subclass tibbles.)

If you build your class using the tools provided by the vctrs package, [will gain this behaviour automatically. You will only need to provide your own [method if you use attributes that depend on the data or want non-standard subsetting behaviour. See ?vctrs::new_vctr for details.

13.6.3 Exercises

1. How does [.Date support subclasses? How does it fail to support subclasses?

2. R has two classes for representing date time data, POSIXct and POSIXlt, which both inherit from POSIXt. Which generics have different behaviours for the two classes? Which generics share the same behaviour?

3. What do you expect this code to return? What does it actually return? Why?

```
generic2 <- function(x) UseMethod("generic2")
generic2.a1 <- function(x) "a1"
generic2.a2 <- function(x) "a2"
generic2.b <- function(x) {
  class(x) <- "a1"
  NextMethod()
}

generic2(structure(list(), class = c("b", "a2")))
```

13.7   Dispatch details

This chapter concludes with a few additional details about method dispatch. It is safe to skip these details if you're new to S3.

13.7.1   S3 and base types

What happens when you call an S3 generic with a base object, i.e. an object with no class? You might think it would dispatch on what `class()` returns:

```
class(matrix(1:5))
#> [1] "matrix"
```

But unfortunately dispatch actually occurs on the **implicit class**, which has three components:

- The string "array" or "matrix" if the object has dimensions
- The result of `typeof()` with a few minor tweaks
- The string "numeric" if object is "integer" or "double"

There is no base function that will compute the implicit class, but you can use `sloop::s3_class()`

```
s3_class(matrix(1:5))
#> [1] "matrix"  "integer" "numeric"
```

This is used by `s3_dispatch()`:

```
s3_dispatch(print(matrix(1:5)))
#>     print.matrix
#>     print.integer
#>     print.numeric
#> => print.default
```

This means that the `class()` of an object does not uniquely determine its dispatch:

```
x1 <- 1:5
class(x1)
#> [1] "integer"
s3_dispatch(mean(x1))
```

```
#>      mean.integer
#>      mean.numeric
#> => mean.default

x2 <- structure(x1, class = "integer")
class(x2)
#> [1] "integer"
s3_dispatch(mean(x2))
#>      mean.integer
#> => mean.default
```

13.7.2 Internal generics

Some base functions, like `[`, `sum()`, and `cbind()`, are called **internal generics** because they don't call `UseMethod()` but instead call the C functions `DispatchGroup()` or `DispatchOrEval()`. `s3_dispatch()` shows internal generics by including the name of the generic followed by `(internal)`:

```
s3_dispatch(Sys.time()[1])
#> => [.POSIXct
#>      [.POSIXt
#>      [.default
#> -> [ (internal)
```

For performance reasons, internal generics do not dispatch to methods unless the class attribute has been set, which means that internal generics do not use the implicit class. Again, if you're ever confused about method dispatch, you can rely on `s3_dispatch()`.

13.7.3 Group generics

Group generics are the most complicated part of S3 method dispatch because they involve both `NextMethod()` and internal generics. Like internal generics, they only exist in base R, and you cannot define your own group generic.

There are four group generics:

- **Math**: `abs()`, `sign()`, `sqrt()`, `floor()`, `cos()`, `sin()`, `log()`, and more (see `?Math` for the complete list).
- **Ops**: `+`, `-`, `*`, `/`, `^`, `%%`, `%/%`, `&`, `|`, `!`, `==`, `!=`, `<`, `<=`, `>=`, and `>`.
- **Summary**: `all()`, `any()`, `sum()`, `prod()`, `min()`, `max()`, and `range()`.

- **Complex**: `Arg()`, `Conj()`, `Im()`, `Mod()`, `Re()`.

Defining a single group generic for your class overrides the default behaviour for all of the members of the group. Methods for group generics are looked for only if the methods for the specific generic do not exist:

```
s3_dispatch(sum(Sys.time()))
#>      sum.POSIXct
#>      sum.POSIXt
#>      sum.default
#> => Summary.POSIXct
#>      Summary.POSIXt
#>      Summary.default
#> -> sum (internal)
```

Most group generics involve a call to `NextMethod()`. For example, take `difftime()` objects. If you look at the method dispatch for `abs()`, you'll see there's a `Math` group generic defined.

```
y <- as.difftime(10, units = "mins")
s3_dispatch(abs(y))
#>      abs.difftime
#>      abs.default
#> => Math.difftime
#>      Math.default
#> -> abs (internal)
```

`Math.difftime` basically looks like this:

```
Math.difftime <- function(x, ...) {
  new_difftime(NextMethod(), units = attr(x, "units"))
}
```

It dispatches to the next method, here the internal default, to perform the actual computation, then restore the class and attributes. (To better support subclasses of `difftime` this would need to call `vec_restore()`, as described in Section 13.6.2.)

Inside a group generic function a special variable `.Generic` provides the actual generic function called. This can be useful when producing error messages, and can sometimes be useful if you need to manually re-call the generic with different arguments.

13.7.4 Double dispatch

Generics in the Ops group, which includes the two-argument arithmetic and Boolean operators like - and &, implement a special type of method dispatch. They dispatch on the type of *both* of the arguments, which is called **double dispatch**. This is necessary to preserve the commutative property of many operators, i.e. a + b should equal b + a. Take the following simple example:

```
date <- as.Date("2017-01-01")
integer <- 1L

date + integer
#> [1] "2017-01-02"
integer + date
#> [1] "2017-01-02"
```

If + dispatched only on the first argument, it would return different values for the two cases. To overcome this problem, generics in the Ops group use a slightly different strategy from usual. Rather than doing a single method dispatch, they do two, one for each input. There are three possible outcomes of this lookup:

- The methods are the same, so it doesn't matter which method is used.

- The methods are different, and R falls back to the internal method with a warning.

- One method is internal, in which case R calls the other method.

This approach is error prone so if you want to implement robust double dispatch for algebraic operators, I recommend using the vctrs package. See ?vctrs::vec_arith for details.

13.7.5 Exercises

1. Explain the differences in dispatch below:

```
length.integer <- function(x) 10

x1 <- 1:5
class(x1)
#> [1] "integer"
s3_dispatch(length(x1))
#>  * length.integer
#>    length.numeric
```

```
#>     length.default
#> => length (internal)

x2 <- structure(x1, class = "integer")
class(x2)
#> [1] "integer"
s3_dispatch(length(x2))
#> => length.integer
#>    length.default
#>  * length (internal)
```

2. What classes have a method for the `Math` group generic in base R? Read the source code. How do the methods work?

3. `Math.difftime()` is more complicated than I described. Why?

14

R6

14.1 Introduction

This chapter describes the R6 OOP system. R6 has two special properties:

- It uses the encapsulated OOP paradigm, which means that methods belong to objects, not generics, and you call them like object$method().

- R6 objects are **mutable**, which means that they are modified in place, and hence have reference semantics.

If you've learned OOP in another programming language, it's likely that R6 will feel very natural, and you'll be inclined to prefer it over S3. Resist the temptation to follow the path of least resistance: in most cases R6 will lead you to non-idiomatic R code. We'll come back to this theme in Section 16.3.

R6 is very similar to a base OOP system called **reference classes**, or RC for short. I describe why I teach R6 and not RC in Section 14.5.

Outline

- Section 14.2 introduces R6::R6class(), the one function that you need to know to create R6 classes. You'll learn about the constructor method, $new(), which allows you to create R6 objects, as well as other important methods like $initialize() and $print().

- Section 14.3 discusses the access mechanisms of R6: private and active fields. Together, these allow you to hide data from the user, or expose private data for reading but not writing.

- Section 14.4 explores the consequences of R6's reference semantics. You'll learn about the use of finalizers to automatically clean up any operations performed in the initializer, and a common gotcha if you use an R6 object as a field in another R6 object.

- Section 14.5 describes why I cover R6, rather than the base RC system.

Prerequisites

Because R6 (`https://r6.r-lib.org`) is not built into base R, you'll need to install and load the R6 package to use it:

```
# install.packages("R6")
library(R6)
```

R6 objects have reference semantics which means that they are modified in-place, not copied-on-modify. If you're not familiar with these terms, brush up your vocab by reading Section 2.5.

14.2   Classes and methods

R6 only needs a single function call to create both the class and its methods: `R6::R6Class()`. This is the only function from the package that you'll ever use![1]

The following example shows the two most important arguments to `R6Class()`:

- The first argument is the `classname`. It's not strictly needed, but it improves error messages and makes it possible to use R6 objects with S3 generics. By convention, R6 classes have `UpperCamelCase` names.

- The second argument, `public`, supplies a list of methods (functions) and fields (anything else) that make up the public interface of the object. By convention, methods and fields use `snake_case`. Methods can access the methods and fields of the current object via `self$`.[2]

```
Accumulator <- R6Class("Accumulator", list(
  sum = 0,
  add = function(x = 1) {
    self$sum <- self$sum + x
    invisible(self)
  })
)
```

---

[1]That means if you're creating R6 in a package, you only need to make sure it's listed in the `Imports` field of the `DESCRIPTION`. There's no need to import the package into the `NAMESPACE`.

[2]Unlike in this in python, the `self` variable is automatically provided by R6, and does not form part of the method signature.

You should always assign the result of R6Class() into a variable with the same name as the class, because R6Class() returns an R6 object that defines the class:

```
Accumulator
#> <Accumulator> object generator
#>   Public:
#>     sum: 0
#>     add: function (x = 1)
#>     clone: function (deep = FALSE)
#>   Parent env: <environment: R_GlobalEnv>
#>   Locked objects: TRUE
#>   Locked class: FALSE
#>   Portable: TRUE
```

You construct a new object from the class by calling the new() method. In R6, methods belong to objects, so you use $ to access new():

```
x <- Accumulator$new()
```

You can then call methods and access fields with $:

```
x$add(4)
x$sum
#> [1] 4
```

In this class, the fields and methods are public, which means that you can get or set the value of any field. Later, we'll see how to use private fields and methods to prevent casual access to the internals of your class.

To make it clear when we're talking about fields and methods as opposed to variables and functions, I'll prefix their names with $. For example, the Accumulate class has field $sum and method $add().

14.2.1 Method chaining

$add() is called primarily for its side-effect of updating $sum.

```
Accumulator <- R6Class("Accumulator", list(
  sum = 0,
  add = function(x = 1) {
    self$sum <- self$sum + x
    invisible(self)
```

```
  })
)
```

Side-effect R6 methods should always return `self` invisibly. This returns the "current" object and makes it possible to chain together multiple method calls:

```
x$add(10)$add(10)$sum
#> [1] 24
```

For, readability, you might put one method call on each line:

```
x$
  add(10)$
  add(10)$
  sum
#> [1] 44
```

This technique is called **method chaining** and is commonly used in languages like Python and JavaScript. Method chaining is deeply related to the pipe, and we'll discuss the pros and cons of each approach in Section 16.3.3.

14.2.2 Important methods

There are two important methods that should be defined for most classes: `$initialize()` and `$print()`. They're not required, but providing them will make your class easier to use.

`$initialize()` overrides the default behaviour of `$new()`. For example, the following code defines an Person class with fields `$name` and `$age`. To ensure that that `$name` is always a single string, and `$age` is always a single number, I placed checks in `$initialize()`.

```
Person <- R6Class("Person", list(
  name = NULL,
  age = NA,
  initialize = function(name, age = NA) {
    stopifnot(is.character(name), length(name) == 1)
    stopifnot(is.numeric(age), length(age) == 1)

    self$name <- name
    self$age <- age
  }
```

```
))

hadley <- Person$new("Hadley", age = "thirty-eight")
#> Error in .subset2(public_bind_env, "initialize")(...):
#> is.numeric(age) is not TRUE

hadley <- Person$new("Hadley", age = 38)
```

If you have more expensive validation requirements, implement them in a separate $validate() and only call when needed.

Defining $print() allows you to override the default printing behaviour. As with any R6 method called for its side effects, $print() should return invisible(self).

```
Person <- R6Class("Person", list(
  name = NULL,
  age = NA,
  initialize = function(name, age = NA) {
    self$name <- name
    self$age <- age
  },
  print = function(...) {
    cat("Person: \n")
    cat("  Name: ", self$name, "\n", sep = "")
    cat("  Age:  ", self$age, "\n", sep = "")
    invisible(self)
  }
))

hadley2 <- Person$new("Hadley")
hadley2
#> Person:
#>   Name: Hadley
#>   Age:  NA
```

This code illustrates an important aspect of R6. Because methods are bound to individual objects, the previously created hadley object does not get this new method:

```
hadley
#> <Person>
#>   Public:
#>     age: 38
```

```
#>     clone: function (deep = FALSE)
#>     initialize: function (name, age = NA)
#>     name: Hadley
```

```
hadley$print
#> NULL
```

From the perspective of R6, there is no relationship between `hadley` and `hadley2`; they just coincidentally share the same class name. This doesn't cause problems when using already developed R6 objects but can make interactive experimentation confusing. If you're changing the code and can't figure out why the results of method calls aren't any different, make sure you've re-constructed R6 objects with the new class.

14.2.3   Adding methods after creation

Instead of continuously creating new classes, it's also possible to modify the fields and methods of an existing class. This is useful when exploring interactively, or when you have a class with many functions that you'd like to break up into pieces. Add new elements to an existing class with `$set()`, supplying the visibility (more on in Section 14.3), the name, and the component.

```
Accumulator <- R6Class("Accumulator")
Accumulator$set("public", "sum", 0)
Accumulator$set("public", "add", function(x = 1) {
  self$sum <- self$sum + x
  invisible(self)
})
```

As above, new methods and fields are only available to new objects; they are not retrospectively added to existing objects.

14.2.4   Inheritance

To inherit behaviour from an existing class, provide the class object to the `inherit` argument:

```
AccumulatorChatty <- R6Class("AccumulatorChatty",
  inherit = Accumulator,
  public = list(
    add = function(x = 1) {
```

```
        cat("Adding ", x, "\n", sep = "")
        super$add(x = x)
      }
    )
  )
)

x2 <- AccumulatorChatty$new()
x2$add(10)$add(1)$sum
#> Adding 10
#> Adding 1
#> [1] 11
```

`$add()` overrides the superclass implementation, but we can still delegate to the superclass implementation by using `super$`. (This is analogous to `NextMethod()` in S3, as discussed in Section 13.6.) Any methods which are not overridden will use the implementation in the parent class.

14.2.5 Introspection

Every R6 object has an S3 class that reflects its hierarchy of R6 classes. This means that the easiest way to determine the class (and all classes it inherits from) is to use `class()`:

```
class(hadley2)
#> [1] "Person" "R6"
```

The S3 hierarchy includes the base "R6" class. This provides common behaviour, including a `print.R6()` method which calls `$print()`, as described above.

You can list all methods and fields with `names()`:

```
names(hadley2)
#> [1] ".__enclos_env__" "age"        "name"
#> [4] "clone"           "print"      "initialize"
```

We defined `$name`, `$age`, `$print`, and `$initialize`. As suggested by the name, `.__enclos_env__` is an internal implementation detail that you shouldn't touch; we'll come back to `$clone()` in Section 14.4.

14.2.6 Exercises

1. Create a bank account R6 class that stores a balance and allows you to deposit and withdraw money. Create a subclass that throws an error if you attempt to go into overdraft. Create another subclass that allows you to go into overdraft, but charges you a fee.

2. Create an R6 class that represents a shuffled deck of cards. You should be able to draw cards from the deck with $draw(n), and return all cards to the deck and reshuffle with $reshuffle(). Use the following code to make a vector of cards.

```
suit <- c(" ", " ", " ", " ")
value <- c("A", 2:10, "J", "Q", "K")
cards <- paste0(rep(value, 4), suit)
```

3. Why can't you model a bank account or a deck of cards with an S3 class?

4. Create an R6 class that allows you to get and set the current time-zone. You can access the current timezone with Sys.timezone() and set it with Sys.setenv(TZ = "newtimezone"). When setting the time zone, make sure the new time zone is in the list provided by Olson-Names().

5. Create an R6 class that manages the current working directory. It should have $get() and $set() methods.

6. Why can't you model the time zone or current working directory with an S3 class?

7. What base type are R6 objects built on top of? What attributes do they have?

14.3 Controlling access

R6Class() has two other arguments that work similarly to public:

- private allows you to create fields and methods that are only available from within the class, not outside of it.

- active allows you to use accessor functions to define dynamic, or active, fields.

These are described in the following sections.

14.3.1 Privacy

With R6 you can define **private** fields and methods, elements that can only be accessed from within the class, not from the outside[3]. There are two things that you need to know to take advantage of private elements:

- The `private` argument to `R6Class` works in the same way as the `public` argument: you give it a named list of methods (functions) and fields (everything else).

- Fields and methods defined in `private` are available within the methods using `private$` instead of `self$`. You cannot access private fields or methods outside of the class.

To make this concrete, we could make `$age` and `$name` fields of the Person class private. With this definition of `Person` we can only set `$age` and `$name` during object creation, and we cannot access their values from outside of the class.

```
Person <- R6Class("Person",
  public = list(
    initialize = function(name, age = NA) {
      private$name <- name
      private$age <- age
    },
    print = function(...) {
      cat("Person: \n")
      cat("  Name: ", private$name, "\n", sep = "")
      cat("  Age:  ", private$age, "\n", sep = "")
    }
  ),
  private = list(
    age = NA,
    name = NULL
  )
)

hadley3 <- Person$new("Hadley")
hadley3
#> Person:
#>   Name: Hadley
#>   Age:  NA
hadley3$name
#> NULL
```

---

[3]Because R is such a flexible language, it's technically still possible to access private values, but you'll have to try much harder, spelunking in to the details of R6's implementation.

The distinction between public and private fields is important when you create complex networks of classes, and you want to make it as clear as possible what it's ok for others to access. Anything that's private can be more easily refactored because you know others aren't relying on it. Private methods tend to be less important in R compared to other programming languages because the object hierarchies in R tend to be simpler.

14.3.2   Active fields

Active fields allow you to define components that look like fields from the outside, but are defined with functions, like methods. Active fields are implemented using **active bindings** (Section 7.2.6). Each active binding is a function that takes a single argument: value. If the argument is missing(), the value is being retrieved; otherwise it's being modified.

For example, you could make an active field random that returns a different value every time you access it:

```
Rando <- R6::R6Class("Rando", active = list(
  random = function(value) {
    if (missing(value)) {
      runif(1)
    } else {
      stop("Can't set `$random`", call. = FALSE)
    }
  }
))
x <- Rando$new()
x$random
#> [1] 0.0808
x$random
#> [1] 0.834
x$random
#> [1] 0.601
```

Active fields are particularly useful in conjunction with private fields, because they make it possible to implement components that look like fields from the outside but provide additional checks. For example, we can use them to make a read-only age field, and to ensure that name is a length 1 character vector.

```
Person <- R6Class("Person",
  private = list(
    .age = NA,
    .name = NULL
```

```
  ),
  active = list(
    age = function(value) {
      if (missing(value)) {
        private$.age
      } else {
        stop("`$age` is read only", call. = FALSE)
      }
    },
    name = function(value) {
      if (missing(value)) {
        private$.name
      } else {
        stopifnot(is.character(value), length(value) == 1)
        private$.name <- value
        self
      }
    }
  ),
  public = list(
    initialize = function(name, age = NA) {
      private$.name <- name
      private$.age <- age
    }
  )
)

hadley4 <- Person$new("Hadley", age = 38)
hadley4$name
#> [1] "Hadley"
hadley4$name <- 10
#> Error in (function (value) : is.character(value) is not TRUE
hadley4$age <- 20
#> Error: `$age` is read only
```

14.3.3   Exercises

1. Create a bank account class that prevents you from directly setting the account balance, but you can still withdraw from and deposit to. Throw an error if you attempt to go into overdraft.

2. Create a class with a write-only $password field. It should have
   $check_password(password) method that returns TRUE or FALSE, but
   there should be no way to view the complete password.

3. Extend the Rando class with another active binding that allows you
   to access the previous random value. Ensure that active binding is
   the only way to access the value.

4. Can subclasses access private fields/methods from their parent? Perform an experiment to find out.

14.4 Reference semantics

One of the big differences between R6 and most other objects is that they
have reference semantics. The primary consequence of reference semantics is
that objects are not copied when modified:

```
y1 <- Accumulator$new()
y2 <- y1

y1$add(10)
c(y1 = y1$sum, y2 = y2$sum)
#> y1 y2
#> 10 10
```

Instead, if you want a copy, you'll need to explicitly $clone() the object:

```
y1 <- Accumulator$new()
y2 <- y1$clone()

y1$add(10)
c(y1 = y1$sum, y2 = y2$sum)
#> y1 y2
#> 10  0
```

($clone() does not recursively clone nested R6 objects. If you want that, you'll
need to use $clone(deep = TRUE).)

There are three other less obvious consequences:

• It is harder to reason about code that uses R6 objects because you need to
  understand more context.

- It makes sense to think about when an R6 object is deleted, and you can write a `$finalize()` to complement the `$initialize()`.

- If one of the fields is an R6 object, you must create it inside `$initialize()`, not `R6Class()`.

These consequences are described in more detail below.

14.4.1 Reasoning

Generally, reference semantics makes code harder to reason about. Take this very simple example:

```
x <- list(a = 1)
y <- list(b = 2)

z <- f(x, y)
```

For the vast majority of functions, you know that the final line only modifies z.

Take a similar example that uses an imaginary `List` reference class:

```
x <- List$new(a = 1)
y <- List$new(b = 2)

z <- f(x, y)
```

The final line is much harder to reason about: if `f()` calls methods of x or y, it might modify them as well as z. This is the biggest potential downside of R6 and you should take care to avoid it by writing functions that either return a value, or modify their R6 inputs, but not both. That said, doing both can lead to substantially simpler code in some cases, and we'll discuss this further in Section 16.3.2.

14.4.2 Finalizer

One useful property of reference semantics is that it makes sense to think about when an R6 object is **finalized**, i.e. when it's deleted. This doesn't make sense for most objects because copy-on-modify semantics mean that there may be many transient versions of an object, as alluded to in Section 2.6. For example, the following creates two factor objects: the second is created when the levels are modified, leaving the first to be destroyed by the garbage collector.

```
x <- factor(c("a", "b", "c"))
levels(x) <- c("c", "b", "a")
```

Since R6 objects are not copied-on-modify they are only deleted once, and it makes sense to think about `$finalize()` as a complement to `$initialize()`. Finalizers usually play a similar role to `on.exit()` (as described in Section 6.7.4), cleaning up any resources created by the initializer. For example, the following class wraps up a temporary file, automatically deleting it when the class is finalized.

```
TemporaryFile <- R6Class("TemporaryFile", list(
  path = NULL,
  initialize = function() {
    self$path <- tempfile()
  },
  finalize = function() {
    message("Cleaning up ", self$path)
    unlink(self$path)
  }
))
```

The finalize method will be run when the object is deleted (or more precisely, by the first garbage collection after the object has been unbound from all names) or when R exits. This means that the finalizer can be called effectively anywhere in your R code, and therefore it's almost impossible to reason about finalizer code that touches shared data structures. Avoid these potential problems by only using the finalizer to clean up private resources allocated by initializer.

```
tf <- TemporaryFile$new()
rm(tf)
#> Cleaning up /tmp/Rtmpk73JdI/file155f31d8424bd
```

14.4.3   R6 fields

A final consequence of reference semantics can crop up where you don't expect it. If you use an R6 class as the default value of a field, it will be shared across all instances of the object! Take the following code: we want to create a temporary database every time we call `TemporaryDatabase$new()`, but the current code always uses the same path.

```
TemporaryDatabase <- R6Class("TemporaryDatabase", list(
  con = NULL,
  file = TemporaryFile$new(),
  initialize = function() {
    self$con <- DBI::dbConnect(RSQLite::SQLite(), path = file$path)
  },
  finalize = function() {
    DBI::dbDisconnect(self$con)
  }
))

db_a <- TemporaryDatabase$new()
db_b <- TemporaryDatabase$new()

db_a$file$path == db_b$file$path
#> [1] TRUE
```

(If you're familiar with Python, this is very similar to the "mutable default argument" problem.)

The problem arises because `TemporaryFile$new()` is called only once when the `TemporaryDatabase` class is defined. To fix the problem, we need to make sure it's called every time that `TemporaryDatabase$new()` is called, i.e. we need to put it in `$initialize()`:

```
TemporaryDatabase <- R6Class("TemporaryDatabase", list(
  con = NULL,
  file = NULL,
  initialize = function() {
    self$file <- TemporaryFile$new()
    self$con <- DBI::dbConnect(RSQLite::SQLite(), path = file$path)
  },
  finalize = function() {
    DBI::dbDisconnect(self$con)
  }
))

db_a <- TemporaryDatabase$new()
db_b <- TemporaryDatabase$new()

db_a$file$path == db_b$file$path
#> [1] FALSE
```

14.4.4 Exercises

1. Create a class that allows you to write a line to a specified file.
   You should open a connection to the file in $initialize(), append
   a line using cat() in $append_line(), and close the connection in
   $finalize().

14.5 Why R6?

R6 is very similar to a built-in OO system called **reference classes**, or RC
for short. I prefer R6 to RC because:

- R6 is much simpler. Both R6 and RC are built on top of environments, but
  while R6 uses S3, RC uses S4. This means to fully understand RC, you need
  to understand how the more complicated S4 works.

- R6 has comprehensive online documentation at https://r6.r-lib.org.

- R6 has a simpler mechanism for cross-package subclassing, which just works
  without you having to think about it. For RC, read the details in the "External Methods; Inter-Package Superclasses" section of ?setRefClass.

- RC mingles variables and fields in the same stack of environments so that you
  get (field) and set (field <<- value) fields like regular values. R6 puts fields
  in a separate environment so you get (self$field) and set (self$field <-
  value) with a prefix. The R6 approach is more verbose but I like it because
  it is more explicit.

- R6 is much faster than RC. Generally, the speed of method dispatch is
  not important outside of microbenchmarks. However, RC is quite slow, and
  switching from RC to R6 led to a substantial performance improvement in
  the shiny package. For more details, see vignette("Performance", "R6").

- RC is tied to R. That means if any bugs are fixed, you can only take advantage of the fixes by requiring a newer version of R. This makes it difficult
  for packages (like those in the tidyverse) that need to work across many R
  versions.

- Finally, because the ideas that underlie R6 and RC are similar, it will only
  require a small amount of additional effort to learn RC if you need to.

15

S4

15.1 Introduction

S4 provides a formal approach to functional OOP. The underlying ideas are similar to S3 (the topic of Chapter 13), but implementation is much stricter and makes use of specialised functions for creating classes (`setClass()`), generics (`setGeneric()`), and methods (`setMethod()`). Additionally, S4 provides both multiple inheritance (i.e. a class can have multiple parents) and multiple dispatch (i.e. method dispatch can use the class of multiple arguments).

An important new component of S4 is the **slot**, a named component of the object that is accessed using the specialised subsetting operator @ (pronounced at). The set of slots, and their classes, forms an important part of the definition of an S4 class.

Outline

- Section 15.2 gives a quick overview of the main components of S4: classes, generics, and methods.

- Section 15.3 dives into the details of S4 classes, including prototypes, constructors, helpers, and validators.

- Section 15.4 shows you how to create new S4 generics, and how to supply those generics with methods. You'll also learn about accessor functions which are designed to allow users to safely inspect and modify object slots.

- Section 15.5 dives into the full details of method dispatch in S4. The basic idea is simple, then it rapidly gets more complex once multiple inheritance and multiple dispatch are combined.

- Section 15.6 discusses the interaction between S4 and S3, showing you how to use them together.

Learning more

Like the other OO chapters, the focus here will be on how S4 works, not how to deploy it most effectively. If you do want to use it in practice, there are two main challenges:

- There is no one reference that will answer all your questions about S4.
- R's built-in documentation sometimes clashes with community best practices.

As you move towards more advanced usage, you will need to piece together needed information by carefully reading the documentation, asking questions on StackOverflow, and performing experiments. Some recommendations:

- The Bioconductor community is a long-term user S4 and has produced much of best material about its effective use. Start with S4 classes and methods (https://bioconductor.org/help/course-materials/2017/Zurich/S4-classes-and-methods.html) taught by Martin Morgan and Hervé Pagès, or check for a newer version at Bioconductor course materials (https://bioconductor.org/help/course-materials/).

  Martin Morgan is a member of R-core and the project lead of Bioconductor. He's a world expert on the practical use of S4, and I recommend reading anything he has written about it, starting with the questions he has answered on stackoverflow (http://stackoverflow.com/search?tab=votes&q=user%3a547331%20%5bs4%5d%20is%3aanswe).

- John Chambers is the author of the S4 system, and provides an overview of its motivation and historical context in *Object-oriented programming, functional programming and R* [Chambers, 2014]. For a fuller exploration of S4, see his book *Software for Data Analysis* [Chambers, 2008].

Prerequisites

All functions related to S4 live in the methods package. This package is always available when you're running R interactively, but may not be available when running R in batch mode, i.e. from Rscript[1]. For this reason, it's a good idea to call library(methods) whenever you use S4. This also signals to the reader that you'll be using the S4 object system.

```
library(methods)
```

---

[1]This is a historical quirk introduced because the methods package used to take a long time to load and Rscript is optimised for fast command line invocation.

15.2 Basics

We'll start with a quick overview of the main components of S4. You define an S4 class by calling `setClass()` with the class name and a definition of its slots, and the names and classes of the class data:

```
setClass("Person",
  slots = c(
    name = "character",
    age = "numeric"
  )
)
```

Once the class is defined, you can construct new objects from it by calling `new()` with the name of the class and a value for each slot:

```
john <- new("Person", name = "John Smith", age = NA_real_)
```

Given an S4 object you can see its class with `is()` and access slots with `@` (equivalent to `$`) and `slot()` (equivalent to `[[`):

```
is(john)
#> [1] "Person"
john@name
#> [1] "John Smith"
slot(john, "age")
#> [1] NA
```

Generally, you should only use `@` in your methods. If you're working with someone else's class, look for **accessor** functions that allow you to safely set and get slot values. As the developer of a class, you should also provide your own accessor functions. Accessors are typically S4 generics allowing multiple classes to share the same external interface.

Here we'll create a setter and getter for the `age` slot by first creating generics with `setGeneric()`:

```
setGeneric("age", function(x) standardGeneric("age"))
setGeneric("age<-", function(x, value) standardGeneric("age<-"))
```

And then defining methods with `setMethod()`:

```
setMethod("age", "Person", function(x) x@age)
setMethod("age<-", "Person", function(x, value) {
  x@age <- value
  x
})

age(john) <- 50
age(john)
#> [1] 50
```

If you're using an S4 class defined in a package, you can get help on it with class?Person. To get help for a method, put ? in front of a call (e.g. ?age(john)) and ? will use the class of the arguments to figure out which help file you need.

Finally, you can use sloop functions to identify S4 objects and generics found in the wild:

```
sloop::otype(john)
#> [1] "S4"
sloop::ftype(age)
#> [1] "S4"        "generic"
```

15.2.1 Exercises

1. lubridate::period() returns an S4 class. What slots does it have? What class is each slot? What accessors does it provide?

2. What other ways can you find help for a method? Read ?"?" and summarise the details.

15.3 Classes

To define an S4 class, call setClass() with three arguments:

• The class **name**. By convention, S4 class names use UpperCamelCase.

• A named character vector that describes the names and classes of the **slots** (fields). For example, a person might be represented by a character name and a numeric age: c(name = "character", age = "numeric"). The pseudo-class ANY allows a slot to accept objects of any type.

- A **prototype**, a list of default values for each slot. Technically, the prototype is optional[2], but you should always provide it.

The code below illustrates the three arguments by creating a `Person` class with character `name` and numeric `age` slots.

```
setClass("Person",
  slots = c(
    name = "character",
    age = "numeric"
  ),
  prototype = list(
    name = NA_character_,
    age = NA_real_
  )
)

me <- new("Person", name = "Hadley")
str(me)
#> Formal class 'Person' [package ".GlobalEnv"] with 2 slots
#>   ..@ name: chr "Hadley"
#>   ..@ age : num NA
```

15.3.1 Inheritance

There is one other important argument to `setClass()`: `contains`. This specifies a class (or classes) to inherit slots and behaviour from. For example, we can create an `Employee` class that inherits from the `Person` class, adding an extra slot that describes their `boss`.

```
setClass("Employee",
  contains = "Person",
  slots = c(
    boss = "Person"
  ),
  prototype = list(
    boss = new("Person")
  )
)
```

---

[2]`?setClass` recommends that you avoid the `prototype` argument, but this is generally considered to be bad advice.

```
str(new("Employee"))
#> Formal class 'Employee' [package ".GlobalEnv"] with 3 slots
#>   ..@ boss:Formal class 'Person' [package ".GlobalEnv"] with 2 slots
#>   .. .. ..@ name: chr NA
#>   .. .. ..@ age : num NA
#>   ..@ name: chr NA
#>   ..@ age : num NA
```

setClass() has 9 other arguments but they are either deprecated or not recommended.

15.3.2 Introspection

To determine what classes an object inherits from, use is():

```
is(new("Person"))
#> [1] "Person"
is(new("Employee"))
#> [1] "Employee" "Person"
```

To test if an object inherits from a specific class, use the second argument of is():

```
is(john, "person")
#> [1] FALSE
```

15.3.3 Redefinition

In most programming languages, class definition occurs at compile-time and object construction occurs later, at run-time. In R, however, both definition and construction occur at run time. When you call setClass(), you are registering a class definition in a (hidden) global variable. As with all state-modifying functions you need to use setClass() with care. It's possible to create invalid objects if you redefine a class after already having instantiated an object:

```
setClass("A", slots = c(x = "numeric"))
a <- new("A", x = 10)

setClass("A", slots = c(a_different_slot = "numeric"))
a
```

```
#> An object of class "A"
#> Slot "a_different_slot":
#> Error in slot(object, what): no slot of name "a_different_slot" for
#> this object of class "A"
```

This can cause confusion during interactive creation of new classes. (R6 classes
have the same problem, as described in Section 14.2.2.)

15.3.4 Helper

new() is a low-level constructor suitable for use by you, the developer. User-
facing classes should always be paired with a user-friendly helper. A helper
should always:

- Have the same name as the class, e.g. myclass().
- Have a thoughtfully crafted user interface with carefully chosen default val-
  ues and useful conversions.
- Create carefully crafted error messages tailored towards an end-user.
- Finish by calling by calling methods::new().

The Person class is so simple so a helper is almost superfluous, but we can
use it to clearly define the contract: age is optional but name is required. We'll
also coerce age to a double so the helper also works when passed an integer.

```
Person <- function(name, age = NA) {
  age <- as.double(age)

  new("Person", name = name, age = age)
}

Person("Hadley")
#> An object of class "Person"
#> Slot "name":
#> [1] "Hadley"
#>
#> Slot "age":
#> [1] NA
```

15.3.5 Validator

The constructor automatically checks that the slots have correct classes:

```
Person(mtcars)
#> Error in validObject(.Object): invalid class "Person" object:
#> invalid object for slot "name" in class "Person": got class
#> "data.frame", should be or extend class "character"
```

You will need to implement more complicated checks (i.e. checks that involve lengths, or multiple slots) yourself. For example, we might want to make it clear that the Person class is a vector class, and can store data about multiple people. That's not currently clear because @name and @age can be different lengths:

```
Person("Hadley", age = c(30, 37))
#> An object of class "Person"
#> Slot "name":
#> [1] "Hadley"
#>
#> Slot "age":
#> [1] 30 37
```

To enforce these additional constraints we write a validator with setValidity(). It takes a class and a function that returns TRUE if the input is valid, and otherwise returns a character vector describing the problem(s):

```
setValidity("Person", function(object) {
  if (length(object@name) != length(object@age)) {
    "@name and @age must be same length"
  } else {
    TRUE
  }
})
```

Now we can no longer create an invalid object:

```
Person("Hadley", age = c(30, 37))
#> Error in validObject(.Object): invalid class "Person" object: @name
#> and @age must be same length
```

NB: The validity method is only called automatically by new(), so you can still create an invalid object by modifying it:

```
alex <- Person("Alex", age = 30)
alex@age <- 1:10
```

You can explicitly check the validity yourself by calling `validObject()`:

```
validObject(alex)
#> Error in validObject(alex): invalid class "Person" object: @name and
#> @age must be same length
```

In Section 15.4.4, we'll use `validObject()` to create accessors that can not create invalid objects.

15.3.6 Exercises

1. Extend the Person class with fields to match `utils::person()`. Think about what slots you will need, what class each slot should have, and what you'll need to check in your validity method.

2. What happens if you define a new S4 class that doesn't have any slots? (Hint: read about virtual classes in `?setClass`.)

3. Imagine you were going to reimplement factors, dates, and data frames in S4. Sketch out the `setClass()` calls that you would use to define the classes. Think about appropriate `slots` and `prototype`.

15.4 Generics and methods

The job of a generic is to perform method dispatch, i.e. find the specific implementation for the combination of classes passed to the generic. Here you'll learn how to define S4 generics and methods, then in the next section we'll explore precisely how S4 method dispatch works.

To create a new S4 generic, call `setGeneric()` with a function that calls standardGeneric():

```
setGeneric("myGeneric", function(x) standardGeneric("myGeneric"))
```

By convention, new S4 generics should use `lowerCamelCase`.

It is bad practice to use `{}` in the generic as it triggers a special case that is more expensive, and generally best avoided.

```
# Don't do this!
setGeneric("myGeneric", function(x) {
```

```
  standardGeneric("myGeneric")
})
```

15.4.1 Signature

Like setClass(), setGeneric() has many other arguments. There is only one
that you need to know about: signature. This allows you to control the arguments that are used for method dispatch. If signature is not supplied, all arguments (apart from ...) are used. It is occasionally useful to remove arguments
from dispatch. This allows you to require that methods provide arguments like
verbose = TRUE or quiet = FALSE, but they don't take part in dispatch.

```
setGeneric("myGeneric",
  function(x, ..., verbose = TRUE) standardGeneric("myGeneric"),
  signature = "x"
)
```

15.4.2 Methods

A generic isn't useful without some methods, and in S4 you define methods with setMethod(). There are three important arguments: the name of the
generic, the name of the class, and the method itself.

```
setMethod("myGeneric", "Person", function(x) {
  # method implementation
})
```

More formally, the second argument to setMethod() is called the **signature**.
In S4, unlike S3, the signature can include multiple arguments. This makes
method dispatch in S4 substantially more complicated, but avoids having to
implement double-dispatch as a special case. We'll talk more about multiple
dispatch in the next section. setMethod() has other arguments, but you should
never use them.

To list all the methods that belong to a generic, or that are associated with a
class, use methods("generic") or methods(class = "class"); to find the implementation of a specific method, use selectMethod("generic", "class").

15.4.3 Show method

The most commonly defined S4 method that controls printing is show(), which controls how the object appears when it is printed. To define a method for an existing generic, you must first determine the arguments. You can get those from the documentation or by looking at the args() of the generic:

```
args(getGeneric("show"))
#> function (object)
#> NULL
```

Our show method needs to have a single argument object:

```
setMethod("show", "Person", function(object) {
  cat(is(object)[[1]], "\n",
      "  Name: ", object@name, "\n",
      "   Age: ", object@age, "\n",
      sep = ""
  )
})
john
#> Person
#>    Name: John Smith
#>    Age:  50
```

15.4.4 Accessors

Slots should be considered an internal implementation detail: they can change without warning and user code should avoid accessing them directly. Instead, all user-accessible slots should be accompanied by a pair of **accessors**. If the slot is unique to the class, this can just be a function:

```
person_name <- function(x) x@name
```

Typically, however, you'll define a generic so that multiple classes can use the same interface:

```
setGeneric("name", function(x) standardGeneric("name"))
setMethod("name", "Person", function(x) x@name)

name(john)
#> [1] "John Smith"
```

If the slot is also writeable, you should provide a setter function. You should always include validObject() in the setter to prevent the user from creating invalid objects.

```
setGeneric("name<-", function(x, value) standardGeneric("name<-"))
setMethod("name<-", "Person", function(x, value) {
  x@name <- value
  validObject(x)

  x
})

name(john) <- "Jon Smythe"
name(john)
#> [1] "Jon Smythe"

name(john) <- letters
#> Error in validObject(x): invalid class "Person" object: @name and
#> @age must be same length
```

(If the name<- notation is unfamiliar, review Section 6.8.)

15.4.5 Exercises

1. Add age() accessors for the Person class.

2. In the definition of the generic, why is it necessary to repeat the name of the generic twice?

3. Why does the show() method defined in Section 15.4.3 use is(object)[[1]]? (Hint: try printing the employee subclass.)

4. What happens if you define a method with different argument names to the generic?

15.5 Method dispatch

S4 dispatch is complicated because S4 has two important features:

- Multiple inheritance, i.e. a class can have multiple parents,
- Multiple dispatch, i.e. a generic can use multiple arguments to pick a method.

These features make S4 very powerful, but can also make it hard to understand which method will get selected for a given combination of inputs. In practice, keep method dispatch as simple as possible by avoiding multiple inheritance, and reserving multiple dispatch only for where it is absolutely necessary.

But it's important to describe the full details, so here we'll start simple with single inheritance and single dispatch, and work our way up to the more complicated cases. To illustrate the ideas without getting bogged down in the details, we'll use an imaginary **class graph** based on emoji:

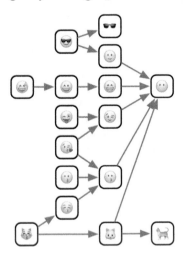

Emoji give us very compact class names that evoke the relationships between the classes. It should be straightforward to remember that 😅 inherits from 😬 which inherits from 🙂, and that 😎 inherits from both 👀 and 🙂.

15.5.1 Single dispatch

Let's start with the simplest case: a generic function that dispatches on a single class with a single parent. The method dispatch here is simple so it's a good place to define the graphical conventions we'll use for the more complex cases.

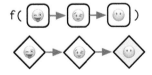

There are two parts to this diagram:

- The top part, f(...), defines the scope of the diagram. Here we have a generic with one argument, that has a class hierarchy that is three levels deep.

- The bottom part is the **method graph** and displays all the possible methods that could be defined. Methods that exist, i.e. that have been defined with setMethod(), have a grey background.

To find the method that gets called, you start with the most specific class of the actual arguments, then follow the arrows until you find a method that exists. For example, if you called the function with an object of class 😎 you would follow the arrow right to find the method defined for the more general 😀 class. If no method is found, method dispatch has failed and an error is thrown. In practice, this means that you should alway define methods defined for the terminal nodes, i.e. those on the far right.

There are two **pseudo-classes** that you can define methods for. These are called pseudo-classes because they don't actually exist, but allow you to define useful behaviours. The first pseudo-class is ANY which matches any class[3]. For technical reasons that we'll get to later, the link to the ANY method is longer than the links between the other classes:

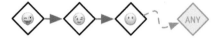

The second pseudo-class is MISSING. If you define a method for this pseudo-class, it will match whenever the argument is missing. It's not useful for single dispatch, but is important for functions like + and - that use double dispatch and behave differently depending on whether they have one or two arguments.

15.5.2 Multiple inheritance

Things get more complicated when the class has multiple parents.

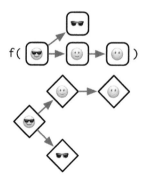

---

[3]The S4 ANY pseudo-class plays the same role as the S3 default pseudo-class.

The basic process remains the same: you start from the actual class supplied to the generic, then follow the arrows until you find a defined method. The wrinkle is that now there are multiple arrows to follow, so you might find multiple methods. If that happens, you pick the method that is closest, i.e. requires travelling the fewest arrows.

NB: while the method graph is a powerful metaphor for understanding method dispatch, implementing it in this way would be rather inefficient, so the actual approach that S4 uses is somewhat different. You can read the details in ?Methods_Details.

What happens if methods are the same distance? For example, imagine we've defined methods for ✌ and ☺, and we call the generic with 😎. Note that no method can be found for the 😶 class, which I'll highlight with a red double outline.

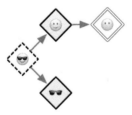

This is called an **ambiguous** method, and in diagrams I'll illustrate it with a thick dotted border. When this happens in R, you'll get a warning, and the method for the class that comes earlier in the alphabet will be picked (this is effectively random and should not be relied upon). When you discover ambiguity you should always resolve it by providing a more precise method:

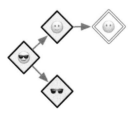

The fallback ANY method still exists but the rules are little more complex. As indicated by the wavy dotted lines, the ANY method is always considered further away than a method for a real class. This means that it will never contribute to ambiguity.

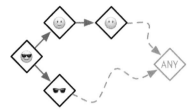

With multiple inheritances it is hard to simultaneously prevent ambiguity, ensure that every terminal method has an implementation, and minimise the number of defined methods (in order to benefit from OOP). For example, of the six ways to define only two methods for this call, only one is free from problems. For this reason, I recommend using multiple inheritance with extreme care: you will need to carefully think about the method graph and plan accordingly.

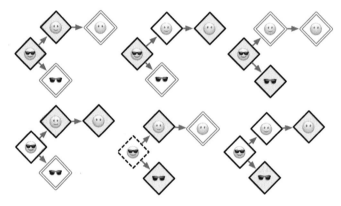

15.5.3 Multiple dispatch

Once you understand multiple inheritance, understanding multiple dispatch is straightforward. You follow multiple arrows in the same way as previously, but now each method is specified by two classes (separated by a comma).

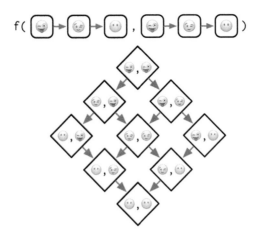

I'm not going to show examples of dispatching on more than two arguments, but you can follow the basic principles to generate your own method graphs.

The main difference between multiple inheritance and multiple dispatch is that there are many more arrows to follow. The following diagram shows four defined methods which produce two ambiguous cases:

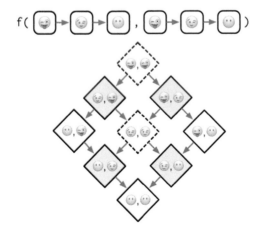

Multiple dispatch tends to be less tricky to work with than multiple inheritance because there are usually fewer terminal class combinations. In this example, there's only one. That means, at a minimum, you can define a single method and have default behaviour for all inputs.

15.5.4 Multiple dispatch and multiple inheritance

Of course you can combine multiple dispatch with multiple inheritance:

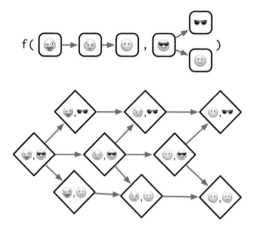

A still more complicated case dispatches on two classes, both of which have multiple inheritance:

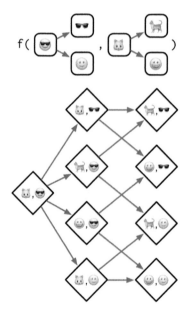

As the method graph gets more and more complicated it gets harder and harder to predict which method will get called given a combination of inputs, and it gets harder and harder to make sure that you haven't introduced ambiguity. If you have to draw diagrams to figure out what method is actually going to be called, it's a strong indication that you should go back and simplify your design.

15.5.5 Exercises

1. Draw the method graph for f(😬, 😵).

2. Draw the method graph for f(😃, 😌, 😵).

3. Take the last example which shows multiple dispatch over two classes that use multiple inheritance. What happens if you define a method for all terminal classes? Why does method dispatch not save us much work here?

15.6 S4 and S3

When writing S4 code, you'll often need to interact with existing S3 classes and generics. This section describes how S4 classes, methods, and generics interact with existing code.

15.6.1 Classes

In slots and contains you can use S4 classes, S3 classes, or the implicit class (Section 13.7.1) of a base type. To use an S3 class, you must first register it with setOldClass(). You call this function once for each S3 class, giving it the class attribute. For example, the following definitions are already provided by base R:

```
setOldClass("data.frame")
setOldClass(c("ordered", "factor"))
setOldClass(c("glm", "lm"))
```

However, it's generally better to be more specific and provide a full S4 definition with slots and a prototype:

```
setClass("factor",
  contains = "integer",
  slots = c(
    levels = "character"
  ),
  prototype = structure(
    integer(),
    levels = character()
```

```
  )
)
setOldClass("factor", S4Class = "factor")
```

Generally, these definitions should be provided by the creator of the S3 class. If you're trying to build an S4 class on top of an S3 class provided by a package, you should request that the package maintainer add this call to their package, rather than adding it to your own code.

If an S4 object inherits from an S3 class or a base type, it will have a special virtual slot called .Data. This contains the underlying base type or S3 object:

```
RangedNumeric <- setClass(
  "RangedNumeric",
  contains = "numeric",
  slots = c(min = "numeric", max = "numeric"),
  prototype = structure(numeric(), min = NA_real_, max = NA_real_)
)
rn <- RangedNumeric(1:10, min = 1, max = 10)
rn@min
#> [1] 1
rn@.Data
#>  [1]  1  2  3  4  5  6  7  8  9 10
```

It is possible to define S3 methods for S4 generics, and S4 methods for S3 generics (provided you've called setOldClass()). However, it's more complicated than it might appear at first glance, so make sure you thoroughly read ?Methods_for_S3.

15.6.2 Generics

As well as creating a new generic from scratch, it's also possible to convert an existing S3 generic to an S4 generic:

```
setGeneric("mean")
```

In this case, the existing function becomes the default (ANY) method:

```
selectMethod("mean", "ANY")
#> Method Definition (Class "derivedDefaultMethod"):
#>
```

```
#> function (x, ...)
#> UseMethod("mean")
#> <bytecode: 0x7fd256468a30>
#> <environment: namespace:base>
#>
#> Signatures:
#>         x
#> target  "ANY"
#> defined "ANY"
```

NB: `setMethod()` will automatically call `setGeneric()` if the first argument isn't already a generic, enabling you to turn any existing function into an S4 generic. It is OK to convert an existing S3 generic to S4, but you should avoid converting regular functions to S4 generics in packages because that requires careful coordination if done by multiple packages.

15.6.3   Exercises

1. What would a full `setOldClass()` definition look like for an ordered factor (i.e. add `slots` and `prototype` the definition above)?

2. Define a `length` method for the `Person` class.

16

Trade-offs

16.1 Introduction

You now know about the three most important OOP toolkits available in R. Now that you understand their basic operation and the principles that underlie them, we can start to compare and contrast the systems in order to understand their strengths and weaknesses. This will help you pick the system that is most likely to solve new problems.

Overall, when picking an OO system, I recommend that you default to S3. S3 is simple, and widely used throughout base R and CRAN. While it's far from perfect, its idiosyncrasies are well understood and there are known approaches to overcome most shortcomings. If you have an existing background in programming you are likely to lean towards R6, because it will feel familiar. I think you should resist this tendency for two reasons. Firstly, if you use R6 it's very easy to create a non-idiomatic API that will feel very odd to native R users, and will have surprising pain points because of the reference semantics. Secondly, if you stick to R6, you'll lose out on learning a new way of thinking about OOP that gives you a new set of tools for solving problems.

Outline

- Section 16.2 compares S3 and S4. In brief, S4 is more formal and tends to require more upfront planning. That makes it more suitable for big projects developed by teams, not individuals.

- Section 16.3 compares S3 and R6. This section is quite long because these two systems are fundamentally different and there are a number of tradeoffs that you need to consider.

Prerequisites

You need to be familiar with S3, S4, and R6, as discussed in the previous three chapters.

16.2   S4 versus S3

Once you've mastered S3, S4 is not too difficult to pick up: the underlying ideas are the same, S4 is just more formal, more strict, and more verbose. The strictness and formality of S4 make it well suited for large teams. Since more structure is provided by the system itself, there is less need for convention, and new contributors don't need as much training. S4 tends to require more upfront design than S3, and this investment is more likely to pay off on larger projects where greater resources are available.

One large team effort where S4 is used to good effect is Bioconductor. Bioconductor is similar to CRAN: it's a way of sharing packages amongst a wider audience. Bioconductor is smaller than CRAN (~1,300 versus ~10,000 packages, July 2017) and the packages tend to be more tightly integrated because of the shared domain and because Bioconductor has a stricter review process. Bioconductor packages are not required to use S4, but most will because the key data structures (e.g. SummarizedExperiment, IRanges, DNAStringSet) are built using S4.

S4 is also a good fit for complex systems of interrelated objects, and it's possible to minimise code duplication through careful implementation of methods. The best example of such a system is the Matrix package [Bates and Maechler, 2018]. It is designed to efficiently store and compute with many different types of sparse and dense matrices. As of version 1.2.15, it defines 102 classes, 21 generic functions, and 1993 methods, and to give you some idea of the complexity, a small subset of the class graph is shown in Figure 16.1.

This domain is a good fit for S4 because there are often computational shortcuts for specific combinations of sparse matrices. S4 makes it easy to provide a general method that works for all inputs, and then provide more specialised methods where the inputs allow a more efficient implementation. This requires careful planning to avoid method dispatch ambiguity, but the planning pays off with higher performance.

The biggest challenge to using S4 is the combination of increased complexity and absence of a single source of documentation. S4 is a complex system and it can be challenging to use effectively in practice. This wouldn't be such a problem if S4 documentation wasn't scattered through R documentation, books, and websites. S4 needs a book length treatment, but that book does not (yet) exist. (The documentation for S3 is no better, but the lack is less painful because S3 is much simpler.)

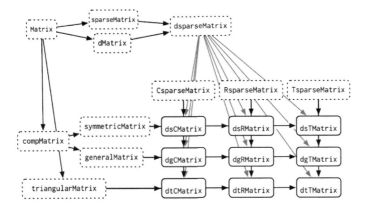

Figure 16.1 A small subset of the Matrix class graph showing the inheritance of sparse matrices. Each concrete class inherits from two virtual parents: one that describes how the data is stored (C = column oriented, R = row oriented, T = tagged) and one that describes any restriction on the matrix (s = symmetric, t = triangle, g = general).

16.3 R6 versus S3

R6 is a profoundly different OO system from S3 and S4 because it is built on encapsulated objects, rather than generic functions. Additionally R6 objects have reference semantics, which means that they can be modified in place. These two big differences have a number of non-obvious consequences which we'll explore here:

- A generic is a regular function so it lives in the global namespace. An R6 method belongs to an object so it lives in a local namespace. This influences how we think about naming.

- R6's reference semantics allow methods to simultaneously return a value and modify an object. This solves a painful problem called "threading state".

- You invoke an R6 method using $, which is an infix operator. If you set up your methods correctly you can use chains of method calls as an alternative to the pipe.

These are general trade-offs between functional and encapsulated OOP, so they also serve as a discussion of system design in R versus Python.

16.3.1   Namespacing

One non-obvious difference between S3 and R6 is the space in which methods are found:

- Generic functions are global: all packages share the same namespace.
- Encapsulated methods are local: methods are bound to a single object.

The advantage of a global namespace is that multiple packages can use the same verbs for working with different types of objects. Generic functions provide a uniform API that makes it easier to perform typical actions with a new object because there are strong naming conventions. This works well for data analysis because you often want to do the same thing to different types of objects. In particular, this is one reason that R's modelling system is so useful: regardless of where the model has been implemented you always work with it using the same set of tools (`summary()`, `predict()`, …).

The disadvantage of a global namespace is that it forces you to think more deeply about naming. You want to avoid multiple generics with the same name in different packages because it requires the user to type `::` frequently. This can be hard because function names are usually English verbs, and verbs often have multiple meanings. Take `plot()` for example:

```
plot(data)        # plot some data
plot(bank_heist)  # plot a crime
plot(land)        # create a new plot of land
plot(movie)       # extract plot of a movie
```

Generally, you should avoid methods that are homonyms of the original generic, and instead define a new generic.

This problem doesn't occur with R6 methods because they are scoped to the object. The following code is fine, because there is no implication that the plot method of two different R6 objects has the same meaning:

```
data$plot()
bank_heist$plot()
land$plot()
movie$plot()
```

These considerations also apply to the arguments to the generic. S3 generics must have the same core arguments, which means they generally have non-specific names like x or .data. S3 generics generally need ... to pass on additional arguments to methods, but this has the downside that misspelled argument names will not create an error. In comparison, R6 methods can vary more widely and use more specific and evocative argument names.

A secondary advantage of local namespacing is that creating an R6 method is very cheap. Most encapsulated OO languages encourage you to create many small methods, each doing one thing well with an evocative name. Creating a new S3 method is more expensive, because you may also have to create a generic, and think about the naming issues described above. That means that the advice to create many small methods does not apply to S3. It's still a good idea to break your code down into small, easily understood chunks, but they should generally just be regular functions, not methods.

16.3.2 Threading state

One challenge of programming with S3 is when you want to both return a value and modify the object. This violates our guideline that a function should either be called for its return value or for its side effects, but is necessary in a handful of cases.

For example, imagine you want to create a **stack** of objects. A stack has two main methods:

- push() adds a new object to the top of the stack.
- pop() returns the top most value, and removes it from the stack.

The implementation of the constructor and the push() method is straightforward. A stack contains a list of items, and pushing an object to the stack simply appends to this list.

```
new_stack <- function(items = list()) {
    structure(list(items = items), class = "stack")
}

push <- function(x, y) {
    x$items <- c(x$items, list(y))
    x
}
```

(I haven't created a real method for push() because making it generic would just make this example more complicated for no real benefit.)

Implementing pop() is more challenging because it has to both return a value (the object at the top of the stack), and have a side-effect (remove that object from that top). Since we can't modify the input object in S3 we need to return two things: the value, and the updated object.

```
pop <- function(x) {
    n <- length(x$items)
```

```
  item <- x$items[[n]]
  x$items <- x$items[-n]

  list(item = item, x = x)
}
```

This leads to rather awkward usage:

```
s <- new_stack()
s <- push(s, 10)
s <- push(s, 20)

out <- pop(s)
out$item
#> [1] 20
s <- out$x
s
#> $items
#> $items[[1]]
#> [1] 10
#>
#>
#> attr(,"class")
#> [1] "stack"
```

This problem is known as **threading state** or **accumulator programming**, because no matter how deeply the pop() is called, you have to thread the modified stack object all the way back to where it lives.

One way that other FP languages deal with this challenge is to provide a **multiple assign** (or destructuring bind) operator that allows you to assign multiple values in a single step. The zeallot package [Teetor, 2018] provides multi-assign for R with %<-%. This makes the code more elegant, but doesn't solve the key problem:

```
library(zeallot)

c(value, s) %<-% pop(s)
value
#> [1] 10
```

An R6 implementation of a stack is simpler because $pop() can modify the object in place, and return only the top-most value:

```
Stack <- R6::R6Class("Stack", list(
  items = list(),
  push = function(x) {
    self$items <- c(self$items, x)
    invisible(self)
  },
  pop = function() {
    item <- self$items[[self$length()]]
    self$items <- self$items[-self$length()]
    item
  },
  length = function() {
    length(self$items)
  }
))
```

This leads to more natural code:

```
s <- Stack$new()
s$push(10)
s$push(20)
s$pop()
#> [1] 20
```

I encountered a real-life example of threading state in ggplot2 scales. Scales are complex because they need to combine data across every facet and every layer. I originally used S3 classes, but it required passing scale data to and from many functions. Switching to R6 made the code substantially simpler. However, it also introduced some problems because I forgot to call to $clone() when modifying a plot. This allowed independent plots to share the same scale data, creating a subtle bug that was hard to track down.

16.3.3   Method chaining

The pipe, %>%, is useful because it provides an infix operator that makes it easy to compose functions from left-to-right. Interestingly, the pipe is not so important for R6 objects because they already use an infix operator: $. This allows the user to chain together multiple method calls in a single expression, a technique known as **method chaining**:

```
s <- Stack$new()
s$
```

```
push(10)$
push(20)$
pop()
#> [1] 20
```

This technique is commonly used in other programming languages, like Python and JavaScript, and is made possible with one convention: any R6 method that is primarily called for its side-effects (usually modifying the object) should return invisible(self).

The primary advantage of method chaining is that you can get useful auto-complete; the primary disadvantage is that only the creator of the class can add new methods (and there's no way to use multiple dispatch).

Part IV

Metaprogramming

Introduction

One of the most intriguing things about R is its ability to do **metaprogramming**. This is the idea that code is data that can be inspected and modified programmatically. This is a powerful idea; one that deeply influences much R code. At the most basic level, it allows you to do things like write `library(purrr)` instead of `library("purrr")` and enable `plot(x, sin(x))` to automatically label the axes with x and sin(x). At a deeper level, it allows you to do things like use `y ~ x1 + x2` to represent a model that predicts the value of y from x1 and x2, to translate `subset(df, x == y)` into `df[df$x == df$y, , drop = FALSE]`, and to use `dplyr::filter(db, is.na(x))` to generate the SQL `WHERE x IS NULL` when db is a remote database table.

Closely related to metaprogramming is **non-standard evaluation**, NSE for short. This term, which is commonly used to describe the behaviour of R functions, is problematic in two ways. Firstly, NSE is actually a property of the argument (or arguments) of a function, so talking about NSE functions is a little sloppy. Secondly, it's confusing to define something by what it's not (standard), so in this book I'll introduce more precise vocabulary.

Specifically, this book focuses on tidy evaluation (sometimes called tidy eval for short). Tidy evaluation is implemented in the rlang package [Henry and Wickham, 2018b], and I'll use rlang extensively in these chapters. This will allow you to focus on the big ideas, without being distracted by the quirks of implementation that arise from R's history. After I introduce each big idea with rlang, I'll then circle back to talk about how those ideas are expressed in base R. This approach may seem backward to some, but it's like learning how to drive using an automatic transmission rather than a stick shift: it allows you to focus on the big picture before having to learn the details. This book focusses on the theoretical side of tidy evaluation, so you can fully understand how it works from the ground up. If you are looking for a more practical introduction, I recommend the tidy evaluation book at `https://tidyeval.`
`tidyverse.org`[1].

You'll learn about metaprogramming and tidy evaluation in the following five chapters:

---

[1] As I write this chapter, the tidy evaluation book is still a work-in-progress, but by the time you read this it will hopefully be finished.

1. Chapter 17 gives a high level description of the whole metaprogramming story, briefly learning about all major components and how they fit together to form a cohesive whole.

2. Chapter 18 shows that that all R code can be described as a tree. You'll learn how to visualise these trees, how the rules of R's grammar convert linear sequences of characters into these trees, and how to use recursive functions to work with code trees.

3. Chapter 19 presents tools from rlang that you can use to capture (quote) unevaluated function arguments. You'll also learn about quasiquotation, which provides a set of techniques to unquote input to make it possible to easily generate new trees from code fragments.

4. Chapter 20 moves on to evaluating captured code. Here you'll learn about an important data structure, the **quosure**, which ensures correct evaluation by capturing both the code to evaluate, and the environment in which to evaluate it. This chapter will show you how to put all the pieces together to understand how NSE works in base R, and how to write functions that work like subset().

5. Chapter 21 finishes up by combining first-class environments, lexical scoping, and metaprogramming to translate R code into other languages, namely HTML and LaTeX.

17

Big picture

17.1 Introduction

Metaprogramming is the hardest topic in this book because it brings together many formerly unrelated topics and forces you grapple with issues that you probably haven't thought about before. You'll also need to learn a lot of new vocabulary, and at first it will seem like every new term is defined by three other terms that you haven't heard of. Even if you're an experienced programmer in another language, your existing skills are unlikely to be much help as few modern popular languages expose the level of metaprogramming that R provides. So don't be surprised if you're frustrated or confused at first; this is a natural part of the process that happens to everyone!

But I think it's easier to learn metaprogramming now than ever before. Over the last few years, the theory and practice have matured substantially, providing a strong foundation paired with tools that allow you to solve common problems. In this chapter, you'll get the big picture of all the main pieces and how they fit together.

Outline

Each section in this chapter introduces one big new idea:

- Section 17.2 shows that code is data and teaches you how to create and modify expressions by capturing code.

- Section 17.3 describes the tree-like structure of code, called an abstract syntax tree.

- Section 17.4 shows how to create new expressions programmatically.

- Section 17.5 shows how to execute expressions by evaluating them in an environment.

- Section 17.6 illustrates how to customise evaluation by supplying custom functions in a new environment.

- Section 17.7 extends that customisation to data masks, which blur the line between environments and data frames.

- Section 17.8 introduces a new data structure called the quosure that makes all this simpler and more correct.

Prerequisites

This chapter introduces the big ideas using rlang; you'll learn the base equivalents in later chapters. We'll also use the lobstr package to explore the tree structure of code.

```
library(rlang)
library(lobstr)
```

Make sure that you're also familiar with the environment (Section 7.2) and data frame (Section 3.6) data structures.

17.2 Code is data

The first big idea is that code is data: you can capture code and compute on as you can with any other type of data. The first way you can capture code is with rlang::expr(). You can think of expr() as returning exactly what you pass in:

```
expr(mean(x, na.rm = TRUE))
#> mean(x, na.rm = TRUE)
expr(10 + 100 + 1000)
#> 10 + 100 + 1000
```

More formally, captured code is called an **expression**. An expression isn't a single type of object, but is a collective term for any of four types (call, symbol, constant, or pairlist), which you'll learn more about in Chapter 18.

expr() lets you capture code that you've typed. You need a different tool to capture code passed to a function because expr() doesn't work:

```
capture_it <- function(x) {
  expr(x)
}
```

```
capture_it(a + b + c)
#> x
```

Here you need to use a function specifically designed to capture user input in a function argument: enexpr(). Think of the "en" in the context of "enrich": enexpr() takes a lazily evaluated argument and turns it into an expression:

```
capture_it <- function(x) {
  enexpr(x)
}
capture_it(a + b + c)
#> a + b + c
```

Because capture_it() uses enexpr() we say that it automatically quotes its first argument. You'll learn more about this term in Section 19.2.1.

Once you have captured an expression, you can inspect and modify it. Complex expressions behave much like lists. That means you can modify them using [[and $:

```
f <- expr(f(x = 1, y = 2))

# Add a new argument
f$z <- 3
f
#> f(x = 1, y = 2, z = 3)

# Or remove an argument:
f[[2]] <- NULL
f
#> f(y = 2, z = 3)
```

The first element of the call is the function to be called, which means the first argument is in the second position. You'll learn the full details in Section 18.3.

17.3 Code is a tree

To do more complex manipulation with expressions, you need to fully understand their structure. Behind the scenes, almost every programming language represents code as a tree, often called the **abstract syntax tree**, or AST for short. R is unusual in that you can actually inspect and manipulate this tree.

A very convenient tool for understanding the tree-like structure is lob-str::ast(). Given some code, this function displays the underlying tree structure. Function calls form the branches of the tree, and are shown by rectangles. The leaves of the tree are symbols (like a) and constants (like "b").

```
lobstr::ast(f(a, "b"))
#> ■─f
#> ├─a
#> └─"b"
```

Nested function calls create more deeply branching trees:

```
lobstr::ast(f1(f2(a, b), f3(1, f4(2))))
#> ■─f1
#> ├─■─f2
#> │ ├─a
#> │ └─b
#> └─■─f3
#>   ├─1
#>   └─■─f4
#>     └─2
```

Because all function forms can be written in prefix form (Section 6.8.2), every R expression can be displayed in this way:

```
lobstr::ast(1 + 2 * 3)
#> ■─`+`
#> ├─1
#> └─■─`*`
#>   ├─2
#>   └─3
```

Displaying the AST in this way is a useful tool for exploring R's grammar, the topic of Section 18.4.

17.4   Code can generate code

As well as seeing the tree from code typed by a human, you can also use code to create new trees. There are two main tools: call2() and unquoting.

rlang::call2() constructs a function call from its components: the function to call, and the arguments to call it with.

```
call2("f", 1, 2, 3)
#> f(1, 2, 3)
call2("+", 1, call2("*", 2, 3))
#> 1 + 2 * 3
```

call2() is often convenient to program with, but is a bit clunky for interactive use. An alternative technique is to build complex code trees by combining simpler code trees with a template. expr() and enexpr() have built-in support for this idea via !! (pronounced bang-bang), the **unquote operator**.

The precise details are the topic of Section 19.4, but basically !!x inserts the code tree stored in x into the expression. This makes it easy to build complex trees from simple fragments:

```
xx <- expr(x + x)
yy <- expr(y + y)

expr(!!xx / !!yy)
#> (x + x)/(y + y)
```

Notice that the output preserves the operator precedence so we get (x + x) / (y + y) not x + x / y + y (i.e. x + (x / y) + y). This is important, particularly if you've been wondering if it wouldn't be easier to just paste strings together.

Unquoting gets even more useful when you wrap it up into a function, first using enexpr() to capture the user's expression, then expr() and !! to create a new expression using a template. The example below shows how you can generate an expression that computes the coefficient of variation:

```
cv <- function(var) {
  var <- enexpr(var)
  expr(sd(!!var) / mean(!!var))
}

cv(x)
#> sd(x)/mean(x)
cv(x + y)
#> sd(x + y)/mean(x + y)
```

(This isn't very useful here, but being able to create this sort of building block is very useful when solving more complex problems.)

Importantly, this works even when given weird variable names:

```
cv(`)`)
#> sd(`)`)/mean(`)`)
```

Dealing with weird names[1] is another good reason to avoid `paste()` when generating R code. You might think this is an esoteric concern, but not worrying about it when generating SQL code in web applications led to SQL injection attacks that have collectively cost billions of dollars.

17.5   Evaluation runs code

Inspecting and modifying code gives you one set of powerful tools. You get another set of powerful tools when you **evaluate**, i.e. execute or run, an expression. Evaluating an expression requires an environment, which tells R what the symbols in the expression mean. You'll learn the details of evaluation in Chapter 20.

The primary tool for evaluating expressions is `base::eval()`, which takes an expression and an environment:

```
eval(expr(x + y), env(x = 1, y = 10))
#> [1] 11
eval(expr(x + y), env(x = 2, y = 100))
#> [1] 102
```

If you omit the environment, `eval` uses the current environment:

```
x <- 10
y <- 100
eval(expr(x + y))
#> [1] 110
```

One of the big advantages of evaluating code manually is that you can tweak the environment. There are two main reasons to do this:

- To temporarily override functions to implement a domain specific language.
- To add a data mask so you can to refer to variables in a data frame as if they are variables in an environment.

---

[1]More technically, these are called non-syntactic names and are the topic of Section 2.2.1.

17.6   Customising evaluation with functions

The above example used an environment that bound x and y to vectors. It's less obvious that you also bind names to functions, allowing you to override the behaviour of existing functions. This is a big idea that we'll come back to in Chapter 21 where I explore generating HTML and LaTeX from R. The example below gives you a taste of the power. Here I evalute code in a special environment where * and + have been overridden to work with strings instead of numbers:

```
string_math <- function(x) {
  e <- env(
    caller_env(),
    `+` = function(x, y) paste0(x, y),
    `*` = function(x, y) strrep(x, y)
  )

  eval(enexpr(x), e)
}

name <- "Hadley"
string_math("Hello " + name)
#> [1] "Hello Hadley"
string_math(("x" * 2 + "-y") * 3)
#> [1] "xx-yxx-yxx-y"
```

dplyr takes this idea to the extreme, running code in an environment that generates SQL for execution in a remote database:

```
library(dplyr)
#>
#> Attaching package: 'dplyr'
#> The following objects are masked from 'package:stats':
#>
#>     filter, lag
#> The following objects are masked from 'package:base':
#>
#>     intersect, setdiff, setequal, union

con <- DBI::dbConnect(RSQLite::SQLite(), filename = ":memory:")
mtcars_db <- copy_to(con, mtcars)
```

```
mtcars_db %>%
  filter(cyl > 2) %>%
  select(mpg:hp) %>%
  head(10) %>%
  show_query()
#> <SQL>
#> SELECT `mpg`, `cyl`, `disp`, `hp`
#> FROM `mtcars`
#> WHERE (`cyl` > 2.0)
#> LIMIT 10

DBI::dbDisconnect(con)
```

17.7   Customising evaluation with data

Rebinding functions is an extremely powerful technique, but it tends to require a lot of investment. A more immediately practical application is modifying evaluation to look for variables in a data frame instead of an environment. This idea powers the base subset() and transform() functions, as well as many tidyverse functions like ggplot2::aes() and dplyr::mutate(). It's possible to use eval() for this, but there are a few potential pitfalls (Section 20.6), so we'll switch to rlang::eval_tidy() instead.

As well as expression and environment, eval_tidy() also takes a **data mask**, which is typically a data frame:

```
df <- data.frame(x = 1:5, y = sample(5))
eval_tidy(expr(x + y), df)
#> [1] 2 6 5 9 8
```

Evaluating with a data mask is a useful technique for interactive analysis because it allows you to write x + y rather than df$x + df$y. However, that convenience comes at a cost: ambiguity. In Section 20.4 you'll learn how to deal with ambiguity using special .data and .env pronouns.

We can wrap this pattern up into a function by using enexpr(). This gives us a function very similar to base::with():

```
with2 <- function(df, expr) {
  eval_tidy(enexpr(expr), df)
```

```
}

with2(df, x + y)
#> [1] 2 6 5 9 8
```

Unfortunately, this function has a subtle bug and we need a new data structure to help deal with it.

17.8   Quosures

To make the problem more obvious, I'm going to modify with2(). The basic problem still occurs without this modification but it's much harder to see.

```
with2 <- function(df, expr) {
  a <- 1000
  eval_tidy(enexpr(expr), df)
}
```

We can see the problem when we use with2() to refer to a variable called a. We want the value of a to come from the binding we can see (10), not the binding internal to the function (1000):

```
df <- data.frame(x = 1:3)
a <- 10
with2(df, x + a)
#> [1] 1001 1002 1003
```

The problem arises because we need to evaluate the captured expression in the environment where it was written (where a is 10), not the environment inside of with2() (where a is 1000).

Fortunately we can solve this problem by using a new data structure: the **quosure** which bundles an expression with an environment. eval_tidy() knows how to work with quosures so all we need to do is switch out enexpr() for enquo():

```
with2 <- function(df, expr) {
  a <- 1000
  eval_tidy(enquo(expr), df)
```

```
}

with2(df, x + a)
#> [1] 11 12 13
```

Whenever you use a data mask, you must always use `enquo()` instead of `en-expr()`. This is the topic of Chapter 20.

18

Expressions

18.1 Introduction

To compute on the language, we first need to understand its structure. That requires some new vocabulary, some new tools, and some new ways of thinking about R code. The first of these is the distinction between an operation and its result. Take the following code, which multiplies a variable x by 10 and saves the result to a new variable called y. It doesn't work because we haven't defined a variable called x:

```
y <- x * 10
#> Error in eval(expr, envir, enclos): object 'x' not found
```

It would be nice if we could capture the intent of the code without executing it. In other words, how can we separate our description of the action from the action itself?

One way is to use `rlang::expr()`:

```
z <- rlang::expr(y <- x * 10)
z
#> y <- x * 10
```

`expr()` returns an expression, an object that captures the structure of the code without evaluating it (i.e. running it). If you have an expression, you can evaluate it with `base::eval()`:

```
x <- 4
eval(z)
y
#> [1] 40
```

The focus of this chapter is the data structures that underlie expressions. Mastering this knowledge will allow you to inspect and modify captured code, and to generate code with code. We'll come back to expr() in Chapter 19, and to eval() in Chapter 20.

Outline

- Section 18.2 introduces the idea of the abstract syntax tree (AST), and reveals the tree like structure that underlies all R code.
- Section 18.3 dives into the details of the data structures that underpin the AST: constants, symbols, and calls, which are collectively known as expressions.
- Section 18.4 covers parsing, the act of converting the linear sequence of character in code into the AST, and uses that idea to explore some details of R's grammar.
- Section 18.5 shows you how you can use recursive functions to compute on the language, writing functions that compute with expressions.
- Section 18.6 circles back to three more specialised data structures: pairlists, missing arguments, and expression vectors.

Prerequisites

Make sure you've read the metaprogramming overview in Chapter 17 to get a broad overview of the motivation and the basic vocabulary. You'll also need the rlang (https://rlang.r-lib.org) package to capture and compute on expressions, and the lobstr (https://lobstr.r-lib.org) package to visualise them.

```
library(rlang)
library(lobstr)
```

18.2 Abstract syntax trees

Expressions are also called **abstract syntax trees** (ASTs) because the structure of code is hierarchical and can be naturally represented as a tree. Understanding this tree structure is crucial for inspecting and modifying expressions (i.e. metaprogramming).

18.2.1 Drawing

We'll start by introducing some conventions for drawing ASTs, beginning with a simple call that shows their main components: f(x, "y", 1). I'll draw trees in two ways[1]:

- By "hand" (i.e. with OmniGraffle):

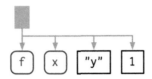

- With lobstr::ast():

```
lobstr::ast(f(x, "y", 1))
#> █─f
#> ├─x
#> ├─"y"
#> └─1
```

Both approaches share conventions as much as possible:

- The leaves of the tree are either symbols, like f and x, or constants, like 1 or "y". Symbols are drawn in purple and have rounded corners. Constants have black borders and square corners. Strings and symbols are easily confused, so strings are always surrounded in quotes.

- The branches of the tree are call objects, which represent function calls, and are drawn as orange rectangles. The first child (f) is the function that gets called; the second and subsequent children (x, "y", and 1) are the arguments to that function.

Colours will be shown when *you* call ast(), but do not appear in the book for complicated technical reasons.

The above example only contained one function call, making for a very shallow tree. Most expressions will contain considerably more calls, creating trees with multiple levels. For example, consider the AST for f(g(1, 2), h(3, 4, i())):

---

[1]For more complex code, you can also use RStudio's tree viewer which doesn't obey quite the same graphical conventions, but allows you to interactively explore large ASTs. Try it out with View(expr(f(x, "y", 1))).

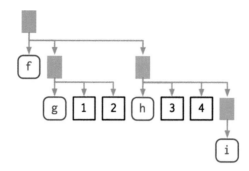

```
lobstr::ast(f(g(1, 2), h(3, 4, i()))))
#> █─f
#> ├─█─g
#> │ ├─1
#> │ └─2
#> └─█─h
#>   ├─3
#>   ├─4
#>   └─█─i
```

You can read the hand-drawn diagrams from left-to-right (ignoring vertical position), and the lobstr-drawn diagrams from top-to-bottom (ignoring horizontal position). The depth within the tree is determined by the nesting of function calls. This also determines evaluation order, as evaluation generally proceeds from deepest-to-shallowest, but this is not guaranteed because of lazy evaluation (Section 6.5). Also note the appearance of i(), a function call with no arguments; it's a branch with a single (symbol) leaf.

18.2.2 Non-code components

You might have wondered what makes these *abstract* syntax trees. They are abstract because they only capture important structural details of the code, not whitespace or comments:

```
ast(
  f(x,  y)  # important!
)
#> █─f
#> ├─x
#> └─y
```

There's only one place where whitespace affects the AST:

```
lobstr::ast(y <- x)
#> █─`<-`
#> ├─y
#> └─x
lobstr::ast(y < -x)
#> █─`<`
#> ├─y
#> └─█─`-`
#>     └─x
```

18.2.3 Infix calls

Every call in R can be written in tree form because any call can be written in prefix form (Section 6.8.1). Take `y <- x * 10` again: what are the functions that are being called? It is not as easy to spot as `f(x, 1)` because this expression contains two infix calls: `<-` and `*`. That means that these two lines of code are equivalent:

```
y <- x * 10
`<-`(y, `*`(x, 10))
```

And they both have this AST[2]:

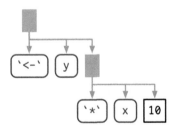

```
lobstr::ast(y <- x * 10)
#> █─`<-`
#> ├─y
#> └─█─`*`
#>     ├─x
#>     └─10
```

---

[2]The names of non-prefix functions are non-syntactic so I surround them with ``` `` ```, as in Section 2.2.1.

There really is no difference between the ASTs, and if you generate an expression with prefix calls, R will still print it in infix form:

```
expr(`<-`(y, `*`(x, 10)))
#> y <- x * 10
```

The order in which infix operators are applied is governed by a set of rules called operator precedence, and we'll use `lobstr::ast()` to explore them in Section 18.4.1.

18.2.4 Exercises

1. Reconstruct the code represented by the trees below:

```
#> █──f
#> └─█──g
#>    └─█──h
#> █──`+`
#> ├─█──`+`
#> │  ├──1
#> │  └──2
#> └──3
#> █──`*`
#> ├─█──`(`
#> │  └─█──`+`
#> │     ├──x
#> │     └──y
#> └──z
```

2. Draw the following trees by hand and then check your answers with `lobstr::ast()`.

   ```
   f(g(h(i(1, 2, 3))))
   f(1, g(2, h(3, i())))
   f(g(1, 2), h(3, i(4, 5)))
   ```

3. What's happening with the ASTs below? (Hint: carefully read ?"^".)

   ```
   lobstr::ast(`x` + `y`)
   #> █──`+`
   #> ├──x
   ```

```
#>    └─y
lobstr::ast(x ** y)
#>  ■─`^`
#>  ├─x
#>  └─y
lobstr::ast(1 -> x)
#>  ■─`<-`
#>  ├─x
#>  └─1
```

4. What is special about the AST below? (Hint: re-read Section 6.2.1.)

```
lobstr::ast(function(x = 1, y = 2) {})
#>  ■─`function`
#>  ├─■─x = 1
#>  │  └─y = 2
#>  ├─■─`{`
#>  └─<inline srcref>
```

5. What does the call tree of an `if` statement with multiple `else if` conditions look like? Why?

18.3   Expressions

Collectively, the data structures present in the AST are called expressions. An **expression** is any member of the set of base types created by parsing code: constant scalars, symbols, call objects, and pairlists. These are the data structures used to represent captured code from `expr()`, and `is_expression(expr(...))` is always true[3]. Constants, symbols and call objects are the most important, and are discussed below. Pairlists and empty symbols are more specialised and we'll come back to them in Sections 18.6.1 and Section 18.6.2.

NB: In base R documentation "expression" is used to mean two things. As well as the definition above, expression is also used to refer to the type of object returned by `expression()` and `parse()`, which are basically lists of expressions

---

[3]It is *possible* to insert any other base object into an expression, but this is unusual and only needed in rare circumstances. We'll come back to that idea in Section 19.4.7.

as defined above. In this book I'll call these **expression vectors**, and I'll come back to them in Section 18.6.3.

18.3.1 Constants

Scalar constants are the simplest component of the AST. More precisely, a **constant** is either NULL or a length-1 atomic vector (or scalar, Section 3.2.1) like TRUE, 1L, 2.5 or "x". You can test for a constant with rlang::is_syntactic_literal().

Constants are self-quoting in the sense that the expression used to represent a constant is the same constant:

```
identical(expr(TRUE), TRUE)
#> [1] TRUE
identical(expr(1), 1)
#> [1] TRUE
identical(expr(2L), 2L)
#> [1] TRUE
identical(expr("x"), "x")
#> [1] TRUE
```

18.3.2 Symbols

A **symbol** represents the name of an object like x, mtcars, or mean. In base R, the terms symbol and name are used interchangeably (i.e. is.name() is identical to is.symbol()), but in this book I used symbol consistently because "name" has many other meanings.

You can create a symbol in two ways: by capturing code that references an object with expr(), or turning a string into a symbol with rlang::sym():

```
expr(x)
#> x
sym("x")
#> x
```

You can turn a symbol back into a string with as.character() or rlang::as_string(). as_string() has the advantage of clearly signalling that you'll get a character vector of length 1.

```
as_string(expr(x))
#> [1] "x"
```

You can recognise a symbol because it's printed without quotes, str() tells you that it's a symbol, and is.symbol() is TRUE:

```
str(expr(x))
#>  symbol x
is.symbol(expr(x))
#> [1] TRUE
```

The symbol type is not vectorised, i.e. a symbol is always length 1. If you want multiple symbols, you'll need to put them in a list, using (e.g.) rlang::syms().

18.3.3 Calls

A **call object** represents a captured function call. Call objects are a special type of list[4] where the first component specifies the function to call (usually a symbol), and the remaining elements are the arguments for that call. Call objects create branches in the AST, because calls can be nested inside other calls.

You can identify a call object when printed because it looks just like a function call. Confusingly typeof() and str() print "language"[5] for call objects, but is.call() returns TRUE:

```
lobstr::ast(read.table("important.csv", row.names = FALSE))
#> █─read.table
#> ├─"important.csv"
#> └─row.names = FALSE
x <- expr(read.table("important.csv", row.names = FALSE))

typeof(x)
#> [1] "language"
is.call(x)
#> [1] TRUE
```

18.3.3.1 Subsetting

Calls generally behave like lists, i.e. you can use standard subsetting tools. The first element of the call object is the function to call, which is usually a symbol:

---

[4]More precisely, they're pairlists, Section 18.6.1, but this distinction rarely matters.

[5]Avoid is.language() which returns TRUE for symbols, calls, and expression vectors.

```
x[[1]]
#> read.table
is.symbol(x[[1]])
#> [1] TRUE
```

The remainder of the elements are the arguments:

```
as.list(x[-1])
#> [[1]]
#> [1] "important.csv"
#>
#> $row.names
#> [1] FALSE
```

You can extract individual arguments with [[or, if named, $:

```
x[[2]]
#> [1] "important.csv"
x$row.names
#> [1] FALSE
```

You can determine the number of arguments in a call object by subtracting 1 from its length:

```
length(x) - 1
#> [1] 2
```

Extracting specific arguments from calls is challenging because of R's flexible rules for argument matching: it could potentially be in any location, with the full name, with an abbreviated name, or with no name. To work around this problem, you can use `rlang::call_standardise()` which standardises all arguments to use the full name:

```
rlang::call_standardise(x)
#> read.table(file = "important.csv", row.names = FALSE)
```

(NB: If the function uses ... it's not possible to standardise all arguments.)

Calls can be modified in the same way as lists:

```
x$header <- TRUE
x
#> read.table("important.csv", row.names = FALSE, header = TRUE)
```

18.3.3.2 Function position

The first element of the call object is the **function position**. This contains
the function that will be called when the object is evaluated, and is usually a
symbol[6]:

```
lobstr::ast(foo())
#> █─foo
```

While R allows you to surround the name of the function with quotes, the
parser converts it to a symbol:

```
lobstr::ast("foo"())
#> █─foo
```

However, sometimes the function doesn't exist in the current environment and
you need to do some computation to retrieve it: for example, if the function is
in another package, is a method of an R6 object, or is created by a function
factory. In this case, the function position will be occupied by another call:

```
lobstr::ast(pkg::foo(1))
#> █─█─`::`
#> │ ├─pkg
#> │ └─foo
#> └─1
lobstr::ast(obj$foo(1))
#> █─█─`$`
#> │ ├─obj
#> │ └─foo
#> └─1
lobstr::ast(foo(1)(2))
#> █─█─foo
#> │ └─1
#> └─2
```

---

[6]Peculiarly, it can also be a number, as in the expression 3(). But this call will always
fail to evaluate because a number is not a function.

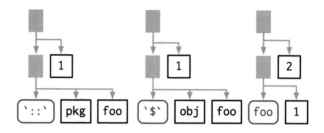

18.3.3.3 Constructing

You can construct a call object from its components using `rlang::call2()`. The first argument is the name of the function to call (either as a string, a symbol, or another call). The remaining arguments will be passed along to the call:

```
call2("mean", x = expr(x), na.rm = TRUE)
#> mean(x = x, na.rm = TRUE)
call2(expr(base::mean), x = expr(x), na.rm = TRUE)
#> base::mean(x = x, na.rm = TRUE)
```

Infix calls created in this way still print as usual.

```
call2("<-", expr(x), 10)
#> x <- 10
```

Using `call2()` to create complex expressions is a bit clunky. You'll learn another technique in Chapter 19.

18.3.4 Summary

The following table summarises the appearance of the different expression subtypes in `str()` and `typeof()`:

	`str()`	`typeof()`
Scalar constant	`logi/int/num/chr`	`logical/integer/double/character`
Symbol	`symbol`	`symbol`
Call object	`language`	`language`
Pairlist	Dotted pair list	`pairlist`
Expression vector	`expression()`	`expression`

Both base R and rlang provide functions for testing for each type of input, although the types covered are slightly different. You can easily tell them apart because all the base functions start with `is.` and the rlang functions start with `is_`.

	base	rlang
Scalar constant	—	is_syntactic_literal()
Symbol	is.symbol()	is_symbol()
Call object	is.call()	is_call()
Pairlist	is.pairlist()	is_pairlist()
Expression vector	is.expression()	—

18.3.5 Exercises

1. Which two of the six types of atomic vector can't appear in an expression? Why? Similarly, why can't you create an expression that contains an atomic vector of length greater than one?

2. What happens when you subset a call object to remove the first element? e.g. `expr(read.csv("foo.csv", header = TRUE))[-1]`. Why?

3. Describe the differences between the following call objects.

```
x <- 1:10

call2(median, x, na.rm = TRUE)
call2(expr(median), x, na.rm = TRUE)
call2(median, expr(x), na.rm = TRUE)
call2(expr(median), expr(x), na.rm = TRUE)
```

4. `rlang::call_standardise()` doesn't work so well for the following calls. Why? What makes `mean()` special?

```
call_standardise(quote(mean(1:10, na.rm = TRUE)))
#> mean(x = 1:10, na.rm = TRUE)
call_standardise(quote(mean(n = T, 1:10)))
#> mean(x = 1:10, n = T)
call_standardise(quote(mean(x = 1:10, , TRUE)))
#> mean(x = 1:10, , TRUE)
```

5. Why does this code not make sense?

```
x <- expr(foo(x = 1))
names(x) <- c("x", "y")
```

6. Construct the expression `if(x > 1) "a" else "b"` using multiple calls to `call2()`. How does the code structure reflect the structure of the AST?

18.4 Parsing and grammar

We've talked a lot about expressions and the AST, but not about how expressions are created from code that you type (like `"x + y"`). The process by which a computer language takes a string and constructs an expression is called **parsing**, and is governed by a set of rules known as a **grammar**. In this section, we'll use `lobstr::ast()` to explore some of the details of R's grammar, and then show how you can transform back and forth between expressions and strings.

18.4.1 Operator precedence

Infix functions introduce two sources of ambiguity[7]. The first source of ambiguity arises from infix functions: what does `1 + 2 * 3` yield? Do you get 9 (i.e. `(1 + 2) * 3`), or 7 (i.e. `1 + (2 * 3)`)? In other words, which of the two possible parse trees below does R use?

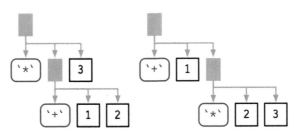

Programming languages use conventions called **operator precedence** to resolve this ambiguity. We can use `ast()` to see what R does:

```
lobstr::ast(1 + 2 * 3)
#> █─`+`
```

---

[7]This ambiguity does not exist in languages with only prefix or postfix calls. It's interesting to compare a simple arithmetic operation in Lisp (prefix) and Forth (postfix). In Lisp you'd write `(* (+ 1 2) 3))`; this avoids ambiguity by requiring parentheses everywhere. In Forth, you'd write `1 2 + 3 *`; this doesn't require any parentheses, but does require more thought when reading.

```
#>  ├─1
#>  └─■─`*`
#>     ├─2
#>     └─3
```

Predicting the precedence of arithmetic operations is usually easy because it's drilled into you in school and is consistent across the vast majority of programming languages.

Predicting the precedence of other operators is harder. There's one particularly surprising case in R: ! has a much lower precedence (i.e. it binds less tightly) than you might expect. This allows you to write useful operations like:

```
lobstr::ast(!x %in% y)
#>  ■─`!`
#>  └─■─`%in%`
#>     ├─x
#>     └─y
```

R has over 30 infix operators divided into 18 precedence groups. While the details are described in ?Syntax, very few people have memorised the complete ordering. If there's any confusion, use parentheses!

```
lobstr::ast((1 + 2) * 3)
#>  ■─`*`
#>  ├─■─`(`
#>  │  └─■─`+`
#>  │     ├─1
#>  │     └─2
#>  └─3
```

Note the appearance of the parentheses in the AST as a call to the (function.

18.4.2 Associativity

The second source of ambiguity is introduced by repeated usage of the same infix function. For example, is 1 + 2 + 3 equivalent to (1 + 2) + 3 or to 1 + (2 + 3)? This normally doesn't matter because x + (y + z) == (x + y) + z, i.e. addition is associative, but is needed because some S3 classes define + in a non-associative way. For example, ggplot2 overloads + to build up a complex plot from simple pieces; this is non-associative because earlier layers are drawn underneath later layers (i.e. geom_point() + geom_smooth() does not yield the same plot as geom_smooth() + geom_point()).

In R, most operators are **left-associative**, i.e. the operations on the left are evaluated first:

```
lobstr::ast(1 + 2 + 3)
#> █─`+`
#> ├─█─`+`
#> │ ├─1
#> │ └─2
#> └─3
```

There are two exceptions: exponentiation and assignment.

```
lobstr::ast(2^2^3)
#> █─`^`
#> ├─2
#> └─█─`^`
#>   ├─2
#>   └─3
lobstr::ast(x <- y <- z)
#> █─`<-`
#> ├─x
#> └─█─`<-`
#>   ├─y
#>   └─z
```

18.4.3 Parsing and deparsing

Most of the time you type code into the console, and R takes care of turning the characters you've typed into an AST. But occasionally you have code stored in a string, and you want to parse it yourself. You can do so using `rlang::parse_expr()`:

```
x1 <- "y <- x + 10"
x1
#> [1] "y <- x + 10"
is.call(x1)
#> [1] FALSE

x2 <- rlang::parse_expr(x1)
x2
#> y <- x + 10
```

```
is.call(x2)
#> [1] TRUE
```

parse_expr() always returns a single expression. If you have multiple expression separated by ; or \n, you'll need to use rlang::parse_exprs(). It returns a list of expressions:

```
x3 <- "a <- 1; a + 1"
rlang::parse_exprs(x3)
#> [[1]]
#> a <- 1
#>
#> [[2]]
#> a + 1
```

If you find yourself working with strings containing code very frequently, you should reconsider your process. Read Chapter 19 and consider whether you can generate expressions using quasiquotation more safely.

In Base R

The base equivalent to parse_exprs() is parse(). It is a little harder to use because it's specialised for parsing R code stored in files. You need to supply your string to the text argument and it returns an expression vector (Section 18.6.3). I recommend turning the output into a list:

```
as.list(parse(text = x1))
#> [[1]]
#> y <- x + 10
```

The inverse of parsing is **deparsing**: given an expression, you want the string that would generate it. This happens automatically when you print an expression, and you can get the string with rlang::expr_text():

```
z <- expr(y <- x + 10)
expr_text(z)
#> [1] "y <- x + 10"
```

Parsing and deparsing are not perfectly symmetric because parsing generates an *abstract* syntax tree. This means we lose backticks around ordinary names, comments, and whitespace:

```
cat(expr_text(expr({
  # This is a comment
  x <-                'x' + 1
})))
#> {
#>      x <- x + 1
#> }
```

In Base R

Be careful when using the base R equivalent, deparse(): it returns a character
vector with one element for each line. Whenever you use it, remember that
the length of the output might be greater than one, and plan accordingly.

18.4.4 Exercises

1. R uses parentheses in two slightly different ways as illustrated by
   these two calls:

   ```
   f((1))
   `(`(1 + 1)
   ```

   Compare and contrast the two uses by referencing the AST.

2. = can also be used in two ways. Construct a simple example that
   shows both uses.

3. Does -2^2 yield 4 or -4? Why?

4. What does !1 + !1 return? Why?

5. Why does x1 <- x2 <- x3 <- 0 work? Describe the two reasons.

6. Compare the ASTs of x + y %+% z and x ^ y %+% z. What have you
   learned about the precedence of custom infix functions?

7. What happens if you call parse_expr() with a string that generates
   multiple expressions? e.g. parse_expr("x + 1; y + 1")

8. What happens if you attempt to parse an invalid expression? e.g.
   "a +" or "f())".

9. deparse() produces vectors when the input is long. For example,
   the following call produces a vector of length two:

```
expr <- expr(g(a + b + c + d + e + f + g + h + i + j + k + l +
  m + n + o + p + q + r + s + t + u + v + w + x + y + z))

deparse(expr)
```

What does `expr_text()` do instead?

10. `pairwise.t.test()` assumes that `deparse()` always returns a length one character vector. Can you construct an input that violates this expectation? What happens?

18.5   Walking AST with recursive functions

To conclude the chapter I'm going to use everything you've learned about ASTs to solve more complicated problems. The inspiration comes from the base codetools package, which provides two interesting functions:

- `findGlobals()` locates all global variables used by a function. This can be useful if you want to check that your function doesn't inadvertently rely on variables defined in their parent environment.

- `checkUsage()` checks for a range of common problems including unused local variables, unused parameters, and the use of partial argument matching.

Getting all of the details of these functions correct is fiddly, so we won't fully develop the ideas. Instead we'll focus on the big underlying idea: recursion on the AST. Recursive functions are a natural fit to tree-like data structures because a recursive function is made up of two parts that correspond to the two parts of the tree:

- The **recursive case** handles the nodes in the tree. Typically, you'll do something to each child of a node, usually calling the recursive function again, and then combine the results back together again. For expressions, you'll need to handle calls and pairlists (function arguments).

- The **base case** handles the leaves of the tree. The base cases ensure that the function eventually terminates, by solving the simplest cases directly. For expressions, you need to handle symbols and constants in the base case.

To make this pattern easier to see, we'll need two helper functions. First we define `expr_type()` which will return "constant" for constant, "symbol" for symbols, "call", for calls, "pairlist" for pairlists, and the "type" of anything else:

```
expr_type <- function(x) {
  if (rlang::is_syntactic_literal(x)) {
    "constant"
  } else if (is.symbol(x)) {
    "symbol"
  } else if (is.call(x)) {
    "call"
  } else if (is.pairlist(x)) {
    "pairlist"
  } else {
    typeof(x)
  }
}

expr_type(expr("a"))
#> [1] "constant"
expr_type(expr(x))
#> [1] "symbol"
expr_type(expr(f(1, 2)))
#> [1] "call"
```

We'll couple this with a wrapper around the switch function:

```
switch_expr <- function(x, ...) {
  switch(expr_type(x),
    ...,
    stop("Don't know how to handle type ", typeof(x), call. = FALSE)
  )
}
```

With these two functions in hand, we can write a basic template for any function that walks the AST using `switch()` (Section 5.2.3):

```
recurse_call <- function(x) {
  switch_expr(x,
    # Base cases
    symbol = ,
    constant = ,

    # Recursive cases
    call = ,
    pairlist =
```

```
  )
}
```

Typically, solving the base case is easy, so we'll do that first, then check the results. The recursive cases are trickier, and will often require some functional programming.

18.5.1 Finding F and T

We'll start with a function that determines whether another function uses the logical abbreviations T and F because using them is often considered to be poor coding practice. Our goal is to return TRUE if the input contains a logical abbreviation, and FALSE otherwise.

Let's first find the type of T versus TRUE:

```
expr_type(expr(TRUE))
#> [1] "constant"

expr_type(expr(T))
#> [1] "symbol"
```

TRUE is parsed as a logical vector of length one, while T is parsed as a name. This tells us how to write our base cases for the recursive function: a constant is never a logical abbreviation, and a symbol is an abbreviation if it's "F" or "T":

```
logical_abbr_rec <- function(x) {
  switch_expr(x,
    constant = FALSE,
    symbol = as_string(x) %in% c("F", "T")
  )
}

logical_abbr_rec(expr(TRUE))
#> [1] FALSE
logical_abbr_rec(expr(T))
#> [1] TRUE
```

I've written logical_abbr_rec() function assuming that the input will be an expression as this will make the recursive operation simpler. However, when writing a recursive function it's common to write a wrapper that provides defaults or makes the function a little easier to use. Here we'll typically make a wrapper that quotes its input (we'll learn more about that in the next chapter), so we don't need to use `expr()` every time.

```
logical_abbr <- function(x) {
  logical_abbr_rec(enexpr(x))
}

logical_abbr(T)
#> [1] TRUE
logical_abbr(FALSE)
#> [1] FALSE
```

Next we need to implement the recursive cases. Here we want to do the same thing for calls and for pairlists: recursively apply the function to each subcomponent, and return `TRUE` if any subcomponent contains a logical abbreviation. This is made easy by `purrr::some()`, which iterates over a list and returns `TRUE` if the predicate function is true for any element.

```
logical_abbr_rec <- function(x) {
  switch_expr(x,
    # Base cases
    constant = FALSE,
    symbol = as_string(x) %in% c("F", "T"),

    # Recursive cases
    call = ,
    pairlist = purrr::some(x, logical_abbr_rec)
  )
}

logical_abbr(mean(x, na.rm = T))
#> [1] TRUE
logical_abbr(function(x, na.rm = T) FALSE)
#> [1] TRUE
```

18.5.2 Finding all variables created by assignment

`logical_abbr()` is relatively simple: it only returns a single `TRUE` or `FALSE`. The next task, listing all variables created by assignment, is a little more complicated. We'll start simply, and then make the function progressively more rigorous.

We start by looking at the AST for assignment:

```
ast(x <- 10)
#> █─`<-`
#> ├─x
#> └─10
```

Assignment is a call object where the first element is the symbol <-, the second is the name of variable, and the third is the value to be assigned.

Next, we need to decide what data structure we're going to use for the results. Here I think it will be easiest if we return a character vector. If we return symbols, we'll need to use a list() and that makes things a little more complicated.

With that in hand we can start by implementing the base cases and providing a helpful wrapper around the recursive function. Here the base cases are straightforward because we know that neither a symbol nor a constant represents assignment.

```
find_assign_rec <- function(x) {
  switch_expr(x,
    constant = ,
    symbol = character()
  )
}
find_assign <- function(x) find_assign_rec(enexpr(x))

find_assign("x")
#> character(0)
find_assign(x)
#> character(0)
```

Next we implement the recursive cases. This is made easier by a function that should exist in purrr, but currently doesn't. flat_map_chr() expects .f to return a character vector of arbitrary length, and flattens all results into a single character vector.

```
flat_map_chr <- function(.x, .f, ...) {
  purrr::flatten_chr(purrr::map(.x, .f, ...))
}

flat_map_chr(letters[1:3], ~ rep(., sample(3, 1)))
#> [1] "a" "b" "b" "b" "c" "c"
```

The recursive case for pairlists is straightforward: we iterate over every element of the pairlist (i.e. each function argument) and combine the results. The case for calls is a little bit more complex: if this is a call to `<-` then we should return the second element of the call:

```
find_assign_rec <- function(x) {
  switch_expr(x,
    # Base cases
    constant = ,
    symbol = character(),

    # Recursive cases
    pairlist = flat_map_chr(as.list(x), find_assign_rec),
    call = {
      if (is_call(x, "<-")) {
        as_string(x[[2]])
      } else {
        flat_map_chr(as.list(x), find_assign_rec)
      }
    }
  )
}

find_assign(a <- 1)
#> [1] "a"
find_assign({
  a <- 1
  {
    b <- 2
  }
})
#> [1] "a" "b"
```

Now we need to make our function more robust by coming up with examples intended to break it. What happens when we assign to the same variable multiple times?

```
find_assign({
  a <- 1
  a <- 2
})
#> [1] "a" "a"
```

It's easiest to fix this at the level of the wrapper function:

```
find_assign <- function(x) unique(find_assign_rec(enexpr(x)))

find_assign({
  a <- 1
  a <- 2
})
#> [1] "a"
```

What happens if we have nested calls to <-? Currently we only return the first. That's because when <- occurs we immediately terminate recursion.

```
find_assign({
  a <- b <- c <- 1
})
#> [1] "a"
```

Instead we need to take a more rigorous approach. I think it's best to keep the recursive function focused on the tree structure, so I'm going to extract out find_assign_call() into a separate function.

```
find_assign_call <- function(x) {
  if (is_call(x, "<-") && is_symbol(x[[2]])) {
    lhs <- as_string(x[[2]])
    children <- as.list(x)[-1]
  } else {
    lhs <- character()
    children <- as.list(x)
  }

  c(lhs, flat_map_chr(children, find_assign_rec))
}

find_assign_rec <- function(x) {
  switch_expr(x,
    # Base cases
    constant = ,
    symbol = character(),

    # Recursive cases
    pairlist = flat_map_chr(x, find_assign_rec),
    call = find_assign_call(x)
  )
}
```

```
find_assign(a <- b <- c <- 1)
#> [1] "a" "b" "c"
find_assign(system.time(x <- print(y <- 5)))
#> [1] "x" "y"
```

The complete version of this function is quite complicated, it's important to remember we wrote it by working our way up by writing simple component parts.

18.5.3 Exercises

1. `logical_abbr()` returns TRUE for T(1, 2, 3). How could you modify `logical_abbr_rec()` so that it ignores function calls that use T or F?

2. `logical_abbr()` works with expressions. It currently fails when you give it a function. Why? How could you modify `logical_abbr()` to make it work? What components of a function will you need to recurse over?

```
logical_abbr(function(x = TRUE) {
    g(x + T)
})
```

3. Modify `find_assign` to also detect assignment using replacement functions, i.e. `names(x) <- y`.

4. Write a function that extracts all calls to a specified function.

18.6 Specialised data structures

There are two data structures and one special symbol that we need to cover for the sake of completeness. They are not usually important in practice.

18.6.1 Pairlists

Pairlists are a remnant of R's past and have been replaced by lists almost everywhere. The only place you are likely to see pairlists in R[8] is when working with calls to the `function` function, as the formal arguments to a function are stored in a pairlist:

```
f <- expr(function(x, y = 10) x + y)

args <- f[[2]]
args
#> $x
#>
#>
#> $y
#> [1] 10
typeof(args)
#> [1] "pairlist"
```

Fortunately, whenever you encounter a pairlist, you can treat it just like a regular list:

```
pl <- pairlist(x = 1, y = 2)
length(pl)
#> [1] 2
pl$x
#> [1] 1
```

Behind the scenes pairlists are implemented using a different data structure, a linked list instead of an array. That makes subsetting a pairlist much slower than subsetting a list, but this has little practical impact.

18.6.2 Missing arguments

The special symbol that needs a little extra discussion is the empty symbol, which is used to represent missing arguments (not missing values!). You only need to care about the missing symbol if you're programmatically creating functions with missing arguments; we'll come back to that in Section 19.4.3.

You can make an empty symbol with `missing_arg()` (or `expr()`):

---

[8]If you're working in C, you'll encounter pairlists more often. For example, call objects are also implemented using pairlists.

```
missing_arg()
typeof(missing_arg())
#> [1] "symbol"
```

An empty symbol doesn't print anything, so you can check if you have one with `rlang::is_missing()`:

```
is_missing(missing_arg())
#> [1] TRUE
```

You'll find them in the wild in function formals:

```
f <- expr(function(x, y = 10) x + y)
args <- f[[2]]
is_missing(args[[1]])
#> [1] TRUE
```

This is particularly important for ... which is always associated with an empty symbol:

```
f <- expr(function(...) list(...))
args <- f[[2]]
is_missing(args[[1]])
#> [1] TRUE
```

The empty symbol has a peculiar property: if you bind it to a variable, then access that variable, you will get an error:

```
m <- missing_arg()
m
#> Error in eval(expr, envir, enclos): argument "m" is missing, with no
#> default
```

But you won't if you store it inside another data structure!

```
ms <- list(missing_arg(), missing_arg())
ms[[1]]
```

If you need to preserve the missingness of a variable, `rlang::maybe_missing()` is often helpful. It allows you to refer to a potentially missing variable without triggering the error. See the documentation for use cases and more details.

18.6.3   Expression vectors

Finally, we need to briefly discuss the expression vector. Expression vectors are only produced by two base functions: `expression()` and `parse()`:

```
exp1 <- parse(text = c("
x <- 4
x
"))
exp2 <- expression(x <- 4, x)

typeof(exp1)
#> [1] "expression"
typeof(exp2)
#> [1] "expression"

exp1
#> expression(x <- 4, x)
exp2
#> expression(x <- 4, x)
```

Like calls and pairlists, expression vectors behave like lists:

```
length(exp1)
#> [1] 2
exp1[[1]]
#> x <- 4
```

Conceptually, an expression vector is just a list of expressions. The only difference is that calling `eval()` on an expression evaluates each individual expression. I don't believe this advantage merits introducing a new data structure, so instead of expression vectors I just use lists of expressions.

19

Quasiquotation

19.1 Introduction

Now that you understand the tree structure of R code, it's time to return to one of the fundamental ideas that make `expr()` and `ast()` work: quotation. In tidy evaluation, all quoting functions are actually quasiquoting functions because they also support unquoting. Where quotation is the act of capturing an unevaluated expression, **unquotation** is the ability to selectively evaluate parts of an otherwise quoted expression. Together, this is called quasiquotation. Quasiquotation makes it easy to create functions that combine code written by the function's author with code written by the function's user. This helps to solve a wide variety of challenging problems.

Quasiquotation is one of the three pillars of tidy evaluation. You'll learn about the other two (quosures and the data mask) in Chapter 20. When used alone, quasiquotation is most useful for programming, particularly for generating code. But when it's combined with the other techniques, tidy evaluation becomes a powerful tool for data analysis.

Outline

- Section 19.2 motivates the development of quasiquotation with a function, `cement()`, that works like `paste()` but automatically quotes its arguments so that you don't have to.

- Section 19.3 gives you the tools to quote expressions, whether they come from you or the user, or whether you use rlang or base R tools.

- Section 19.4 introduces the biggest difference between rlang quoting functions and base quoting function: unquoting with `!!` and `!!!`.

- Section 19.5 discusses the three main non-quoting techniques that base R functions uses to disable quoting behaviour.

- Section 19.6 explores another place that you can use `!!!`, functions that take `....`. It also introduces the special `:=` operator, which allows you to dynamically change argument names.

- Section 19.7 shows a few practical uses of quoting to solve problems that naturally require some code generation.

- Section 19.8 finishes up with a little history of quasiquotation for those who are interested.

Prerequisites

Make sure you've read the metaprogramming overview in Chapter 17 to get a broad overview of the motivation and the basic vocabulary, and that you're familiar with the tree structure of expressions as described in Section 18.3.

Code-wise, we'll mostly be using the tools from rlang (`https://rlang.r-lib.org`), but at the end of the chapter you'll also see some powerful applications in conjunction with purrr (`https://purrr.tidyverse.org`).

```r
library(rlang)
library(purrr)
```

Related work

Quoting functions have deep connections to Lisp **macros**. But macros are usually run at compile-time, which doesn't exist in R, and they always input and output ASTs. See Lumley [2001] for one approach to implementing them in R. Quoting functions are more closely related to the more esoteric Lisp **fexprs** (`http://en.wikipedia.org/wiki/Fexpr`), functions where all arguments are quoted by default. These terms are useful to know when looking for related work in other programming languages.

19.2   Motivation

We'll start with a concrete example that helps motivate the need for unquoting, and hence quasiquotation. Imagine you're creating a lot of strings by joining together words:

```r
paste("Good", "morning", "Hadley")
#> [1] "Good morning Hadley"
paste("Good", "afternoon", "Alice")
#> [1] "Good afternoon Alice"
```

You are sick and tired of writing all those quotes, and instead you just want to use bare words. To that end, you've written the following function. (Don't worry about the implementation for now; you'll learn about the pieces later.)

```
cement <- function(...) {
  args <- ensyms(...)
  paste(purrr::map(args, as_string), collapse = " ")
}

cement(Good, morning, Hadley)
#> [1] "Good morning Hadley"
cement(Good, afternoon, Alice)
#> [1] "Good afternoon Alice"
```

Formally, this function quotes all of its inputs. You can think of it as automatically putting quotation marks around each argument. That's not precisely true as the intermediate objects it generates are expressions, not strings, but it's a useful approximation, and the root meaning of the term "quote".

This function is nice because we no longer need to type quotation marks. The problem comes when we want to use variables. It's easy to use variables with `paste()`: just don't surround them with quotation marks.

```
name <- "Hadley"
time <- "morning"

paste("Good", time, name)
#> [1] "Good morning Hadley"
```

Obviously this doesn't work with `cement()` because every input is automatically quoted:

```
cement(Good, time, name)
#> [1] "Good time name"
```

We need some way to explicitly *unquote* the input to tell `cement()` to remove the automatic quote marks. Here we need `time` and `name` to be treated differently to `Good`. Quasiquotation gives us a standard tool to do so: `!!`, called "unquote", and pronounced bang-bang. `!!` tells a quoting function to drop the implicit quotes:

```
cement(Good, !!time, !!name)
#> [1] "Good morning Hadley"
```

It's useful to compare `cement()` and `paste()` directly. `paste()` evaluates its arguments, so we must quote where needed; `cement()` quotes its arguments, so we must unquote where needed.

```
paste("Good", time, name)
cement(Good, !!time, !!name)
```

19.2.1 Vocabulary

The distinction between quoted and evaluated arguments is important:

- An **evaluated** argument obeys R's usual evaluation rules.
- A **quoted** argument is captured by the function, and is processed in some custom way.

`paste()` evaluates all its arguments; `cement()` quotes all its arguments.

If you're ever unsure about whether an argument is quoted or evaluated, try executing the code outside of the function. If it doesn't work or does something different, then that argument is quoted. For example, you can use this technique to determine that the first argument to `library()` is quoted:

```
# works
library(MASS)

# fails
MASS
#> Error in eval(expr, envir, enclos): object 'MASS' not found
```

Talking about whether an argument is quoted or evaluated is a more precise way of stating whether or not a function uses non-standard evaluation (NSE). I will sometimes use "quoting function" as short-hand for a function that quotes one or more arguments, but generally, I'll talk about quoted arguments since that is the level at which the difference applies.

19.2.2 Exercises

1. For each function in the following base R code, identify which arguments are quoted and which are evaluated.

    ```
    library(MASS)
    ```

```
mtcars2 <- subset(mtcars, cyl == 4)

with(mtcars2, sum(vs))
sum(mtcars2$am)

rm(mtcars2)
```

2. For each function in the following tidyverse code, identify which arguments are quoted and which are evaluated.

```
library(dplyr)
library(ggplot2)

by_cyl <- mtcars %>%
  group_by(cyl) %>%
  summarise(mean = mean(mpg))

ggplot(by_cyl, aes(cyl, mean)) + geom_point()
```

19.3 Quoting

The first part of quasiquotation is quotation: capturing an expression without evaluating it. We'll need a pair of functions because the expression can be supplied directly or indirectly, via lazily-evaluated function argument. I'll start with the rlang quoting functions, then circle back to those provided by base R.

19.3.1 Capturing expressions

There are four important quoting functions. For interactive exploration, the most important is expr(), which captures its argument exactly as provided:

```
expr(x + y)
#> x + y
expr(1 / 2 / 3)
#> 1/2/3
```

(Remember that white space and comments are not part of the expression, so will not be captured by a quoting function.)

`expr()` is great for interactive exploration, because it captures what you, the developer, typed. It's not so useful inside a function:

```
f1 <- function(x) expr(x)
f1(a + b + c)
#> x
```

We need another function to solve this problem: `enexpr()`. This captures what the caller supplied to the function by looking at the internal promise object that powers lazy evaluation (Section 6.5.1).

```
f2 <- function(x) enexpr(x)
f2(a + b + c)
#> a + b + c
```

(It's called "en"-`expr()` by analogy to enrich. Enriching someone makes them richer; `enexpr()`ing a argument makes it an expression.)

To capture all arguments in ..., use `enexprs()`.

```
f <- function(...) enexprs(...)
f(x = 1, y = 10 * z)
#> $x
#> [1] 1
#>
#> $y
#> 10 * z
```

Finally, `exprs()` is useful interactively to make a list of expressions:

```
exprs(x = x ^ 2, y = y ^ 3, z = z ^ 4)
# shorthand for
# list(x = expr(x ^ 2), y = expr(y ^ 3), z = expr(z ^ 4))
```

In short, use `enexpr()` and `enexprs()` to capture the expressions supplied as arguments *by the user*. Use `expr()` and `exprs()` to capture expressions that *you* supply.

19.3.2 Capturing symbols

Sometimes you only want to allow the user to specify a variable name, not an arbitrary expression. In this case, you can use `ensym()` or `ensyms()`. These are variants of `enexpr()` and `enexprs()` that check the captured expression is either symbol or a string (which is converted to a symbol[1]). `ensym()` and `ensyms()` throw an error if given anything else.

```
f <- function(...) ensyms(...)
f(x)
#> [[1]]
#> x
f("x")
#> [[1]]
#> x
```

19.3.3 With base R

Each rlang function described above has an equivalent in base R. Their primary difference is that the base equivalents do not support unquoting (which we'll talk about very soon). This make them quoting functions, rather than quasiquoting functions.

The base equivalent of `expr()` is `quote()`:

```
quote(x + y)
#> x + y
```

The base function closest to `enexpr()` is `substitute()`:

```
f3 <- function(x) substitute(x)
f3(x + y)
#> x + y
```

The base equivalent to `exprs()` is `alist()`:

```
alist(x = 1, y = x + 2)
#> $x
#> [1] 1
#>
#> $y
#> x + 2
```

---

[1]This is for compatibility with base R, which allows you to provide a string instead of a symbol in many places: `"x" <- 1`, `"foo"(x, y)`, `c("x" = 1)`.

The equivalent to `enexprs()` is an undocumented feature of `substitute()`[2]:

```
f <- function(...) as.list(substitute(...()))
f(x = 1, y = 10 * z)
#> $x
#> [1] 1
#>
#> $y
#> 10 * z
```

There are two other important base quoting functions that we'll cover elsewhere:

- `bquote()` provides a limited form of quasiquotation, and is discussed in Section 19.5.

- `~`, the formula, is a quoting function that also captures the environment. It's the inspiration for quosures, the topic of the next chapter, and is discussed in Section 20.3.4.

19.3.4 Substitution

You'll most often see `substitute()` used to capture unevaluated arguments. However, as well as quoting, `substitute()` also does substitution (as its name suggests!). If you give it an expression, rather than a symbol, it will substitute in the values of symbols defined in the current environment.

```
f4 <- function(x) substitute(x * 2)
f4(a + b + c)
#> (a + b + c) * 2
```

I think this makes code hard to understand, because if it is taken out of context, you can't tell if the goal of `substitute(x + y)` is to replace x, y, or both. If you do want to use `substitute()` for substitution, I recommend that you use the second argument to make your goal clear:

```
substitute(x * y * z, list(x = 10, y = quote(a + b)))
#> 10 * (a + b) * z
```

---

[2]Discovered by Peter Meilstrup and described in R-devel on 2018-08-13 (http://r.789695.n4.nabble.com/substitute-on-arguments-in-ellipsis-quot-dot-dot-dot-quot-td4751658.html).

19.3.5 Summary

When quoting (i.e. capturing code), there are two important distinctions:

- Is it supplied by the developer of the code or the user of the code? In other words, is it fixed (supplied in the body of the function) or varying (supplied via an argument)?

- Do you want to capture a single expression or multiple expressions?

This leads to a 2 × 2 table of functions for rlang, Table 19.1, and for base R, Table 19.2.

Table 19.1: rlang quasiquoting functions

	Developer	User
One	expr()	enexpr()
Many	exprs()	enexprs()

Table 19.2: base R quoting functions

	Developer	User
One	quote()	substitute()
Many	alist()	as.list(substitute(...()))

19.3.6 Exercises

1. How is expr() implemented? Look at its source code.

2. Compare and contrast the following two functions. Can you predict the output before running them?

```
f1 <- function(x, y) {
  exprs(x = x, y = y)
}
f2 <- function(x, y) {
  enexprs(x = x, y = y)
}
f1(a + b, c + d)
f2(a + b, c + d)
```

3. What happens if you try to use enexpr() with an expression (i.e.
   enexpr(x + y) ? What happens if enexpr() is passed a missing argument?

4. How are exprs(a) and exprs(a =) different? Think about both the
   input and the output.

5. What are other differences between exprs() and alist()? Read the
   documentation for the named arguments of exprs() to find out.

6. The documentation for substitute() says:

   > Substitution takes place by examining each compo-
   > nent of the parse tree as follows:
   >
   > - *If it is not a bound symbol in env, it is un-
   >   changed.*
   > - *If it is a promise object (i.e., a formal argument
   >   to a function) the expression slot of the promise
   >   replaces the symbol.*
   > - *If it is an ordinary variable, its value is substi-
   >   tuted, unless env is .GlobalEnv in which case
   >   the symbol is left unchanged.*

   Create examples that illustrate each of the above cases.

19.4   Unquoting

So far, you've only seen relatively small advantages of the rlang quoting functions over the base R quoting functions: they have a more consistent naming scheme. The big difference is that rlang quoting functions are actually
quasiquoting functions because they can also unquote.

Unquoting allows you to selectively evaluate parts of the expression that would
otherwise be quoted, which effectively allows you to merge ASTs using a template AST. Since base functions don't use unquoting, they instead use a variety
of other techniques, which you'll learn about in Section 19.5.

Unquoting is one inverse of quoting. It allows you to selectively evaluate code
inside expr(), so that expr(!!x) is equivalent to x. In Chapter 20, you'll
learn about another inverse, evaluation. This happens outside expr(), so that
eval(expr(x)) is equivalent to x.

19.4.1 Unquoting one argument

Use `!!` to unquote a single argument in a function call. `!!` takes a single expression, evaluates it, and inlines the result in the AST.

```
x <- expr(-1)
expr(f(!!x, y))
#> f(-1, y)
```

I think this is easiest to understand with a diagram. `!!` introduces a placeholder in the AST, shown with dotted borders. Here the placeholder x is replaced by an AST, illustrated by a dotted connection.

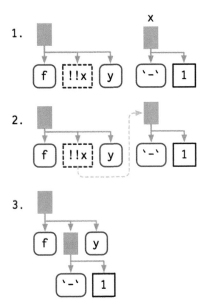

As well as call objects, `!!` also works with symbols and constants:

```
a <- sym("y")
b <- 1
expr(f(!!a, !!b))
#> f(y, 1)
```

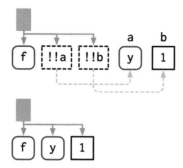

If the right-hand side of !! is a function call, !! will evaluate it and insert the results:

```
mean_rm <- function(var) {
  var <- ensym(var)
  expr(mean(!!var, na.rm = TRUE))
}
expr(!!mean_rm(x) + !!mean_rm(y))
#> mean(x, na.rm = TRUE) + mean(y, na.rm = TRUE)
```

!! preserves operator precedence because it works with expressions.

```
x1 <- expr(x + 1)
x2 <- expr(x + 2)

expr(!!x1 / !!x2)
#> (x + 1)/(x + 2)
```

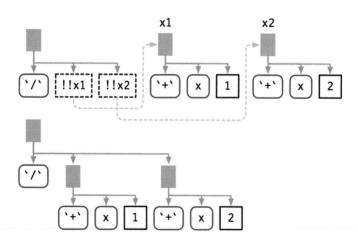

If we simply pasted the text of the expressions together, we'd end up with x
+ 1 / x + 2, which has a very different AST:

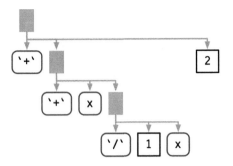

19.4.2 Unquoting a function

!! is most commonly used to replace the arguments to a function, but you
can also use it to replace the function. The only challenge here is operator
precedence: expr(!!f(x, y)) unquotes the result of f(x, y), so you need an
extra pair of parentheses.

```
f <- expr(foo)
expr((!!f)(x, y))
#> foo(x, y)
```

This also works when f is a call:

```
f <- expr(pkg::foo)
expr((!!f)(x, y))
#> pkg::foo(x, y)
```

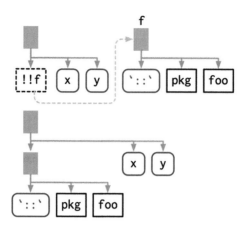

Because of the large number of parentheses involved, it can be clearer to use
rlang::call2():

```
f <- expr(pkg::foo)
call2(f, expr(x), expr(y))
#> pkg::foo(x, y)
```

19.4.3 Unquoting a missing argument

Very occasionally it is useful to unquote a missing argument (Section 18.6.2),
but the naive approach doesn't work:

```
arg <- missing_arg()
expr(foo(!!arg, !!arg))
#> Error in enexpr(expr): argument "arg" is missing, with no default
```

You can work around this with the rlang::maybe_missing() helper:

```
expr(foo(!!maybe_missing(arg), !!maybe_missing(arg)))
#> foo(, )
```

19.4.4 Unquoting in special forms

There are a few special forms where unquoting is a syntax error. Take $ for
example: it must always be followed by the name of a variable, not another
expression. This means attempting to unquote with $ will fail with a syntax
error:

```
expr(df$!!x)
#> Error: unexpected '!' in "expr(df$!"
```

To make unquoting work, you'll need to use the prefix form (Section 6.8.1):

```
x <- expr(x)
expr(`$`(df, !!x))
#> df$x
```

19.4.5 Unquoting many arguments

!! is a one-to-one replacement. !!! (called "unquote-splice", and pronounced bang-bang-bang) is a one-to-many replacement. It takes a list of expressions and inserts them at the location of the !!!:

```
xs <- exprs(1, a, -b)
expr(f(!!!xs, y))
#> f(1, a, -b, y)

# Or with names
ys <- set_names(xs, c("a", "b", "c"))
expr(f(!!!ys, d = 4))
#> f(a = 1, b = a, c = -b, d = 4)
```

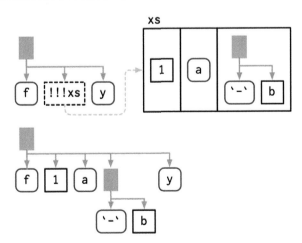

!!! can be used in any rlang function that takes ... regardless of whether or not ... is quoted or evaluated. We'll come back to this in Section 19.6; for now note that this can be useful in call2().

```
call2("f", !!!xs, expr(y))
#> f(1, a, -b, y)
```

19.4.6   The polite fiction of !!

So far we have acted as if !! and !!! are regular prefix operators like + , -,
and !. They're not. From R's perspective, !! and !!! are simply the repeated
application of !:

```
!!TRUE
#> [1] TRUE
!!!TRUE
#> [1] FALSE
```

!! and !!! behave specially inside all quoting functions powered by rlang,
where they behave like real operators with precedence equivalent to unary +
and -. This requires considerable work inside rlang, but means that you can
write !!x + !!y instead of (!!x) + (!!y).

The biggest downside[3] to using a fake operator is that you might get silent
errors when misusing !! outside of quasiquoting functions. Most of the time
this is not an issue because !! is typically used to unquote expressions or
quosures. Since expressions are not supported by the negation operator, you
will get an argument type error in this case:

```
x <- quote(variable)
!!x
#> Error in !x: invalid argument type
```

But you can get silently incorrect results when working with numeric values:

```
df <- data.frame(x = 1:5)
y <- 100
with(df, x + !!y)
#> [1] 2 3 4 5 6
```

Given these drawbacks, you might wonder why we introduced new syntax
instead of using regular function calls. Indeed, early versions of tidy evaluation

---

[3]Prior to R 3.5.1, there was another major downside: the R deparser treated !!x as !(!x).
This is why in old versions of R you might see extra parentheses when printing expressions.
The good news is that these parentheses are not real and can be safely ignored most of the
time. The bad news is that they will become real if you reparse that printed output to R
code. These roundtripped functions will not work as expected since !(!x) does not unquote.

used function calls like UQ() and UQS(). However, they're not really function calls, and pretending they are leads to a misleading mental mode. We chose !! and !!! as the least-bad solution:

- They are visually strong and don't look like existing syntax. When you see !!x or !!!x it's clear that something unusual is happening.

- They override a rarely used piece of syntax, as double negation is not a common pattern in R[4]. If you do need it, you can just add parentheses !(!x).

19.4.7 Non-standard ASTs

With unquoting, it's easy to create non-standard ASTs, i.e. ASTs that contain components that are not expressions. (It is also possible to create non-standard ASTs by directly manipulating the underlying objects, but it's harder to do so accidentally.) These are valid, and occasionally useful, but their correct use is beyond the scope of this book. However, it's important to learn about them, because they can be deparsed, and hence printed, in misleading ways.

For example, if you inline more complex objects, their attributes are not printed. This can lead to confusing output:

```
x1 <- expr(class(!!data.frame(x = 10)))
x1
#> class(list(x = 10))
eval(x1)
#> [1] "data.frame"
```

You have two main tools to reduce this confusion: rlang::expr_print() and lobstr::ast():

```
expr_print(x1)
#> class(<data.frame>)
lobstr::ast(!!x1)
#> █—class
#> └—<inline data.frame>
```

Another confusing case arises if you inline an integer sequence:

```
x2 <- expr(f(!!c(1L, 2L, 3L, 4L, 5L)))
x2
```

---

[4]Unlike, say, Javascript, where !!x is a commonly used shortcut to convert an integer into a logical.

```
#> f(1:5)
expr_print(x2)
#> f(<int: 1L, 2L, 3L, 4L, 5L>)
lobstr::ast(!!x2)
#> ▮—f
#> └—<inline integer>
```

It's also possible to create regular ASTs that can not be generated from code because of operator precedence. In this case, R will print parentheses that do not exist in the AST:

```
x3 <- expr(1 + !!expr(2 + 3))
x3
#> 1 + (2 + 3)

lobstr::ast(!!x3)
#> ▮—`+`
#> ├—1
#> └—▮—`+`
#>    ├—2
#>    └—3
```

19.4.8   Exercises

1. Given the following components:

   ```
   xy <- expr(x + y)
   xz <- expr(x + z)
   yz <- expr(y + z)
   abc <- exprs(a, b, c)
   ```

   Use quasiquotation to construct the following calls:

   ```
   (x + y) / (y + z)
   -(x + z) ^ (y + z)
   (x + y) + (y + z) - (x + y)
   atan2(x + y, y + z)
   sum(x + y, x + y, y + z)
   sum(a, b, c)
   ```

```
mean(c(a, b, c), na.rm = TRUE)
foo(a = x + y, b = y + z)
```

2. The following two calls print the same, but are actually different:

```
(a <- expr(mean(1:10)))
#> mean(1:10)
(b <- expr(mean(!!(1:10))))
#> mean(1:10)
identical(a, b)
#> [1] FALSE
```

What's the difference? Which one is more natural?

19.5   Non-quoting

Base R has one function that implements quasiquotation: bquote(). It uses .() for unquoting:

```
xyz <- bquote((x + y + z))
bquote(-.(xyz) / 2)
#> -(x + y + z)/2
```

bquote() isn't used by any other function in base R, and has had relatively little impact on how R code is written. There are three challenges to effective use of bquote():

- It is only easily used with your code; it is hard to apply it to arbitrary code supplied by a user.

- It does not provide an unquote-splice operator that allows you to unquote multiple expressions stored in a list.

- It lacks the ability to handle code accompanied by an environment, which is crucial for functions that evaluate code in the context of a data frame, like subset() and friends.

Base functions that quote an argument use some other technique to allow indirect specification. Base R approaches selectively turn quoting off, rather than using unquoting, so I call them **non-quoting** techniques.

There are four basic forms seen in base R:

- A pair of quoting and non-quoting functions. For example, `$` has two arguments, and the second argument is quoted. This is easier to see if you write in prefix form: `mtcars$cyl` is equivalent to `` `$`(mtcars, cyl) ``. If you want to refer to a variable indirectly, you use `[[`, as it takes the name of a variable as a string.

```
x <- list(var = 1, y = 2)
var <- "y"

x$var
#> [1] 1
x[[var]]
#> [1] 2
```

There are three other quoting functions closely related to `$`: `subset()`, `transform()`, and `with()`. These are seen as wrappers around `$` only suitable for interactive use so they all have the same non-quoting alternative:
`[`

`<-`/`assign()` and `::`/`getExportedValue()` work similarly to `$`/`[`.

- A pair of quoting and non-quoting arguments. For example, `rm()` allows you to provide bare variable names in ..., or a character vector of variable names in `list`:

```
x <- 1
rm(x)

y <- 2
vars <- c("y", "vars")
rm(list = vars)
```

`data()` and `save()` work similarly.

- An argument that controls whether a different argument is quoting or non-quoting. For example, in `library()`, the `character.only` argument controls the quoting behaviour of the first argument, `package`:

```
library(MASS)

pkg <- "MASS"
library(pkg, character.only = TRUE)
```

`demo()`, `detach()`, `example()`, and `require()` work similarly.

- Quoting if evaluation fails. For example, the first argument to help() is non-quoting if it evaluates to a string; if evaluation fails, the first argument is quoted.

```
# Shows help for var
help(var)

var <- "mean"
# Shows help for mean
help(var)

var <- 10
# Shows help for var
help(var)
```

ls(), page(), and match.fun() work similarly.

Another important class of quoting functions are the base modelling and plotting functions, which follow the so-called standard non-standard evaluation rules: http://developer.r-project.org/nonstandard-eval.pdf. For example, lm() quotes the weight and subset arguments, and when used with a formula argument, the plotting function quotes the aesthetic arguments (col, cex, etc). Take the following code: we only need col = Species rather than col = iris$Species.

```
palette(RColorBrewer::brewer.pal(3, "Set1"))
plot(
  Sepal.Length ~ Petal.Length,
  data = iris,
  col = Species,
  pch = 20,
  cex = 2
)
```

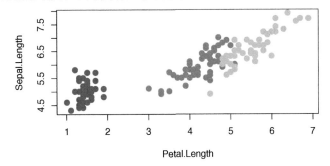

These functions have no built-in options for indirect specification, but you'll learn how to simulate unquoting in Section 20.6.

19.6   ... (dot-dot-dot)

!!! is useful because it's not uncommon to have a list of expressions that you want to insert into a call. It turns out that this pattern is common elsewhere. Take the following two motivating problems:

- What do you do if the elements you want to put in ... are already stored in a list? For example, imagine you have a list of data frames that you want to rbind() together:

```
dfs <- list(
    a = data.frame(x = 1, y = 2),
    b = data.frame(x = 3, y = 4)
)
```

You could solve this specific case with rbind(dfsa, dfsb), but how do you generalise that solution to a list of arbitrary length?

- What do you do if you want to supply the argument name indirectly? For example, imagine you want to create a single column data frame where the name of the column is specified in a variable:

```
var <- "x"
val <- c(4, 3, 9)
```

In this case, you could create a data frame and then change names (i.e. setNames(data.frame(val), var)), but this feels inelegant. How can we do better?

One way to think about these problems is to draw explicit parallels to quasiquotation:

- Row-binding multiple data frames is like unquote-splicing: we want to inline individual elements of the list into the call:

```
dplyr::bind_rows(!!!dfs)
#>    x y
#> 1  1 2
#> 2  3 4
```

When used in this context, the behaviour of !!! is known as "spatting" in Ruby, Go, PHP, and Julia. It is closely related to *args (star-args) and **kwarg (star-star-kwargs) in Python, which are sometimes called argument unpacking.

- The second problem is like unquoting the left-hand side of =: rather than interpreting var literally, we want to use the value stored in the variable called var:

```
tibble::tibble(!!var := val)
#> # A tibble: 3 x 1
#>       x
#>   <dbl>
#> 1     4
#> 2     3
#> 3     9
```

Note the use of := (pronounced colon-equals) rather than =. Unfortunately we need this new operation because R's grammar does not allow expressions as argument names:

```
tibble::tibble(!!var = value)
#> Error: unexpected '=' in "tibble::tibble(!!var ="
```

:= is like a vestigial organ: it's recognised by R's parser, but it doesn't have any code associated with it. It looks like an = but allows expressions on either side, making it a more flexible alternative to =. It is used in data.table for similar reasons.

Base R takes a different approach, which we'll come back to in Section 19.6.4.

We say functions that support these tools, without quoting arguments, have **tidy dots**[5]. To gain tidy dots behaviour in your own function, all you need to do is use list2().

19.6.1 Examples

One place we could use list2() is to create a wrapper around attributes() that allows us to set attributes flexibly:

```
set_attr <- function(.x, ...) {
  attr <- rlang::list2(...)
```

---

[5]This is admittedly not the most creative of names, but it clearly suggests it's something that has been added to R after the fact.

```
  attributes(.x) <- attr
  .x
}

attrs <- list(x = 1, y = 2)
attr_name <- "z"

1:10 %>%
  set_attr(w = 0, !!!attrs, !!attr_name := 3) %>%
  str()
#>  int [1:10] 1 2 3 4 5 6 7 8 9 10
#>  - attr(*, "w")= num 0
#>  - attr(*, "x")= num 1
#>  - attr(*, "y")= num 2
#>  - attr(*, "z")= num 3
```

19.6.2 exec()

What if you want to use this technique with a function that doesn't have
tidy dots? One option is to use rlang::exec() to call a function with some
arguments supplied directly (in ...) and others indirectly (in a list):

```
# Directly
exec("mean", x = 1:10, na.rm = TRUE, trim = 0.1)
#> [1] 5.5

# Indirectly
args <- list(x = 1:10, na.rm = TRUE, trim = 0.1)
exec("mean", !!!args)
#> [1] 5.5

# Mixed
params <- list(na.rm = TRUE, trim = 0.1)
exec("mean", x = 1:10, !!!params)
#> [1] 5.5
```

rlang::exec() also makes it possible to supply argument names indirectly:

```
arg_name <- "na.rm"
arg_val <- TRUE
exec("mean", 1:10, !!arg_name := arg_val)
#> [1] 5.5
```

And finally, it's useful if you have a vector of function names or a list of functions that you want to call with the same arguments:

```
x <- c(runif(10), NA)
funs <- c("mean", "median", "sd")

purrr::map_dbl(funs, exec, x, na.rm = TRUE)
#> [1] 0.444 0.482 0.298
```

exec() is closely related to call2(); where call2() returns an expression, exec() evaluates it.

19.6.3 dots_list()

list2() provides one other handy feature: by default it will ignore any empty arguments at the end. This is useful in functions like tibble::tibble() because it means that you can easily change the order of variables without worrying about the final comma:

```
# Can easily move x to first entry:
tibble::tibble(
  y = 1:5,
  z = 3:-1,
  x = 5:1,
)

# Need to remove comma from z and add comma to x
data.frame(
  y = 1:5,
  z = 3:-1,
  x = 5:1
)
```

list2() is a wrapper around rlang::dots_list() with defaults set to the most commonly used settings. You can get more control by calling dots_list() directly:

- .ignore_empty allows you to control exactly which arguments are ignored. The default ignores a single trailing argument to get the behaviour described above, but you can choose to ignore all missing arguments, or no missing arguments.

- .homonoyms controls what happens if multiple arguments use the same name:

```
str(dots_list(x = 1, x = 2))
#> List of 2
#>  $ x: num 1
#>  $ x: num 2
str(dots_list(x = 1, x = 2, .homonyms = "first"))
#> List of 1
#>  $ x: num 1
str(dots_list(x = 1, x = 2, .homonyms = "last"))
#> List of 1
#>  $ x: num 2
str(dots_list(x = 1, x = 2, .homonyms = "error"))
#> Error: Arguments can't have the same name.
#> We found multiple arguments named `x` at positions 1 and 2
```

- If there are empty arguments that are not ignored, .preserve_empty controls what to do with them. The default throws an error; setting .preserve_empty = TRUE instead returns missing symbols. This is useful if you're using dots_list() to generate function calls.

19.6.4 With base R

Base R provides a Swiss army knife to solve these problems: do.call(). do.call() has two main arguments. The first argument, what, gives a function to call. The second argument, args, is a list of arguments to pass to that function, and so do.call("f", list(x, y, z)) is equivalent to f(x, y, z).

- do.call() gives a straightforward solution to rbind()ing together many data frames:

```
do.call("rbind", dfs)
#>   x y
#> a 1 2
#> b 3 4
```

- With a little more work, we can use do.call() to solve the second problem. We first create a list of arguments, then name that, then use do.call():

```
args <- list(val)
names(args) <- var

do.call("data.frame", args)
```

```
#>    x
#> 1 4
#> 2 3
#> 3 9
```

Some base functions (including interaction(), expand.grid(), options(), and par()) use a trick to avoid do.call(): if the first component of ... is a list, they'll take its components instead of looking at the other elements of .... The implementation looks something like this:

```
f <- function(...) {
  dots <- list(...)
  if (length(dots) == 1 && is.list(dots[[1]])) {
    dots <- dots[[1]]
  }

  # Do something
  ...
}
```

Another approach to avoiding do.call() is found in the RCurl::getURL() function written by Duncan Temple Lang. getURL() takes both ... and .dots which are concatenated together and looks something like this:

```
f <- function(..., .dots) {
  dots <- c(list(...), .dots)
  # Do something
}
```

At the time I discovered it, I found this technique particularly compelling so you can see it used throughout the tidyverse. Now, however, I prefer the approach described previously.

19.6.5 Exercises

1. One way to implement exec() is shown below. Describe how it works. What are the key ideas?

   ```
   exec <- function(f, ..., .env = caller_env()) {
     args <- list2(...)
   ```

```
      do.call(f, args, envir = .env)
    }
```

2. Carefully read the source code for interaction(), expand.grid(), and par(). Compare and contrast the techniques they use for switching between dots and list behaviour.

3. Explain the problem with this definition of set_attr()

```
set_attr <- function(x, ...) {
  attr <- rlang::list2(...)
  attributes(x) <- attr

  x
}
set_attr(1:10, x = 10)
#> Error in attributes(x) <- attr: attributes must be named
```

19.7 Case studies

To make the ideas of quasiquotation concrete, this section contains a few small case studies that use it to solve real problems. Some of the case studies also use purrr: I find the combination of quasiquotation and functional programming to be particularly elegant.

19.7.1 lobstr::ast()

Quasiquotation allows us to solve an annoying problem with lobstr::ast(): what happens if we've already captured the expression?

```
z <- expr(foo(x, y))
lobstr::ast(z)
#> z
```

Because ast() quotes its first argument, we can use !!:

```
lobstr::ast(!!z)
#> █—foo
```

```
#> ├─x
#> └─y
```

19.7.2 Map-reduce to generate code

Quasiquotation gives us powerful tools for generating code, particularly when combined with purrr::map() and purr::reduce(). For example, assume you have a linear model specified by the following coefficients:

```
intercept <- 10
coefs <- c(x1 = 5, x2 = -4)
```

And you want to convert it into an expression like 10 + (x1 * 5) + (x2 * -4). The first thing we need to do is turn the character names vector into a list of symbols. rlang::syms() is designed precisely for this case:

```
coef_sym <- syms(names(coefs))
coef_sym
#> [[1]]
#> x1
#>
#> [[2]]
#> x2
```

Next we need to combine each variable name with its coefficient. We can do this by combining rlang::expr() with purrr::map2():

```
summands <- map2(coef_sym, coefs, ~ expr((!!.x * !!.y)))
summands
#> [[1]]
#> (x1 * 5)
#>
#> [[2]]
#> (x2 * -4)
```

In this case, the intercept is also a part of the sum, although it doesn't involve a multiplication. We can just add it to the start of the summands vector:

```
summands <- c(intercept, summands)
summands
#> [[1]]
```

```
#> [1] 10
#>
#> [[2]]
#> (x1 * 5)
#>
#> [[3]]
#> (x2 * -4)
```

Finally, we need to reduce (Section 9.5) the individual terms into a single sum by adding the pieces together:

```
eq <- reduce(summands, ~ expr(!!.x + !!.y))
eq
#> 10 + (x1 * 5) + (x2 * -4)
```

We could make this even more general by allowing the user to supply the name of the coefficient, and instead of assuming many different variables, index into a single one.

```
var <- expr(y)
coef_sym <- map(seq_along(coefs), ~ expr((!!var)[[!!.x]]))
coef_sym
#> [[1]]
#> y[[1L]]
#>
#> [[2]]
#> y[[2L]]
```

And finish by wrapping this up in a function:

```
linear <- function(var, val) {
  var <- ensym(var)
  coef_name <- map(seq_along(val[-1]), ~ expr((!!var)[[!!.x]]))

  summands <- map2(val[-1], coef_name, ~ expr((!!.x * !!.y)))
  summands <- c(val[[1]], summands)

  reduce(summands, ~ expr(!!.x + !!.y))
}

linear(x, c(10, 5, -4))
#> 10 + (5 * x[[1L]]) + (-4 * x[[2L]])
```

Note the use of `ensym()`: we want the user to supply the name of a single variable, not a more complex expression.

19.7.3 Slicing an array

An occasionally useful tool missing from base R is the ability to extract a slice of an array given a dimension and an index. For example, we'd like to write `slice(x, 2, 1)` to extract the first slice along the second dimension, i.e. `x[, 1, ]`. This is a moderately challenging problem because it requires working with missing arguments.

We'll need to generate a call with multiple missing arguments. We first generate a list of missing arguments with `rep()` and `missing_arg()`, then unquote-splice them into a call:

```
indices <- rep(list(missing_arg()), 3)
expr(x[!!!indices])
#> x[, , ]
```

Then we use subset-assignment to insert the index in the desired position:

```
indices[[2]] <- 1
expr(x[!!!indices])
#> x[, 1, ]
```

We then wrap this into a function, using a couple of `stopifnot()`s to make the interface clear:

```
slice <- function(x, along, index) {
  stopifnot(length(along) == 1)
  stopifnot(length(index) == 1)

  nd <- length(dim(x))
  indices <- rep(list(missing_arg()), nd)
  indices[[along]] <- index

  expr(x[!!!indices])
}

x <- array(sample(30), c(5, 2, 3))
slice(x, 1, 3)
#> x[3, , ]
slice(x, 2, 2)
```

```
#> x[, 2, ]
slice(x, 3, 1)
#> x[, , 1]
```

A real `slice()` would evaluate the generated call (Chapter 20), but here I think it's more illuminating to see the code that's generated, as that's the hard part of the challenge.

19.7.4 Creating functions

Another powerful application of quotation is creating functions "by hand", using `rlang::new_function()`. It's a function that creates a function from its three components (Section 6.2.1): arguments, body, and (optionally) an environment:

```
new_function(
  exprs(x = , y = ),
  expr({x + y})
)
#> function (x, y)
#> {
#>     x + y
#> }
```

NB: the empty arguments in `exprs()` generates arguments with no defaults.

One use of `new_function()` is as an alternative to function factories with scalar or symbol arguments. For example, we could write a function that generates functions that raise a function to the power of a number.

```
power <- function(exponent) {
  new_function(
    exprs(x = ),
    expr({
      x ^ !!exponent
    }),
    caller_env()
  )
}
power(0.5)
#> function (x)
#> {
```

```
#>      x^0.5
#> }
```

Another application of `new_function()` is for functions that work like `graphics::curve()`, which allows you to plot a mathematical expression without creating a function:

```
curve(sin(exp(4 * x)), n = 1000)
```

In this code, x is a pronoun: it doesn't represent a single concrete value, but is instead a placeholder that varies over the range of the plot. One way to implement `curve()` is to turn that expression into a function with a single argument, x, then call that function:

```
curve2 <- function(expr, xlim = c(0, 1), n = 100) {
  expr <- enexpr(expr)
  f <- new_function(exprs(x = ), expr)

  x <- seq(xlim[1], xlim[2], length = n)
  y <- f(x)

  plot(x, y, type = "l", ylab = expr_text(expr))
}
curve2(sin(exp(4 * x)), n = 1000)
```

Functions like `curve()` that use an expression containing a pronoun are known as **anaphoric** functions[6].

---

[6] Anaphoric comes from the linguistics term "anaphora", an expression that is context dependent. Anaphoric functions are found in Arc (http://www.arcfn.com/doc/anaphoric.html) (a Lisp like language), Perl (http://www.perlmonks.org/index.pl?node_id=666047), and Clojure (http://amalloy.hubpages.com/hub/Unhygenic-anaphoric-Clojure-macros-for-fun-and-profit).

19.7.5 Exercises

1. In the linear-model example, we could replace the expr()
   in   reduce(summands, ~ expr(!!.x + !!.y))   with   call2():
   reduce(summands, call2, "+"). Compare and contrast the two
   approaches. Which do you think is easier to read?

2. Re-implement the Box-Cox transform defined below using unquoting and new_function():

```r
bc <- function(lambda) {
  if (lambda == 0) {
    function(x) log(x)
  } else {
    function(x) (x ^ lambda - 1) / lambda
  }
}
```

3. Re-implement the simple compose() defined below using quasiquotation and new_function():

```r
compose <- function(f, g) {
  function(...) f(g(...))
}
```

19.8 History

The idea of quasiquotation is an old one. It was first developed by the philoso-
pher Willard van Orman Quine[7] in the early 1940s. It's needed in philosophy
because it helps when precisely delineating the use and mention of words,
i.e. distinguishing between the object and the words we use to refer to that
object.

Quasiquotation was first used in a programming language, Lisp, in the mid-
1970s [Bawden, 1999]. Lisp has one quoting function `, and uses , for un-
quoting. Most languages with a Lisp heritage behave similarly. For example,

---

[7]You might be familiar with the name Quine from "quines", computer programs that
return a copy of their own source when run.

Racket (` and @), Clojure (` and ~), and Julia (: and @) all have quasiquotation tools that differ only slightly from Lisp. These languages have a single quoting function and you must call it explicitly.

In R, however, many functions quote one or more inputs. This introduces ambiguity (because you need to read the documentation to determine if an argument is quoted or not), but allows for concise and elegant data exploration code. In base R, only one function supports quasiquotation: `bquote()`, written in 2003 by Thomas Lumley. However, `bquote()` has some major limitations which prevented it from having a wide impact on R code (Section 19.5).

My attempt to resolve these limitations led to the lazyeval package (2014-2015). Unfortunately, my analysis of the problem was incomplete and while lazyeval solved some problems, it created others. It was not until I started working with Lionel Henry on the problem that all the pieces finally fell into place and we created the full tidy evaluation framework (2017). Despite the newness of tidy evaluation, I teach it here because it is a rich and powerful theory that, once mastered, makes many hard problems much easier.

20

Evaluation

20.1 Introduction

The user-facing inverse of quotation is unquotation: it gives the *user* the ability to selectively evaluate parts of an otherwise quoted argument. The developer-facing complement of quotation is evaluation: this gives the *developer* the ability to evaluate quoted expressions in custom environments to achieve specific goals.

This chapter begins with a discussion of evaluation in its purest form. You'll learn how `eval()` evaluates an expression in an environment, and then how it can be used to implement a number of important base R functions. Once you have the basics under your belt, you'll learn extensions to evaluation that are needed for robustness. There are two big new ideas:

- The quosure: a data structure that captures an expression along with its associated environment, as found in function arguments.

- The data mask, which makes it easier to evaluate an expression in the context of a data frame. This introduces potential evaluation ambiguity which we'll then resolve with data pronouns.

Together, quasiquotation, quosures, and data masks form what we call **tidy evaluation**, or tidy eval for short. Tidy eval provides a principled approach to non-standard evaluation that makes it possible to use such functions both interactively and embedded with other functions. Tidy evaluation is the most important practical implication of all this theory so we'll spend a little time exploring the implications. The chapter finishes off with a discussion of the closest related approaches in base R, and how you can program around their drawbacks.

Outline

- Section 20.2 discusses the basics of evaluation using `eval()`, and shows how you can use it to implement key functions like `local()` and `source()`.

- Section 20.3 introduces a new data structure, the quosure, which combines an expression with an environment. You'll learn how to capture quosures from promises, and evaluate them using `rlang::eval_tidy()`.

- Section 20.4 extends evaluation with the data mask, which makes it trivial to intermingle symbols bound in an environment with variables found in a data frame.

- Section 20.5 shows how to use tidy evaluation in practice, focussing on the common pattern of quoting and unquoting, and how to handle ambiguity with pronouns.

- Section 20.6 circles back to evaluation in base R, discusses some of the downsides, and shows how to use quasiquotation and evaluation to wrap functions that use NSE.

Prerequisites

You'll need to be familiar with the content of Chapter 18 and Chapter 19, as well as the environment data structure (Section 7.2) and the caller environment (Section 7.5).

We'll continue to use rlang (`https://rlang.r-lib.org`) and purrr (`https://purrr.tidyverse.org`).

```
library(rlang)
library(purrr)
```

20.2   Evaluation basics

Here we'll explore the details of `eval()` which we briefly mentioned in the last chapter. It has two key arguments: `expr` and `envir`. The first argument, `expr`, is the object to evaluate, typically a symbol or expression[1]. None of the evaluation functions quote their inputs, so you'll usually use them with `expr()` or similar:

```
x <- 10
eval(expr(x))
```

---

[1]All other objects yield themselves when evaluated; i.e. `eval(x)` yields x, except when x is a symbol or expression.

```
#> [1] 10

y <- 2
eval(expr(x + y))
#> [1] 12
```

The second argument, env, gives the environment in which the expression should be evaluated, i.e. where to look for the values of x, y, and +. By default, this is the current environment, i.e. the calling environment of eval(), but you can override it if you want:

```
eval(expr(x + y), env(x = 1000))
#> [1] 1002
```

The first argument is evaluated, not quoted, which can lead to confusing results once if you use a custom environment and forget to manually quote:

```
eval(print(x + 1), env(x = 1000))
#> [1] 11
#> [1] 11

eval(expr(print(x + 1)), env(x = 1000))
#> [1] 1001
```

Now that you've seen the basics, let's explore some applications. We'll focus primarily on base R functions that you might have used before, reimplementing the underlying principles using rlang.

20.2.1 Application: `local()`

Sometimes you want to perform a chunk of calculation that creates some intermediate variables. The intermediate variables have no long-term use and could be quite large, so you'd rather not keep them around. One approach is to clean up after yourself using rm(); another is to wrap the code in a function and just call it once. A more elegant approach is to use local():

```
# Clean up variables created earlier
rm(x, y)

foo <- local({
  x <- 10
  y <- 200
```

```
  x + y
})
```

```
foo
#> [1] 210
x
#> Error in eval(expr, envir, enclos): object 'x' not found
y
#> Error in eval(expr, envir, enclos): object 'y' not found
```

The essence of local() is quite simple and re-implemented below. We capture the input expression, and create a new environment in which to evaluate it. This is a new environment (so assignment doesn't affect the existing environment) with the caller environment as parent (so that expr can still access variables in that environment). This effectively emulates running expr as if it was inside a function (i.e. it's lexically scoped, Section 6.4).

```
local2 <- function(expr) {
  env <- env(caller_env())
  eval(enexpr(expr), env)
}
```

```
foo <- local2({
  x <- 10
  y <- 200
  x + y
})
```

```
foo
#> [1] 210
x
#> Error in eval(expr, envir, enclos): object 'x' not found
y
#> Error in eval(expr, envir, enclos): object 'y' not found
```

Understanding how base::local() works is harder, as it uses eval() and substitute() together in rather complicated ways. Figuring out exactly what's going on is good practice if you really want to understand the subtleties of substitute() and the base eval() functions, so they are included in the exercises below.

20.2.2 Application: source()

We can create a simple version of source() by combining eval() with parse_expr() from Section 18.4.3. We read in the file from disk, use parse_expr() to parse the string into a list of expressions, and then use eval() to evaluate each element in turn. This version evaluates the code in the caller environment, and invisibly returns the result of the last expression in the file just like base::source().

```
source2 <- function(path, env = caller_env()) {
  file <- paste(readLines(path, warn = FALSE), collapse = "\n")
  exprs <- parse_exprs(file)

  res <- NULL
  for (i in seq_along(exprs)) {
    res <- eval(exprs[[i]], env)
  }

  invisible(res)
}
```

The real source() is considerably more complicated because it can echo input and output, and has many other settings that control its behaviour.

Expression vectors

base::eval() has special behaviour for expression *vectors*, evaluating each component in turn. This makes for a very compact implementation of source2() because base::parse() also returns an expression object:

```
source3 <- function(file, env = parent.frame()) {
  lines <- parse(file)
  res <- eval(lines, envir = env)
  invisible(res)
}
```

While source3() is considerably more concise than source2(), this is the only advantage to expression vectors. Overall I don't believe this benefit outweighs the cost of introducing a new data structure, and hence this book avoids the use of expression vectors.

20.2.3   Gotcha: `function()`

There's one small gotcha that you should be aware of if you're using `eval()` and `expr()` to generate functions:

```
x <- 10
y <- 20
f <- eval(expr(function(x, y) !!x + !!y))
f
#> function(x, y) !!x + !!y
```

This function doesn't look like it will work, but it does:

```
f()
#> [1] 30
```

This is because, if available, functions print their `srcref` attribute (Section 6.2.1), and because `srcref` is a base R feature it's unaware of quasiquotation.

To work around this problem, either use `new_function()` (Section 19.7.4) or remove the `srcref` attribute:

```
attr(f, "srcref") <- NULL
f
#> function (x, y)
#> 10 + 20
```

20.2.4   Exercises

1. Carefully read the documentation for `source()`. What environment does it use by default? What if you supply `local = TRUE`? How do you provide a custom environment?

2. Predict the results of the following lines of code:

   ```
   eval(expr(eval(expr(eval(expr(2 + 2))))))
   eval(eval(expr(eval(expr(eval(expr(2 + 2)))))))
   expr(eval(expr(eval(expr(eval(expr(2 + 2)))))))
   ```

3. Fill in the function bodies below to re-implement `get()` using `sym()` and `eval()`, and`assign()` using `sym()`, `expr()`, and `eval()`. Don't worry about the multiple ways of choosing an environment that

get() and assign() support; assume that the user supplies it explicitly.

```
# name is a string
get2 <- function(name, env) {}
assign2 <- function(name, value, env) {}
```

4. Modify source2() so it returns the result of *every* expression, not just the last one. Can you eliminate the for loop?

5. We can make base::local() slightly easier to understand by spreading out over multiple lines:

```
local3 <- function(expr, envir = new.env()) {
  call <- substitute(eval(quote(expr), envir))
  eval(call, envir = parent.frame())
}
```

Explain how local() works in words. (Hint: you might want to print(call) to help understand what substitute() is doing, and read the documentation to remind yourself what environment new.env() will inherit from.)

20.3 Quosures

Almost every use of eval() involves both an expression and environment. This coupling is so important that we need a data structure that can hold both pieces. Base R does not have such a structure[2] so rlang fills the gap with the **quosure**, an object that contains an expression and an environment. The name is a portmanteau of quoting and closure, because a quosure both quotes the expression and encloses the environment. Quosures reify the internal promise object (Section 6.5.1) into something that you can program with.

In this section, you'll learn how to create and manipulate quosures, and a little about how they are implemented.

---

[2] Technically a formula combines an expression and environment, but formulas are tightly coupled to modelling so a new data structure makes sense.

20.3.1 Creating

There are three ways to create quosures:

- Use `enquo()` and `enquos()` to capture user-supplied expressions. The vast majority of quosures should be created this way.

```
foo <- function(x) enquo(x)
foo(a + b)
#> <quosure>
#> expr: ^a + b
#> env:  global
```

- `quo()` and `quos()` exist to match to `expr()` and `exprs()`, but they are included only for the sake of completeness and are needed very rarely. If you find yourself using them, think carefully if `expr()` and careful unquoting can eliminate the need to capture the environment.

```
quo(x + y + z)
#> <quosure>
#> expr: ^x + y + z
#> env:  global
```

- `new_quosure()` create a quosure from its components: an expression and an environment. This is rarely needed in practice, but is useful for learning, so is used a lot in this chapter.

```
new_quosure(expr(x + y), env(x = 1, y = 10))
#> <quosure>
#> expr: ^x + y
#> env:  0x7f87b9cc5fb8
```

20.3.2 Evaluating

Quosures are paired with a new evaluation function `eval_tidy()` that takes a single quosure instead of a expression-environment pair. It is straightforward to use:

```
q1 <- new_quosure(expr(x + y), env(x = 1, y = 10))
eval_tidy(q1)
#> [1] 11
```

For this simple case, `eval_tidy(q1)` is basically a shortcut for `eval(get_expr(q1), get_env(q2))`. However, it has two important features that you'll learn about later in the chapter: it supports nested quosures (Section 20.3.5) and pronouns (Section 20.4.2).

20.3.3 Dots

Quosures are typically just a convenience: they make code cleaner because you only have one object to pass around, instead of two. They are, however, essential when it comes to working with ... because it's possible for each argument passed to ... to be associated with a different environment. In the following example note that both quosures have the same expression, x, but a different environment:

```
f <- function(...) {
  x <- 1
  g(..., f = x)
}
g <- function(...) {
  enquos(...)
}

x <- 0
qs <- f(global = x)
qs
#> <list_of<quosure>>
#>
#> $global
#> <quosure>
#> expr:  ^x
#> env:   global
#>
#> $f
#> <quosure>
#> expr:  ^x
#> env:   0x7f87bab6aa10
```

That means that when you evaluate them, you get the correct results:

```
map_dbl(qs, eval_tidy)
#> global      f
#>      0      1
```

Correctly evaluating the elements of ... was one of the original motivations for the development of quosures.

20.3.4 Under the hood

Quosures were inspired by R's formulas, because formulas capture an expression and an environment:

```
f <- ~runif(3)
str(f)
#> Class 'formula'  language ~runif(3)
#>    ..- attr(*, ".Environment")=<environment: R_GlobalEnv>
```

An early version of tidy evaluation used formulas instead of quosures, as an attractive feature of ~ is that it provides quoting with a single keystroke. Unfortunately, however, there is no clean way to make ~ a quasiquoting function.

Quosures are a subclass of formulas:

```
q4 <- new_quosure(expr(x + y + z))
class(q4)
#> [1] "quosure" "formula"
```

which means that under the hood, quosures, like formulas, are call objects:

```
is_call(q4)
#> [1] TRUE

q4[[1]]
#> `~`
q4[[2]]
#> x + y + z
```

with an attribute that stores the environment:

```
attr(q4, ".Environment")
#> <environment: R_GlobalEnv>
```

If you need to extract the expression or environment, don't rely on these implementation details. Instead use get_expr() and get_env():

```
get_expr(q4)
#> x + y + z
get_env(q4)
#> <environment: R_GlobalEnv>
```

20.3.5 Nested quosures

It's possible to use quasiquotation to embed a quosure in an expression. This is an advanced tool, and most of the time you don't need to think about it because it just works, but I talk about it here so you can spot nested quosures in the wild and not be confused. Take this example, which inlines two quosures into an expression:

```
q2 <- new_quosure(expr(x), env(x = 1))
q3 <- new_quosure(expr(x), env(x = 10))

x <- expr(!!q2 + !!q3)
```

It evaluates correctly with `eval_tidy()`:

```
eval_tidy(x)
#> [1] 11
```

However, if you print it, you only see the xs, with their formula heritage leaking through:

```
x
#> (~x) + ~x
```

You can get a better display with `rlang::expr_print()` (Section 19.4.7):

```
expr_print(x)
#> (^x) + (^x)
```

When you use `expr_print()` in the console, quosures are coloured according to their environment, making it easier to spot when symbols are bound to different variables.

20.3.6 Exercises

1. Predict what each of the following quosures will return if evaluated.

```
q1 <- new_quosure(expr(x), env(x = 1))
q1
#> <quosure>
#> expr: ^x
#> env:  0x7f87b70f83c8

q2 <- new_quosure(expr(x + !!q1), env(x = 10))
q2
#> <quosure>
#> expr: ^x + (^x)
#> env:  0x7f87baa5a908

q3 <- new_quosure(expr(x + !!q2), env(x = 100))
q3
#> <quosure>
#> expr: ^x + (^x + (^x))
#> env:  0x7f87bb0a5fb8
```

2. Write an `enenv()` function that captures the environment associated with an argument. (Hint: this should only require two function calls.)

20.4 Data masks

In this section, you'll learn about the **data mask**, a data frame where the evaluated code will look first for variable definitions. The data mask is the key idea that powers base functions like `with()`, `subset()` and `transform()`, and is used throughout the tidyverse in packages like dplyr and ggplot2.

20.4.1 Basics

The data mask allows you to mingle variables from an environment and a data frame in a single expression. You supply the data mask as the second argument to `eval_tidy()`:

```
q1 <- new_quosure(expr(x * y), env(x = 100))
df <- data.frame(y = 1:10)
```

```
eval_tidy(q1, df)
#>  [1]  100  200  300  400  500  600  700  800  900 1000
```

This code is a little hard to follow because there's so much syntax as we're creating every object from scratch. It's easier to see what's going on if we make a little wrapper. I call this with2() because it's equivalent to base::with().

```
with2 <- function(data, expr) {
  expr <- enquo(expr)
  eval_tidy(expr, data)
}
```

We can now rewrite the code above as below:

```
x <- 100
with2(df, x * y)
#>  [1]  100  200  300  400  500  600  700  800  900 1000
```

base::eval() has similar functionality, although it doesn't call it a data mask. Instead you can supply a data frame to the second argument and an environment to the third. That gives the following implementation of with():

```
with3 <- function(data, expr) {
  expr <- substitute(expr)
  eval(expr, data, caller_env())
}
```

20.4.2 Pronouns

Using a data mask introduces ambiguity. For example, in the following code you can't know whether x will come from the data mask or the environment, unless you know what variables are found in df.

```
with2(df, x)
```

That makes code harder to reason about (because you need to know more context), which can introduce bugs. To resolve that issue, the data mask provides two pronouns: .data and .env.

- .data$x always refers to x in the data mask.
- .env$x always refers to x in the environment.

```
x <- 1
df <- data.frame(x = 2)

with2(df, .data$x)
#> [1] 2
with2(df, .env$x)
#> [1] 1
```

You can also subset .data and .env using [[, e.g. .data[["x"]]. Otherwise the pronouns are special objects and you shouldn't expect them to behave like data frames or environments. In particular, they throw an error if the object isn't found:

```
with2(df, .data$y)
#> Error: Column `y` not found in `.data`
```

20.4.3 Application: subset()

We'll explore tidy evaluation in the context of base::subset(), because it's a simple yet powerful function that makes a common data manipulation challenge easier. If you haven't used it before, subset(), like dplyr::filter(), provides a convenient way of selecting rows of a data frame. You give it some data, along with an expression that is evaluated in the context of that data. This considerably reduces the number of times you need to type the name of the data frame:

```
sample_df <- data.frame(a = 1:5, b = 5:1, c = c(5, 3, 1, 4, 1))

# Shorthand for sample_df[sample_df$a >= 4, ]
subset(sample_df, a >= 4)
#>   a b c
#> 4 4 2 4
#> 5 5 1 1

# Shorthand for sample_df[sample_df$b == sample_df$c, ]
subset(sample_df, b == c)
#>   a b c
#> 1 1 5 5
#> 5 5 1 1
```

The core of our version of subset(), subset2(), is quite simple. It takes two arguments: a data frame, data, and an expression, rows. We evaluate rows using `df` as a data mask, then use the results to subset the data frame with `[`. I've included a very simple check to ensure the result is a logical vector; real code would do more to create an informative error.

```
subset2 <- function(data, rows) {
  rows <- enquo(rows)
  rows_val <- eval_tidy(rows, data)
  stopifnot(is.logical(rows_val))

  data[rows_val, , drop = FALSE]
}

subset2(sample_df, b == c)
#>   a b c
#> 1 1 5 5
#> 5 5 1 1
```

20.4.4 Application: transform

A more complicated situation is `base::transform()` which allows you to add new variables to a data frame, evaluating their expressions in the context of the existing variables:

```
df <- data.frame(x = c(2, 3, 1), y = runif(3))
transform(df, x = -x, y2 = 2 * y)
#>    x      y    y2
#> 1 -2 0.0808 0.162
#> 2 -3 0.8343 1.669
#> 3 -1 0.6008 1.202
```

Again, our own `transform2()` requires little code. We capture the unevaluated `...` with `enquos(...)`, and then evaluate each expression using a for loop. Real code would do more error checking to ensure that each input is named and evaluates to a vector the same length as `data`.

```
transform2 <- function(.data, ...) {
  dots <- enquos(...)

  for (i in seq_along(dots)) {
    name <- names(dots)[[i]]
    dot <- dots[[i]]
```

```
    .data[[name]] <- eval_tidy(dot, .data)
  }

  .data
}

transform2(df, x2 = x * 2, y = -y)
#>   x        y x2
#> 1 2 -0.0808  4
#> 2 3 -0.8343  6
#> 3 1 -0.6008  2
```

NB: I named the first argument .data to avoid problems if users tried to create
a variable called data. They will still have problems if they attempt to create
a variable called .data, but this is much less likely. This is the same reasoning
that leads to the .x and .f arguments to map() (Section 9.2.4).

20.4.5 Application: select()

A data mask will typically be a data frame, but it's sometimes useful to
provide a list filled with more exotic contents. This is basically how the select
argument in base::subset() works. It allows you to refer to variables as if they
were numbers:

```
df <- data.frame(a = 1, b = 2, c = 3, d = 4, e = 5)
subset(df, select = b:d)
#>   b c d
#> 1 2 3 4
```

The key idea is to create a named list where each component gives the position
of the corresponding variable:

```
vars <- as.list(set_names(seq_along(df), names(df)))
str(vars)
#> List of 5
#>  $ a: int 1
#>  $ b: int 2
#>  $ c: int 3
#>  $ d: int 4
#>  $ e: int 5
```

Then implementation is again only a few lines of code:

```
select2 <- function(data, ...) {
  dots <- enquos(...)

  vars <- as.list(set_names(seq_along(data), names(data)))
  cols <- unlist(map(dots, eval_tidy, vars))

  df[, cols, drop = FALSE]
}
select2(df, b:d)
#>   b c d
#> 1 2 3 4
```

dplyr::select() takes this idea and runs with it, providing a number of helpers that allow you to select variables based on their names (e.g. starts_with("x") or ends_with("_a")).

20.4.6 Exercises

1. Why did I use a for loop in transform2() instead of map()? Consider transform2(df, x = x * 2, x = x * 2).

2. Here's an alternative implementation of subset2():

    ```
    subset3 <- function(data, rows) {
      rows <- enquo(rows)
      eval_tidy(expr(data[!!rows, , drop = FALSE]), data = data)
    }

    df <- data.frame(x = 1:3)
    subset3(df, x == 1)
    ```

    Compare and contrast subset3() to subset2(). What are its advantages and disadvantages?

3. The following function implements the basics of dplyr::arrange(). Annotate each line with a comment explaining what it does. Can you explain why !!.na.last is strictly correct, but omitting the !! is unlikely to cause problems?

    ```
    arrange2 <- function(.df, ..., .na.last = TRUE) {
      args <- enquos(...)
    ```

```
    order_call <- expr(order(!!!args, na.last = !!.na.last))

    ord <- eval_tidy(order_call, .df)
    stopifnot(length(ord) == nrow(.df))

    .df[ord, , drop = FALSE]
  }
```

20.5   Using tidy evaluation

While it's important to understand how eval_tidy() works, most of the time
you won't call it directly. Instead, you'll usually use it indirectly by calling a
function that uses eval_tidy(). This section will give a few practical examples
of wrapping functions that use tidy evaluation.

20.5.1   Quoting and unquoting

Imagine we have written a function that resamples a dataset:

```
resample <- function(df, n) {
  idx <- sample(nrow(df), n, replace = TRUE)
  df[idx, , drop = FALSE]
}
```

We want to create a new function that allows us to resample and subset in a
single step. Our naive approach doesn't work:

```
subsample <- function(df, cond, n = nrow(df)) {
  df <- subset2(df, cond)
  resample(df, n)
}

df <- data.frame(x = c(1, 1, 1, 2, 2), y = 1:5)
subsample(df, x == 1)
#> Error in eval_tidy(rows, data): object 'x' not found
```

subsample() doesn't quote any arguments so cond is evaluated normally (not
in a data mask), and we get an error when it tries to find a binding for x. To

fix this problem we need to quote cond, and then unquote it when we pass it on ot subset2():

```
subsample <- function(df, cond, n = nrow(df)) {
  cond <- enquo(cond)

  df <- subset2(df, !!cond)
  resample(df, n)
}

subsample(df, x == 1)
#>     x y
#> 1   1 1
#> 1.1 1 1
#> 2   1 2
```

This is a very common pattern; whenever you call a quoting function with arguments from the user, you need to quote them and then unquote.

20.5.2 Handling ambiguity

In the case above, we needed to think about tidy evaluation because of quasiquotation. We also need to think about tidy evaluation even when the wrapper doesn't need to quote any arguments. Take this wrapper around subset2():

```
threshold_x <- function(df, val) {
  subset2(df, x >= val)
}
```

This function can silently return an incorrect result in two situations:

- When x exists in the calling environment, but not in df:

  ```
  x <- 10
  no_x <- data.frame(y = 1:3)
  threshold_x(no_x, 2)
  #>   y
  #> 1 1
  #> 2 2
  #> 3 3
  ```

- When val exists in df:

```
has_val <- data.frame(x = 1:3, val = 9:11)
threshold_x(has_val, 2)
#> [1] x    val
#> <0 rows> (or 0-length row.names)
```

These failure modes arise because tidy evaluation is ambiguous: each variable can be found in **either** the data mask **or** the environment. To make this function safe we need to remove the ambiguity using the .data and .env pronouns:

```
threshold_x <- function(df, val) {
  subset2(df, .data$x >= .env$val)
}


x <- 10
threshold_x(no_x, 2)
#> Error: Column `x` not found in `.data`
threshold_x(has_val, 2)
#>   x val
#> 2 2  10
#> 3 3  11
```

Generally, whenever you use the .env pronoun, you can use unquoting instead:

```
threshold_x <- function(df, val) {
  subset2(df, .data$x >= !!val)
}
```

There are subtle differences in when val is evaluated. If you unquote, val will be early evaluated by enquo(); if you use a pronoun, val will be lazily evaluated by eval_tidy(). These differences are usually unimportant, so pick the form that looks most natural.

20.5.3   Quoting and ambiguity

To finish our discussion let's consider the case where we have both quoting and potential ambiguity. I'll generalise threshold_x() slightly so that the user can pick the variable used for thresholding. Here I used .data[[var]] because it makes the code a little simpler; in the exercises you'll have a chance to explore how you might use $ instead.

```
threshold_var <- function(df, var, val) {
  var <- as_string(ensym(var))
```

```
  subset2(df, .data[[var]] >= !!val)
}

df <- data.frame(x = 1:10)
threshold_var(df, x, 8)
#>     x
#> 8   8
#> 9   9
#> 10 10
```

It is not always the responsibility of the function author to avoid ambiguity. Imagine we generalise further to allow thresholding based on any expression:

```
threshold_expr <- function(df, expr, val) {
  expr <- enquo(expr)
  subset2(df, !!expr >= !!val)
}
```

It's not possible to evaluate `expr` only in the data mask, because the data mask doesn't include any functions like + or ==. Here, it's the user's responsibility to avoid ambiguity. As a general rule of thumb, as a function author it's your responsibility to avoid ambiguity with any expressions that you create; it's the user's responsibility to avoid ambiguity in expressions that they create.

20.5.4 Exercises

1. I've included an alternative implementation of `threshold_var()` below. What makes it different to the approach I used above? What makes it harder?

```
threshold_var <- function(df, var, val) {
  var <- ensym(var)
  subset2(df, `$`(.data, !!var) >= !!val)
}
```

20.6   Base evaluation

Now that you understand tidy evaluation, it's time to come back to the alternative approaches taken by base R. Here I'll explore the two most common uses in base R:

- substitute() and evaluation in the caller environment, as used by subset().
  I'll use this technique to demonstrate why this technique is not programming
  friendly, as warned about in the subset() documentation.

- match.call(), call manipulation, and evaluation in the caller environment,
  as used by write.csv() and lm(). I'll use this technique to demonstrate how
  quasiquotation and (regular) evaluation can help you write wrappers around
  such functions.

These two approaches are common forms of non-standard evaluation (NSE).

20.6.1   substitute()

The most common form of NSE in base R is substitute() + eval(). The following code shows how you might write the core of subset() in this style using substitute() and eval() rather than enquo() and eval_tidy(). I repeat the code introduced in Section 20.4.3 so you can compare easily. The main difference is the evaluation environment: in subset_base() the argument is evaluated in the caller environment, while in subset_tidy(), it's evaluated in the environment where it was defined.

```r
subset_base <- function(data, rows) {
  rows <- substitute(rows)
  rows_val <- eval(rows, data, caller_env())
  stopifnot(is.logical(rows_val))

  data[rows_val, , drop = FALSE]
}

subset_tidy <- function(data, rows) {
  rows <- enquo(rows)
  rows_val <- eval_tidy(rows, data)
  stopifnot(is.logical(rows_val))

  data[rows_val, , drop = FALSE]
}
```

20.6.1.1 Programming with subset()

The documentation of subset() includes the following warning:

> *This is a convenience function intended for use interactively.*
> *For programming it is better to use the standard subsetting*
> *functions like* [, *and in particular the non-standard evaluation*
> *of argument* subset *can have unanticipated consequences.*

There are three main problems:

- base::subset() always evaluates rows in the calling environment, but if ...
  has been used, then the expression might need to be evaluated elsewhere:

```
f1 <- function(df, ...) {
  xval <- 3
  subset_base(df, ...)
}

my_df <- data.frame(x = 1:3, y = 3:1)
xval <- 1
f1(my_df, x == xval)
#>   x y
#> 3 3 1
```

This may seems like an esoteric concern, but it means that subset_base()
cannot reliably work with functionals like map() or lapply():

```
local({
  zzz <- 2
  dfs <- list(data.frame(x = 1:3), data.frame(x = 4:6))
  lapply(dfs, subset_base, x == zzz)
})
#> Error in eval(rows, data, caller_env()): object 'zzz' not found
```

- Calling subset() from another function requires some care: you have to
  use substitute() to capture a call to subset() complete expression, and
  then evaluate. I think this code is hard to understand because substitute()
  doesn't use a syntactic marker for unquoting. Here I print the generated call
  to make it a little easier to see what's happening.

```
f2 <- function(df1, expr) {
  call <- substitute(subset_base(df1, expr))
  expr_print(call)
  eval(call, caller_env())
```

```
}
my_df <- data.frame(x = 1:3, y = 3:1)
f2(my_df, x == 1)
#> subset_base(my_df, x == 1)
#>   x y
#> 1 1 3
```

- eval() doesn't provide any pronouns so there's no way to require part of the expression to come from the data. As far as I can tell, there's no way to make the following function safe except by manually checking for the presence of z variable in df.

```
f3 <- function(df) {
  call <- substitute(subset_base(df, z > 0))
  expr_print(call)
  eval(call, caller_env())
}

my_df <- data.frame(x = 1:3, y = 3:1)
z <- -1
f3(my_df)
#> subset_base(my_df, z > 0)
#> [1] x y
#> <0 rows> (or 0-length row.names)
```

20.6.1.2 What about [?

Given that tidy evaluation is quite complex, why not simply use [as ?subset recommends? Primarily, it seems unappealing to me to have functions that can only be used interactively, and never inside another function.

Additionally, even the simple subset() function provides two useful features compared to [:

- It sets drop = FALSE by default, so it's guaranteed to return a data frame.

- It drops rows where the condition evaluates to NA.

That means subset(df, x == y) is not equivalent to df[x == y,] as you might expect. Instead, it is equivalent to df[x == y & !is.na(x == y), , drop = FALSE]: that's a lot more typing! Real-life alternatives to subset(), like dplyr::filter(), do even more. For example, dplyr::filter() can translate R expressions to SQL so that they can be executed in a database. This makes programming with filter() relatively more important.

20.6.2 `match.call()`

Another common form of NSE is to capture the complete call with `match.call()`, modify it, and evaluate the result. `match.call()` is similar to `substitute()`, but instead of capturing a single argument, it captures the complete call. It doesn't have an equivalent in rlang.

```
g <- function(x, y, z) {
  match.call()
}
g(1, 2, z = 3)
#> g(x = 1, y = 2, z = 3)
```

One prominent user of `match.call()` is `write.csv()`, which basically works by transforming the call into a call to `write.table()` with the appropriate arguments set. The following code shows the heart of `write.csv()`:

```
write.csv <- function(...) {
  call <- match.call(write.table, expand.dots = TRUE)

  call[[1]] <- quote(write.table)
  call$sep <- ","
  call$dec <- "."

  eval(call, parent.frame())
}
```

I don't think this technique is a good idea because you can achieve the same result without NSE:

```
write.csv <- function(...) {
  write.table(..., sep = ",", dec = ".")
}
```

Nevertheless, it's important to understand this technique because it's commonly used in modelling functions. These functions also prominently print the captured call, which poses some special challenges, as you'll see next.

20.6.2.1 Wrapping modelling functions

To begin, consider the simplest possible wrapper around `lm()`:

```
lm2 <- function(formula, data) {
  lm(formula, data)
}
```

This wrapper works, but is suboptimal because lm() captures its call and displays it when printing.

```
lm2(mpg ~ disp, mtcars)
#>
#> Call:
#> lm(formula = formula, data = data)
#>
#> Coefficients:
#> (Intercept)              disp
#>     29.5999           -0.0412
```

Fixing this is important because this call is the chief way that you see the model specification when printing the model. To overcome this problem, we need to capture the arguments, create the call to lm() using unquoting, then evaluate that call. To make it easier to see what's going on, I'll also print the expression we generate. This will become more useful as the calls get more complicated.

```
lm3 <- function(formula, data, env = caller_env()) {
  formula <- enexpr(formula)
  data <- enexpr(data)

  lm_call <- expr(lm(!!formula, data = !!data))
  expr_print(lm_call)
  eval(lm_call, env)
}
```

```
lm3(mpg ~ disp, mtcars)
#> lm(mpg ~ disp, data = mtcars)
#>
#> Call:
#> lm(formula = mpg ~ disp, data = mtcars)
#>
#> Coefficients:
#> (Intercept)              disp
#>     29.5999           -0.0412
```

There are three pieces that you'll use whenever wrapping a base NSE function in this way:

- You capture the unevaluated arguments using `enexpr()`, and capture the caller environment using `caller_env()`.

- You generate a new expression using `expr()` and unquoting.

- You evaluate that expression in the caller environment. You have to accept that the function will not work correctly if the arguments are not defined in the caller environment. Providing the `env` argument at least provides a hook that experts can use if the default environment isn't correct.

The use of `enexpr()` has a nice side-effect: we can use unquoting to generate formulas dynamically:

```
resp <- expr(mpg)
disp1 <- expr(vs)
disp2 <- expr(wt)
lm3(!!resp ~ !!disp1 + !!disp2, mtcars)
#> lm(mpg ~ vs + wt, data = mtcars)
#>
#> Call:
#> lm(formula = mpg ~ vs + wt, data = mtcars)
#>
#> Coefficients:
#> (Intercept)            vs            wt
#>       33.00          3.15         -4.44
```

20.6.2.2 Evaluation environment

What if you want to mingle objects supplied by the user with objects that you create in the function? For example, imagine you want to make an auto-resampling version of `lm()`. You might write it like this:

```
resample_lm0 <- function(formula, data, env = caller_env()) {
  formula <- enexpr(formula)
  resample_data <- resample(data, n = nrow(data))

  lm_call <- expr(lm(!!formula, data = resample_data))
  expr_print(lm_call)
  eval(lm_call, env)
}

df <- data.frame(x = 1:10, y = 5 + 3 * (1:10) + round(rnorm(10), 2))
resample_lm0(y ~ x, data = df)
#> lm(y ~ x, data = resample_data)
#> Error in is.data.frame(data): object 'resample_data' not found
```

Why doesn't this code work? We're evaluating lm_call in the caller environment, but resample_data exists in the execution environment. We could instead evaluate in the execution environment of resample_lm0(), but there's no guarantee that formula could be evaluated in that environment.

There are two basic ways to overcome this challenge:

1. Unquote the data frame into the call. This means that no lookup has to occur, but has all the problems of inlining expressions (Section 19.4.7). For modelling functions this means that the captured call is suboptimal:

```
resample_lm1 <- function(formula, data, env = caller_env()) {
  formula <- enexpr(formula)
  resample_data <- resample(data, n = nrow(data))

  lm_call <- expr(lm(!!formula, data = !!resample_data))
  expr_print(lm_call)
  eval(lm_call, env)
}
resample_lm1(y ~ x, data = df)$call
#> lm(y ~ x, data = <data.frame>)
#> lm(formula = y ~ x, data = list(x = c(3L, 7L, 4L, 4L,
#> 2L, 7L, 2L, 1L, 8L, 9L), y = c(13.21, 27.04, 18.63,
#> 18.63, 10.99, 27.04, 10.99, 7.83, 28.14, 32.72)))
```

2. Alternatively you can create a new environment that inherits from the caller, and bind variables that you've created inside the function to that environment.

```
resample_lm2 <- function(formula, data, env = caller_env()) {
  formula <- enexpr(formula)
  resample_data <- resample(data, n = nrow(data))

  lm_env <- env(env, resample_data = resample_data)
  lm_call <- expr(lm(!!formula, data = resample_data))
  expr_print(lm_call)
  eval(lm_call, lm_env)
}
resample_lm2(y ~ x, data = df)
#> lm(y ~ x, data = resample_data)
#>
#> Call:
```

```
#> lm(formula = y ~ x, data = resample_data)
#>
#> Coefficients:
#> (Intercept)              x
#>        5.17           3.00
```

This is more work, but gives the cleanest specification.

20.6.3 Exercises

1. Why does this function fail?

```
lm3a <- function(formula, data) {
  formula <- enexpr(formula)

  lm_call <- expr(lm(!!formula, data = data))
  eval(lm_call, caller_env())
}
lm3a(mpg ~ disp, mtcars)$call
#> Error in as.data.frame.default(data, optional = TRUE):
#> cannot coerce class '"function"' to a data.frame
```

2. When model building, typically the response and data are relatively constant while you rapidly experiment with different predictors. Write a small wrapper that allows you to reduce duplication in the code below.

```
lm(mpg ~ disp, data = mtcars)
lm(mpg ~ I(1 / disp), data = mtcars)
lm(mpg ~ disp * cyl, data = mtcars)
```

3. Another way to write resample_lm() would be to include the resample expression (data[sample(nrow(data), replace = TRUE), , drop = FALSE]) in the data argument. Implement that approach. What are the advantages? What are the disadvantages?

21

Translating R code

21.1 Introduction

The combination of first-class environments, lexical scoping, and metaprogramming gives us a powerful toolkit for translating R code into other languages. One fully-fledged example of this idea is dbplyr, which powers the database backends for dplyr, allowing you to express data manipulation in R and automatically translate it into SQL. You can see the key idea in `translate_sql()` which takes R code and returns the equivalent SQL:

```
library(dbplyr)
translate_sql(x ^ 2)
#> <SQL> POWER(`x`, 2.0)
translate_sql(x < 5 & !is.na(x))
#> <SQL> `x` < 5.0 AND NOT(((`x`) IS NULL))
translate_sql(!first %in% c("John", "Roger", "Robert"))
#> <SQL> NOT(`first` IN ('John', 'Roger', 'Robert'))
translate_sql(select == 7)
#> <SQL> `select` = 7.0
```

Translating R to SQL is complex because of the many idiosyncrasies of SQL dialects, so here I'll develop two simple, but useful, domain specific languages (DSL): one to generate HTML, and the other to generate mathematical equations in LaTeX.

If you're interested in learning more about domain specific languages in general, I highly recommend *Domain Specific Languages* [Fowler, 2010]. It discusses many options for creating a DSL and provides many examples of different languages.

Outline

- Section 21.2 creates a DSL for generating HTML, using quasiquotation and purrr to generate a function for each HTML tag, then tidy evaluation to easily access them.

- Section 21.3 transforms mathematically R code into its LaTeX equivalent using a combination of tidy evaluation and expression walking.

Prerequisites

This chapter pulls together many techniques discussed elsewhere in the book. In particular, you'll need to understand environments, expressions, tidy evaluation, and a little functional programming and S3. We'll use rlang (https://rlang.r-lib.org) for metaprogramming tools, and purrr (https://purrr.tidyverse.org) for functional programming.

```r
library(rlang)
library(purrr)
```

21.2 HTML

HTML (HyperText Markup Language) underlies the majority of the web. It's a special case of SGML (Standard Generalised Markup Language), and it's similar but not identical to XML (eXtensible Markup Language). HTML looks like this:

```html
<body>
  <h1 id='first'>A heading</h1>
  <p>Some text &amp; <b>some bold text.</b></p>
  <img src='myimg.png' width='100' height='100' />
</body>
```

Even if you've never looked at HTML before, you can still see that the key component of its coding structure is tags, which look like <tag></tag> or <tag />. Tags can be nested within other tags and intermingled with text. There are over 100 HTML tags, but in this chapter we'll focus on just a handful:

- <body> is the top-level tag that contains all content.
- <h1> defines a top level heading.

- `<p>` defines a paragraph.
- `<b>` emboldens text.
- `<img>` embeds an image.

Tags can have named **attributes** which look like `<tag name1='value1'`
`name2='value2'></tag>`. Two of the most important attributes are `id` and
`class`, which are used in conjunction with CSS (Cascading Style Sheets) to
control the visual appearance of the page.

Void tags, like `<img>`, don't have any children, and are written `<img />`, not
`<img></img>`. Since they have no content, attributes are more important, and
`img` has three that are used with almost every image: `src` (where the image
lives), `width`, and `height`.

Because `<` and `>` have special meanings in HTML, you can't write them directly. Instead you have to use the HTML **escapes**: `&gt;` and `&lt;`. And since
those escapes use `&`, if you want a literal ampersand you have to escape it as
`&amp;`.

21.2.1 Goal

Our goal is to make it easy to generate HTML from R. To give a concrete
example, we want to generate the following HTML:

```
<body>
  <h1 id='first'>A heading</h1>
  <p>Some text &amp; <b>some bold text.</b></p>
  <img src='myimg.png' width='100' height='100' />
</body>
```

Using the following code that matches the structure of the HTML as closely
as possible:

```
with_html(
  body(
    h1("A heading", id = "first"),
    p("Some text &", b("some bold text.")),
    img(src = "myimg.png", width = 100, height = 100)
  )
)
```

This DSL has the following three properties:

- The nesting of function calls matches the nesting of tags.

- Unnamed arguments become the content of the tag, and named arguments become their attributes.

- & and other special characters are automatically escaped.

21.2.2   Escaping

Escaping is so fundamental to translation that it'll be our first topic. There are two related challenges:

- In user input, we need to automatically escape &, < and >.

- At the same time we need to make sure that the &, < and > we generate are not double-escaped (i.e. that we don't accidentally generate &amp;, &lt; and &gt;).

The easiest way to do this is to create an S3 class (Section 13.3) that distinguishes between regular text (that needs escaping) and HTML (that doesn't).

```r
html <- function(x) structure(x, class = "advr_html")

print.advr_html <- function(x, ...) {
  out <- paste0("<HTML> ", x)
  cat(paste(strwrap(out), collapse = "\n"), "\n", sep = "")
}
```

We then write an escape generic. It has two important methods:

- escape.character() takes a regular character vector and returns an HTML vector with special characters (&, <, >) escaped.

- escape.advr_html() leaves already escaped HTML alone.

```r
escape <- function(x) UseMethod("escape")

escape.character <- function(x) {
  x <- gsub("&", "&amp;", x)
  x <- gsub("<", "&lt;", x)
  x <- gsub(">", "&gt;", x)

  html(x)
}

escape.advr_html <- function(x) x
```

Now we check that it works

```
escape("This is some text.")
#> <HTML> This is some text.
escape("x > 1 & y < 2")
#> <HTML> x &gt; 1 &amp; y &lt; 2

# Double escaping is not a problem
escape(escape("This is some text. 1 > 2"))
#> <HTML> This is some text. 1 &gt; 2

# And text we know is HTML doesn't get escaped.
escape(html("<hr />"))
#> <HTML> <hr />
```

Conveniently, this also allows a user to opt out of our escaping if they know the content is already escaped.

21.2.3   Basic tag functions

Next, we'll write a one-tag function by hand, then figure out how to generalise it so we can generate a function for every tag with code.

Let's start with <p>. HTML tags can have both attributes (e.g., id or class) and children (like <b> or <i>). We need some way of separating these in the function call. Given that attributes are named and children are not, it seems natural to use named and unnamed arguments for them respectively. For example, a call to p() might look like:

```
p("Some text. ", b(i("some bold italic text")), class = "mypara")
```

We could list all the possible attributes of the <p> tag in the function definition, but that's hard because there are many attributes, and because it's possible to use custom attributes (http://html5doctor.com/html5-custom-data-attributes/). Instead, we'll use ... and separate the components based on whether or not they are named. With this in mind, we create a helper function that wraps around rlang::list2() (Section 19.6) and returns named and unnamed components separately:

```
dots_partition <- function(...) {
  dots <- list2(...)

  if (is.null(names(dots))) {
    is_named <- rep(FALSE, length(dots))
  } else {
```

```
  is_named <- names(dots) != ""
}

  list(
    named = dots[is_named],
    unnamed = dots[!is_named]
  )
}

str(dots_partition(a = 1, 2, b = 3, 4))
#> List of 2
#>  $ named   :List of 2
#>   ..$ a: num 1
#>   ..$ b: num 3
#>  $ unnamed:List of 2
#>   ..$ : num 2
#>   ..$ : num 4
```

We can now create our `p()` function. Notice that there's one new function here: `html_attributes()`. It takes a named list and returns the HTML attribute specification as a string. It's a little complicated (in part, because it deals with some idiosyncrasies of HTML that I haven't mentioned here), but it's not that important and doesn't introduce any new programming ideas, so I won't discuss it in detail. You can find the source online (`https://github.com/hadley/adv-r/blob/master/dsl-html-attributes.r`) if you want to work through it yourself.

```
source("dsl-html-attributes.r")
p <- function(...) {
  dots <- dots_partition(...)
  attribs <- html_attributes(dots$named)
  children <- map_chr(dots$unnamed, escape)

  html(paste0(
    "<p", attribs, ">",
    paste(children, collapse = ""),
    "</p>"
  ))
}

p("Some text")
#> <HTML> <p>Some text</p>
p("Some text", id = "myid")
```

```
#> <HTML> <p id='myid'>Some text</p>
p("Some text", class = "important", `data-value` = 10)
#> <HTML> <p class='important' data-value='10'>Some text</p>
```

21.2.4 Tag functions

It's straightforward to adapt p() to other tags: we just need to replace "p"
with the name of the tag. One elegant way to do that is to create a function
with rlang::new_function() (Section 19.7.4), using unquoting and paste0()
to generate the starting and ending tags.

```
tag <- function(tag) {
  new_function(
    exprs(... = ),
    expr({
      dots <- dots_partition(...)
      attribs <- html_attributes(dots$named)
      children <- map_chr(dots$unnamed, escape)

      html(paste0(
        !!paste0("<", tag), attribs, ">",
        paste(children, collapse = ""),
        !!paste0("</", tag, ">")
      ))
    }),
    caller_env()
  )
}
tag("b")
#> function (...)
#> {
#>     dots <- dots_partition(...)
#>     attribs <- html_attributes(dots$named)
#>     children <- map_chr(dots$unnamed, escape)
#>     html(paste0("<b", attribs, ">", paste(children, collapse = ""),
#>         "</b>"))
#> }
```

We need the weird exprs(... =) syntax to generate the empty ... argument
in the tag function. See Section 18.6.2 for more details.

Now we can run our earlier example:

```
p <- tag("p")
b <- tag("b")
i <- tag("i")
p("Some text. ", b(i("some bold italic text")), class = "mypara")
#> <HTML> <p class='mypara'>Some text. <b><i>some bold italic
#> text</i></b></p>
```

Before we generate functions for every possible HTML tag, we need to create
a variant that handles void tags. void_tag() is quite similar to tag(), but it
throws an error if there are any unnamed tags, and the tag itself looks a little
different.

```
void_tag <- function(tag) {
  new_function(
    exprs(... = ),
    expr({
      dots <- dots_partition(...)
      if (length(dots$unnamed) > 0) {
        abort(!!paste0("<", tag, "> must not have unnamed arguments"))
      }
      attribs <- html_attributes(dots$named)

      html(paste0(!!paste0("<", tag), attribs, " />"))
    }),
    caller_env()
  )
}

img <- void_tag("img")
img
#> function (...)
#> {
#>     dots <- dots_partition(...)
#>     if (length(dots$unnamed) > 0) {
#>         abort("<img> must not have unnamed arguments")
#>     }
#>     attribs <- html_attributes(dots$named)
#>     html(paste0("<img", attribs, " />"))
#> }
img(src = "myimage.png", width = 100, height = 100)
#> <HTML> <img src='myimage.png' width='100' height='100' />
```

21.2.5 Processing all tags

Next we need to generate these functions for every tag. We'll start with a list of all HTML tags:

```
tags <- c("a", "abbr", "address", "article", "aside", "audio",
  "b","bdi", "bdo", "blockquote", "body", "button", "canvas",
  "caption","cite", "code", "colgroup", "data", "datalist",
  "dd", "del","details", "dfn", "div", "dl", "dt", "em",
  "eventsource","fieldset", "figcaption", "figure", "footer",
  "form", "h1", "h2", "h3", "h4", "h5", "h6", "head", "header",
  "hgroup", "html", "i","iframe", "ins", "kbd", "label",
  "legend", "li", "mark", "map","menu", "meter", "nav",
  "noscript", "object", "ol", "optgroup", "option", "output",
  "p", "pre", "progress", "q", "ruby", "rp","rt", "s", "samp",
  "script", "section", "select", "small", "span", "strong",
  "style", "sub", "summary", "sup", "table", "tbody", "td",
  "textarea", "tfoot", "th", "thead", "time", "title", "tr",
  "u", "ul", "var", "video"
)

void_tags <- c("area", "base", "br", "col", "command", "embed",
  "hr", "img", "input", "keygen", "link", "meta", "param",
  "source", "track", "wbr"
)
```

If you look at this list carefully, you'll see there are quite a few tags that have the same name as base R functions (body, col, q, source, sub, summary, table). This means we don't want to make all the functions available by default, either in the global environment or in a package. Instead, we'll put them in a list (like in Section 10.5) and then provide a helper to make it easy to use them when desired. First, we make a named list containing all the tag functions:

```
html_tags <- c(
  tags %>% set_names() %>% map(tag),
  void_tags %>% set_names() %>% map(void_tag)
)
```

This gives us an explicit (but verbose) way to create HTML:

```
html_tags$p(
  "Some text. ",
  html_tags$b(html_tags$i("some bold italic text")),
  class = "mypara"
```

```
)
#> <HTML> <p class='mypara'>Some text. <b><i>some bold italic
#> text</i></b></p>
```

We can then finish off our HTML DSL with a function that allows us to evaluate code in the context of that list. Here we slightly abuse the data mask, passing it a list of functions rather than a data frame. This is quick hack to mingle the execution environment of code with the functions in html_tags.

```
with_html <- function(code) {
  code <- enquo(code)
  eval_tidy(code, html_tags)
}
```

This gives us a succinct API which allows us to write HTML when we need it but doesn't clutter up the namespace when we don't.

```
with_html(
  body(
    h1("A heading", id = "first"),
    p("Some text &", b("some bold text.")),
    img(src = "myimg.png", width = 100, height = 100)
  )
)
#> <HTML> <body><h1 id='first'>A heading</h1><p>Some text
#> &amp;<b>some bold text.</b></p><img src='myimg.png'
#> width='100' height='100' /></body>
```

If you want to access the R function overridden by an HTML tag with the same name inside with_html(), you can use the full package::function specification.

21.2.6 Exercises

1. The escaping rules for <script> tags are different because they contain JavaScript, not HTML. Instead of escaping angle brackets or ampersands, you need to escape </script> so that the tag isn't closed too early. For example, script("'</script>'"), shouldn't generate this:

   ```
   <script>'</script>'</script>
   ```

But

```
<script>'<\/script>'</script>
```

Adapt the `escape()` to follow these rules when a new argument `script` is set to `TRUE`.

2. The use of ... for all functions has some big downsides. There's no input validation and there will be little information in the documentation or autocomplete about how they are used in the function. Create a new function that, when given a named list of tags and their attribute names (like below), creates tag functions with named arguments.

```
list(
  a = c("href"),
  img = c("src", "width", "height")
)
```

All tags should get `class` and `id` attributes.

3. Reason about the following code that calls `with_html()` referencing objects from the environment. Will it work or fail? Why? Run the code to verify your predictions.

```
greeting <- "Hello!"
with_html(p(greeting))

p <- function() "p"
address <- "123 anywhere street"
with_html(p(address))
```

4. Currently the HTML doesn't look terribly pretty, and it's hard to see the structure. How could you adapt `tag()` to do indenting and formatting? (You may need to do some research into block and inline tags.)

21.3 LaTeX

The next DSL will convert R expressions into their LaTeX math equivalents. (This is a bit like `?plotmath`, but for text instead of plots.) LaTeX is the lingua franca of mathematicians and statisticians: it's common to use LaTeX notation whenever you want to express an equation in text, like in email. Since many reports are produced using both R and LaTeX, it might be useful to be able to automatically convert mathematical expressions from one language to the other.

Because we need to convert both functions and names, this mathematical DSL will be more complicated than the HTML DSL. We'll also need to create a default conversion, so that symbols that we don't know about get a standard conversion. This means that we can no longer use just evaluation: we also need to walk the abstract syntax tree (AST).

21.3.1 LaTeX mathematics

Before we begin, let's quickly cover how formulas are expressed in LaTeX. The full standard is very complex, but fortunately is well documented (http:// en.wikibooks.org/wiki/LaTeX/Mathematics), and the most common commands have a fairly simple structure:

- Most simple mathematical equations are written in the same way you'd type them in R: `x * y`, `z ^ 5`. Subscripts are written using `_` (e.g., `x_1`).
- Special characters start with a `\`: `\pi = π`, `\pm = ±`, and so on. There are a huge number of symbols available in LaTeX: searching online for `latex math symbols` returns many lists (http://www.sunilpatel.co.uk/latex-type/ latex-math-symbols/). There's even a service (http://detexify.kirelabs. org/classify.html) that will look up the symbol you sketch in the browser.
- More complicated functions look like `\name{arg1}{arg2}`. For example, to write a fraction you'd use `\frac{a}{b}`. To write a square root, you'd use `\sqrt{a}`.
- To group elements together use `{}`: i.e., `x ^ a + b` versus `x ^ {a + b}`.
- In good math typesetting, a distinction is made between variables and functions. But without extra information, LaTeX doesn't know whether `f(a * b)` represents calling the function `f` with input `a * b`, or is shorthand for `f * (a * b)`. If `f` is a function, you can tell LaTeX to typeset it using an upright font with `\textrm{f}(a * b)`. (The `rm` stands for "Roman", the opposite of italics.)

21.3.2 Goal

Our goal is to use these rules to automatically convert an R expression to its appropriate LaTeX representation. We'll tackle this in four stages:

- Convert known symbols: `pi` → `\pi`
- Leave other symbols unchanged: `x` → `x`, `y` → `y`
- Convert known functions to their special forms: `sqrt(frac(a, b))` → `\sqrt{\frac{a, b}}`
- Wrap unknown functions with `\textrm`: `f(a)` → `\textrm{f}(a)`

We'll code this translation in the opposite direction of what we did with the HTML DSL. We'll start with infrastructure, because that makes it easy to experiment with our DSL, and then work our way back down to generate the desired output.

21.3.3 `to_math()`

To begin, we need a wrapper function that will convert R expressions into LaTeX math expressions. This will work like `to_html()` by capturing the unevaluated expression and evaluating it in a special environment. There are two main differences:

- The evaluation environment is no longer constant, as it has to vary depending on the input. This is necessary to handle unknown symbols and functions.
- We never evaluate in the argument environment because we're translating every function to a LaTeX expression. The user will need to use explicitly `!!` in order to evaluate normally.

This gives us:

```r
to_math <- function(x) {
  expr <- enexpr(x)
  out <- eval_bare(expr, latex_env(expr))

  latex(out)
}

latex <- function(x) structure(x, class = "advr_latex")
print.advr_latex <- function(x) {
  cat("<LATEX> ", x, "\n", sep = "")
}
```

Next we'll build up `latex_env()`, starting simply and getting progressively more complex.

21.3.4 Known symbols

Our first step is to create an environment that will convert the special LaTeX symbols used for Greek characters, e.g., `pi` to `\pi`. We'll use the trick from Section 20.4.3 to bind the symbol `pi` to the value `"\pi"`.

```
greek <- c(
  "alpha", "theta", "tau", "beta", "vartheta", "pi", "upsilon",
  "gamma", "varpi", "phi", "delta", "kappa", "rho",
  "varphi", "epsilon", "lambda", "varrho", "chi", "varepsilon",
  "mu", "sigma", "psi", "zeta", "nu", "varsigma", "omega", "eta",
  "xi", "Gamma", "Lambda", "Sigma", "Psi", "Delta", "Xi",
  "Upsilon", "Omega", "Theta", "Pi", "Phi"
)
greek_list <- set_names(paste0("\\", greek), greek)
greek_env <- as_environment(greek_list)
```

We can then check it:

```
latex_env <- function(expr) {
  greek_env
}

to_math(pi)
#> <LATEX> \pi
to_math(beta)
#> <LATEX> \beta
```

Looks good so far!

21.3.5 Unknown symbols

If a symbol isn't Greek, we want to leave it as is. This is tricky because we don't know in advance what symbols will be used, and we can't possibly generate them all. Instead, we'll use the approach described in Section 18.5: walking the AST to find all symbols. This gives us `all_names_rec()` and helper `all_names()`:

```r
all_names_rec <- function(x) {
  switch_expr(x,
    constant = character(),
    symbol =   as.character(x),
    call =     flat_map_chr(as.list(x[-1]), all_names)
  )
}

all_names <- function(x) {
  unique(all_names_rec(x))
}

all_names(expr(x + y + f(a, b, c, 10)))
#> [1] "x" "y" "a" "b" "c"
```

We now want to take that list of symbols and convert it to an environment so that each symbol is mapped to its corresponding string representation (e.g., so eval(quote(x), env) yields "x"). We again use the pattern of converting a named character vector to a list, then converting the list to an environment.

```r
latex_env <- function(expr) {
  names <- all_names(expr)
  symbol_env <- as_environment(set_names(names))

  symbol_env
}

to_math(x)
#> <LATEX> x
to_math(longvariablename)
#> <LATEX> longvariablename
to_math(pi)
#> <LATEX> pi
```

This works, but we need to combine it with the Greek symbols environment. Since we want to give preference to Greek over defaults (e.g., to_math(pi) should give "\\pi", not "pi"), symbol_env needs to be the parent of greek_env. To do that, we need to make a copy of greek_env with a new parent. This gives us a function that can convert both known (Greek) and unknown symbols.

```r
latex_env <- function(expr) {
  # Unknown symbols
  names <- all_names(expr)
```

```r
  symbol_env <- as_environment(set_names(names))

  # Known symbols
  env_clone(greek_env, parent = symbol_env)
}

to_math(x)
#> <LATEX> x
to_math(longvariablename)
#> <LATEX> longvariablename
to_math(pi)
#> <LATEX> \pi
```

21.3.6 Known functions

Next we'll add functions to our DSL. We'll start with a couple of helpers that
make it easy to add new unary and binary operators. These functions are very
simple: they only assemble strings.

```r
unary_op <- function(left, right) {
  new_function(
    exprs(e1 = ),
    expr(
      paste0(!!left, e1, !!right)
    ),
    caller_env()
  )
}

binary_op <- function(sep) {
  new_function(
    exprs(e1 = , e2 = ),
    expr(
      paste0(e1, !!sep, e2)
    ),
    caller_env()
  )
}

unary_op("\\sqrt{", "}")
#> function (e1)
#> paste0("\\sqrt{", e1, "}")
```

```
binary_op("+")
#> function (e1, e2)
#> paste0(e1, "+", e2)
```

Using these helpers, we can map a few illustrative examples of converting R to LaTeX. Note that with R's lexical scoping rules helping us, we can easily provide new meanings for standard functions like +, -, and *, and even (and {.

```
# Binary operators
f_env <- child_env(
  .parent = empty_env(),
  `+` = binary_op(" + "),
  `-` = binary_op(" - "),
  `*` = binary_op(" * "),
  `/` = binary_op(" / "),
  `^` = binary_op("^"),
  `[` = binary_op("_"),

  # Grouping
  `{` = unary_op("\\left{ ", " \\right}"),
  `(` = unary_op("\\left( ", " \\right)"),
  paste = paste,

  # Other math functions
  sqrt = unary_op("\\sqrt{", "}"),
  sin  = unary_op("\\sin(", ")"),
  log  = unary_op("\\log(", ")"),
  abs  = unary_op("\\left| ", "\\right| "),
  frac = function(a, b) {
    paste0("\\frac{", a, "}{", b, "}")
  },

  # Labelling
  hat   = unary_op("\\hat{", "}"),
  tilde = unary_op("\\tilde{", "}")
)
```

We again modify `latex_env()` to include this environment. It should be the last environment R looks for names in so that expressions like `sin(sin)` will work.

```
latex_env <- function(expr) {
  # Known functions
  f_env

  # Default symbols
  names <- all_names(expr)
  symbol_env <- as_environment(set_names(names), parent = f_env)

  # Known symbols
  greek_env <- env_clone(greek_env, parent = symbol_env)

  greek_env
}

to_math(sin(x + pi))
#> <LATEX> \sin(x + \pi)
to_math(log(x[i]^2))
#> <LATEX> \log(x_i^2)
to_math(sin(sin))
#> <LATEX> \sin(sin)
```

21.3.7 Unknown functions

Finally, we'll add a default for functions that we don't yet know about. We can't know in advance what the unknown funtions will be so we again walk the AST to find them:

```
all_calls_rec <- function(x) {
  switch_expr(x,
    constant = ,
    symbol =   character(),
    call = {
      fname <- as.character(x[[1]])
      children <- flat_map_chr(as.list(x[-1]), all_calls)
      c(fname, children)
    }
  )
}
all_calls <- function(x) {
  unique(all_calls_rec(x))
}
```

```
all_calls(expr(f(g + b, c, d(a))))
#> [1] "f" "+" "d"
```

We need a closure that will generate the functions for each unknown call:

```
unknown_op <- function(op) {
  new_function(
    exprs(... = ),
    expr({
      contents <- paste(..., collapse = ", ")
      paste0(!!paste0("\\mathrm{", op, "}("), contents, ")")
    })
  )
}
unknown_op("foo")
#> function (...)
#> {
#>     contents <- paste(..., collapse = ", ")
#>     paste0("\\mathrm{foo}(", contents, ")")
#> }
#> <environment: 0x7f98e739d128>
```

And again we update `latex_env()`:

```
latex_env <- function(expr) {
  calls <- all_calls(expr)
  call_list <- map(set_names(calls), unknown_op)
  call_env <- as_environment(call_list)

  # Known functions
  f_env <- env_clone(f_env, call_env)

  # Default symbols
  names <- all_names(expr)
  symbol_env <- as_environment(set_names(names), parent = f_env)

  # Known symbols
  greek_env <- env_clone(greek_env, parent = symbol_env)
  greek_env
}
```

This completes our original requirements:

```
to_math(sin(pi) + f(a))
#> <LATEX> \sin(\pi) + \mathrm{f}(a)
```

You could certainly take this idea further and translate types of mathematical expression, but you should not need any additional metaprogramming tools.

21.3.8 Exercises

1. Add escaping. The special symbols that should be escaped by adding a backslash in front of them are \, $, and %. Just as with HTML, you'll need to make sure you don't end up double-escaping. So you'll need to create a small S3 class and then use that in function operators. That will also allow you to embed arbitrary LaTeX if needed.

2. Complete the DSL to support all the functions that plotmath supports.

Part V

Techniques

Introduction

The final four chapters cover two general programming techniques: finding and fixing bugs, and finding and fixing performance issues. Tools to measure and improve performance are particularly important because R is not a fast language. This is not an accident: R was purposely designed to make interactive data analysis easier for humans, not to make computers as fast as possible. While R is slow compared to other programming languages, for most purposes, it's fast enough. These chapters help you handle the cases where R is no longer fast enough, either by improving the performance of your R code, or by switching to a language, C++, that is designed for performance.

1. Chapter 22 talks about debugging, because finding the root cause of error can be extremely frustrating. Fortunately R has some great tools for debugging, and when they're coupled with a solid strategy, you should be able to find the root cause for most problems rapidly and relatively painlessly.

2. Chapter 23 focuses on measuring performance.

3. Chapter 24 then shows how to improve performance.

22

Debugging

22.1 Introduction

What do you do when R code throws an unexpected error? What tools do you have to find and fix the problem? This chapter will teach you the art and science of debugging, starting with a general strategy, then following up with specific tools.

I'll show the tools provided by both R and the RStudio IDE. I recommend using RStudio's tools if possible, but I'll also show you the equivalents that work everywhere. You may also want to refer to the official RStudio debugging documentation (https://support.rstudio.com/hc/en-us/articles/205612627-Debugging-with-RStudio) which always reflects the latest version of RStudio.

NB: You shouldn't need to use these tools when writing *new* functions. If you find yourself using them frequently with new code, reconsider your approach. Instead of trying to write one big function all at once, work interactively on small pieces. If you start small, you can quickly identify why something doesn't work, and don't need sophisticated debugging tools.

Outline

- Section 22.2 outlines a general strategy for finding and fixing errors.

- Section 22.3 introduces you to the `traceback()` function which helps you locate exactly where an error occurred.

- Section 22.4 shows you how to pause the execution of a function and launch environment where you can interactively explore what's happening.

- Section 22.5 discusses the challenging problem of debugging when you're running code non-interactively.

- Section 22.6 discusses a handful of non-error problems that occassionally also need debugging.

22.2   Overall approach

Finding your bug is a process of confirming the many things that you believe are true — until you find one which is not true.

—Norm Matloff

Finding the root cause of a problem is always challenging. Most bugs are subtle and hard to find because if they were obvious, you would've avoided them in the first place. A good strategy helps. Below I outline a four step process that I have found useful:

1. **Google!**

    Whenever you see an error message, start by googling it. If you're lucky, you'll discover that it's a common error with a known solution. When googling, improve your chances of a good match by removing any variable names or values that are specific to your problem.

    You can automate this process with the errorist [Balamuta, 2018a] and searcher [Balamuta, 2018b] packages. See their websites for more details.

2. **Make it repeatable**

    To find the root cause of an error, you're going to need to execute the code many times as you consider and reject hypotheses. To make that iteration as quick possible, it's worth some upfront investment to make the problem both easy and fast to reproduce.

    Start by creating a reproducible example (Section 1.7). Next, make the example minimal by removing code and simplifying data. As you do this, you may discover inputs that don't trigger the error. Make note of them: they will be helpful when diagnosing the root cause.

    If you're using automated testing, this is also a good time to create an automated test case. If your existing test coverage is low, take the opportunity to add some nearby tests to ensure that existing good behaviour is preserved. This reduces the chances of creating a new bug.

3. **Figure out where it is**

    If you're lucky, one of the tools in the following section will help you to quickly identify the line of code that's causing the bug. Usually, however, you'll have to think a bit more about the problem. It's a great idea to adopt the scientific method. Generate hypotheses, design experiments to test them, and record your results. This may seem like a lot of work, but a systematic approach will end up saving you time. I often waste a lot of time relying on my intuition to solve a bug ("oh, it must be an off-by-one error, so I'll just subtract 1 here"), when I would have been better off taking a systematic approach.

If this fails, you might need to ask help from someone else. If you've followed the previous step, you'll have a small example that's easy to share with others. That makes it much easier for other people to look at the problem, and more likely to help you find a solution.

4. **Fix it and test it**

   Once you've found the bug, you need to figure out how to fix it and to check that the fix actually worked. Again, it's very useful to have automated tests in place. Not only does this help to ensure that you've actually fixed the bug, it also helps to ensure you haven't introduced any new bugs in the process. In the absence of automated tests, make sure to carefully record the correct output, and check against the inputs that previously failed.

22.3 Locating errors

Once you've made the error repeatable, the next step is to figure out where it comes from. The most important tool for this part of the process is traceback(), which shows you the sequence of calls (also known as the call stack, Section 7.5) that lead to the error.

Here's a simple example: you can see that f() calls g() calls h() calls i(), which checks if its argument is numeric:

```
f <- function(a) g(a)
g <- function(b) h(b)
h <- function(c) i(c)
i <- function(d) {
  if (!is.numeric(d)) {
    stop("`d` must be numeric", call. = FALSE)
  }
  d + 10
}
```

When we run `f("a")` code in RStudio we see:

```
> f("a")
```

```
Error: `d` must be numeric                                   ↱ Show Traceback
                                                             ⟳ Rerun with Debug
```

Two options appear to the right of the error message: "Show Traceback" and "Rerun with Debug". If you click "Show traceback" you see:

```
> f("a")
```

```
Error: `d` must be numeric                                   ↱ Hide Traceback
                                                             ⟳ Rerun with Debug

  5. stop("`d` must be numeric", call. = FALSE) at debugging.R#6
  4. i(c) at debugging.R#3
  3. h(b) at debugging.R#2
  2. g(a) at debugging.R#1
  1. f("a")
```

If you're not using RStudio, you can use `traceback()` to get the same information (sans pretty formatting):

```
traceback()
#> 5: stop("`d` must be numeric", call. = FALSE) at debugging.R#6
#> 4: i(c) at debugging.R#3
#> 3: h(b) at debugging.R#2
#> 2: g(a) at debugging.R#1
#> 1: f("a")
```

NB: You read the `traceback()` output from bottom to top: the initial call is `f()`, which calls `g()`, then `h()`, then `i()`, which triggers the error. If you're calling code that you `source()`d into R, the traceback will also display the location of the function, in the form `filename.r#linenumber`. These are clickable in RStudio, and will take you to the corresponding line of code in the editor.

22.3.1 Lazy evaluation

One drawback to `traceback()` is that it always linearises the call tree, which can be confusing if there is much lazy evaluation involved (Section 7.5.2). For example, take the following example where the error happens when evaluating the first argument to `f()`:

```
j <- function() k()
k <- function() stop("Oops!", call. = FALSE)
```

```
f(j())
#> Error: Oops!
```

```
traceback()
#> 7: stop("Oops!") at #1
#> 6: k() at #1
#> 5: j() at debugging.R#1
#> 4: i(c) at debugging.R#3
#> 3: h(b) at debugging.R#2
#> 2: g(a) at debugging.R#1
#> 1: f(j())
```

You can using `rlang::with_abort()` and `rlang::last_trace()` to see the call tree. Here, I think it makes it much easier to see the source of the problem. Look at the last branch of the call tree to see that the error comes from `j()` calling `k()`.

```
rlang::with_abort(f(j()))
#> Error: Oops!
rlang::last_trace()
#>
#> 1.   ─rlang::with_abort(f(j()))
#> 2.   │ └─base::withCallingHandlers(...)
#> 3.   ─global::f(j())
#> 4.   │ └─global::g(a) debugging.R:1:5
#> 5.   │   └─global::h(b) debugging.R:2:5
#> 6.   │     └─global::i(c) debugging.R:3:5
#> 7.   └─global::j() debugging.R:1:5
#> 8.     └─global::k()
```

NB: `rlang::last_trace()` is ordered in the opposite way to `traceback()`. We'll come back to that issue in Section 22.4.2.4.

22.4 Interactive debugger

Sometimes, the precise location of the error is enough to let you track it down and fix it. Frequently, however, you need more information, and the easiest way to get it is with the interactive debugger which allows you to pause execution of a function and interactively explore its state.

If you're using RStudio, the easiest way to enter the interactive debugger is through RStudio's "Rerun with Debug" tool. This reruns the command that created the error, pausing execution where the error occurred. Otherwise, you can insert a call to browser() where you want to pause, and re-run the function. For example, we could insert a call browser() in g():

```r
g <- function(b) {
  browser()
  h(b)
}
f(10)
```

browser() is just a regular function call which means that you can run it conditionally by wrapping it in an if statement:

```r
g <- function(b) {
  if (b < 0) {
    browser()
  }
  h(b)
}
```

In either case, you'll end up in an interactive environment *inside* the function where you can run arbitrary R code to explore the current state. You'll know when you're in the interactive debugger because you get a special prompt:

```
Browse[1]>
```

In RStudio, you'll see the corresponding code in the editor (with the statement that will be run next highlighted), objects in the current environment in the Environment pane, and the call stack in the Traceback pane.

22.4.1 browser() commands

As well as allowing you to run regular R code, browser() provides a few special commands. You can use them by either typing short text commands, or by clicking a button in the RStudio toolbar, Figure 22.1:

Figure 22.1 RStudio debugging toolbar

- Next, n: executes the next step in the function. If you have a variable named n, you'll need print(n) to display its value.

- Step into, ![] or s: works like next, but if the next step is a function, it will step into that function so you can explore it interactively.

- Finish, ⇐ or f: finishes execution of the current loop or function.

- Continue, c: leaves interactive debugging and continues regular execution of the function. This is useful if you've fixed the bad state and want to check that the function proceeds correctly.

- Stop, Q: stops debugging, terminates the function, and returns to the global workspace. Use this once you've figured out where the problem is, and you're ready to fix it and reload the code.

There are two other slightly less useful commands that aren't available in the toolbar:

- Enter: repeats the previous command. I find this too easy to activate accidentally, so I turn it off using `options(browserNLdisabled = TRUE)`.

- `where`: prints stack trace of active calls (the interactive equivalent of `traceback`).

22.4.2   Alternatives

There are three alternatives to using `browser()`: setting breakpoints in RStudio, `option(error = recover)`, and `debug()` and other related functions.

22.4.2.1   Breakpoints

In RStudio, you can set a breakpoint by clicking to the left of the line number, or pressing `Shift + F9`. Breakpoints behave similarly to `browser()` but they are easier to set (one click instead of nine key presses), and you don't run the risk of accidentally including a `browser()` statement in your source code. There are two small downsides to breakpoints:

- There are a few unusual situations in which breakpoints will not work. Read breakpoint troubleshooting (`http://www.rstudio.com/ide/docs/debugging/breakpoint-troubleshooting`) for more details.

- RStudio currently does not support conditional breakpoints.

22.4.2.2   `recover()`

Another way to activate `browser()` is to use `options(error = recover)`. Now when you get an error, you'll get an interactive prompt that displays the traceback and gives you the ability to interactively debug inside any of the frames:

```
options(error = recover)
f("x")
#> Error: `d` must be numeric
#>
#> Enter a frame number, or 0 to exit
#>
#> 1: f("x")
#> 2: debugging.R#1: g(a)
#> 3: debugging.R#2: h(b)
#> 4: debugging.R#3: i(c)
#>
#> Selection:
```

You can return to default error handling with options(error = NULL).

22.4.2.3 debug()

Another approach is to call a function that inserts the browser() call for you:

- debug() inserts a browser statement in the first line of the specified function. undebug() removes it. Alternatively, you can use debugonce() to browse only on the next run.

- utils::setBreakpoint() works similarly, but instead of taking a function name, it takes a file name and line number and finds the appropriate function for you.

These two functions are both special cases of trace(), which inserts arbitrary code at any position in an existing function. trace() is occasionally useful when you're debugging code that you don't have the source for. To remove tracing from a function, use untrace(). You can only perform one trace per function, but that one trace can call multiple functions.

22.4.2.4 Call stack

Unfortunately, the call stacks printed by traceback(), browser() & where, and recover() are not consistent. The following table shows how the call stacks from a simple nested set of calls are displayed by the three tools. The numbering is different between traceback() and where, and recover() displays calls in the opposite order.

traceback()	where	recover()	rlang functions
5: stop("...")			
4: i(c)	where 1: i(c)	1: f()	1. └─global::f(10)
3: h(b)	where 2: h(b)	2: g(a)	2. └─global::g(a)

traceback()	where	recover()		rlang functions
2: g(a)	where 3: g(a)	3: h(b)	3.	└─global::h(b)
1: f("a")	where 4: f("a")	4: i("a")	4.	└─global::i("a")

RStudio displays calls in the same order as traceback(). rlang functions use the same ordering and numbering as recover(), but also use indenting to reinforce the hierarchy of calls.

22.4.3 Compiled code

It is also possible to use an interactive debugger (gdb or lldb) for compiled code (like C or C++). Unfortunately that's beyond the scope of this book, but there are a few resources that you might find useful:

- http://r-pkgs.had.co.nz/src.html#src-debugging
- https://github.com/wch/r-debug/blob/master/debugging-r.md
- http://kevinushey.github.io/blog/2015/04/05/debugging-with-valgrind/
- https://www.jimhester.com/2018/08/22/debugging-rstudio/

22.5 Non-interactive debugging

Debugging is most challenging when you can't run code interactively, typically because it's part of some pipeline run automatically (possibly on another computer), or because the error doesn't occur when you run same code interactively. This can be extremely frustrating!

This section will give you some useful tools, but don't forget the general strategy in Section 22.2. When you can't explore interactively, it's particularly important to spend some time making the problem as small as possible so you can iterate quickly. Sometimes callr::r(f, list(1, 2)) can be useful; this calls f(1, 2) in a fresh session, and can help to reproduce the problem.

You might also want to double check for these common issues:

- Is the global environment different? Have you loaded different packages? Are objects left from previous sessions causing differences?

- Is the working directory different?

- Is the PATH environment variable, which determines where external commands (like git) are found, different?

- Is the R_LIBS environment variable, which determines where library() looks for packages, different?

22.5.1 dump.frames()

dump.frames() is the equivalent to recover() for non-interactive code; it saves a last.dump.rda file in the working directory. Later, an interactive session, you can load("last.dump.rda"); debugger() to enter an interactive debugger with the same interface as recover(). This lets you "cheat", interactively debugging code that was run non-interactively.

```r
# In batch R process ----
dump_and_quit <- function() {
  # Save debugging info to file last.dump.rda
  dump.frames(to.file = TRUE)
  # Quit R with error status
  q(status = 1)
}
options(error = dump_and_quit)

# In a later interactive session ----
load("last.dump.rda")
debugger()
```

22.5.2 Print debugging

If dump.frames() doesn't help, a good fallback is **print debugging**, where you insert numerous print statements to precisely locate the problem, and see the values of important variables. Print debugging is slow and primitive, but it always works, so it's particularly useful if you can't get a good traceback. Start by inserting coarse-grained markers, and then make them progressively more fine-grained as you determine exactly where the problem is.

```r
f <- function(a) {
  cat("f()\n")
  g(a)
}
g <- function(b) {
  cat("g()\n")
  cat("b =", b, "\n")
  h(b)
```

```
}
h <- function(c) {
  cat("i()\n")
  i(c)
}

f(10)
#> f()
#> g()
#> b = 10
#> i()
#> [1] 20
```

Print debugging is particularly useful for compiled code because it's not uncommon for the compiler to modify your code to such an extent you can't figure out the root problem even when inside an interactive debugger.

22.5.3 RMarkdown

Debugging code inside RMarkdown files requires some special tools. First, if you're knitting the file using RStudio, switch to calling rmarkdown::render("path/to/file.Rmd") instead. This runs the code in the current session, which makes it easier to debug. If doing this makes the problem go away, you'll need to figure out what makes the environments different.

If the problem persists, you'll need to use your interactive debugging skills. Whatever method you use, you'll need an extra step: in the error handler, you'll need to call sink(). This removes the default sink that knitr uses to capture all output, and ensures that you can see the results in the console. For example, to use recover() with RMarkdown, you'd put the following code in your setup block:

```
options(error = function() {
  sink()
  recover()
})
```

This will generate a "no sink to remove" warning when knitr completes; you can safely ignore this warning.

If you simply want a traceback, the easiest option is to use rlang::trace_back(), taking advantage of the rlang_trace_top_env option. This ensures that you only see the traceback from your code, instead of all the functions called by RMarkdown and knitr.

```
options(rlang_trace_top_env = rlang::current_env())
options(error = function() {
  sink()
  print(rlang::trace_back(bottom = sys.frame(-1)), simplify = "none")
})
```

22.6 Non-error failures

There are other ways for a function to fail apart from throwing an error:

- A function may generate an unexpected warning. The easiest way to track down warnings is to convert them into errors with options(warn = 2) and use the the call stack, like doWithOneRestart(), withOneRestart(), regular debugging tools. When you do this you'll see some extra calls withRestarts(), and .signalSimpleWarning(). Ignore these: they are internal functions used to turn warnings into errors.

- A function may generate an unexpected message. You can use rlang::with_abort() to turn these messages into errors:

```
f <- function() g()
g <- function() message("Hi!")
f()
#> Hi!

rlang::with_abort(f(), "message")
#> Error: Hi!
rlang::last_trace()
#>
#>  1. ─rlang::with_abort(f(), "message")
#>  2. │ └─base::withCallingHandlers(...)
#>  3. └─global::f()
#>  4.     └─global::g()
```

- A function might never return. This is particularly hard to debug automatically, but sometimes terminating the function and looking at the traceback() is informative. Otherwise, use use print debugging, as in Section 22.5.2.

- The worst scenario is that your code might crash R completely, leaving you with no way to interactively debug your code. This indicates a bug in compiled (C or C++) code.

If the bug is in your compiled code, you'll need to follow the links in Section 22.4.3 and learn how to use an interactive C debugger (or insert many print statements).

If the bug is in a package or base R, you'll need to contact the package maintainer. In either case, work on making the smallest possible reproducible example (Section 1.7) to help the developer help you.

23

Measuring performance

23.1 Introduction

> *Programmers waste enormous amounts of time thinking about, or worrying about, the speed of noncritical parts of their programs, and these attempts at efficiency actually have a strong negative impact when debugging and maintenance are considered.*
>
> — *Donald Knuth.*

Before you can make your code faster, you first need to figure out what's making it slow. This sounds easy, but it's not. Even experienced programmers have a hard time identifying bottlenecks in their code. So instead of relying on your intuition, you should **profile** your code: measure the run-time of each line of code using realistic inputs.

Once you've identified bottlenecks you'll need to carefully experiment with alternatives to find faster code that is still equivalent. In Chapter 24 you'll learn a bunch of ways to speed up code, but first you need to learn how to **microbenchmark** so that you can precisely measure the difference in performance.

Outline

- Section 23.2 shows you how to use profiling tools to dig into exactly what is making code slow.

- Section 23.3 shows how to use microbenchmarking to explore alternative implementations and figure out exactly which one is fastest.

Prerequisites

We'll use profvis (https://rstudio.github.io/profvis/) for profiling, and bench (https://bench.r-lib.org/) for microbenchmarking.

```
library(profvis)
library(bench)
```

23.2   Profiling

Across programming languages, the primary tool used to understand code performance is the profiler. There are a number of different types of profilers, but R uses a fairly simple type called a sampling or statistical profiler. A sampling profiler stops the execution of code every few milliseconds and records the call stack (i.e. which function is currently executing, and the function that called the function, and so on). For example, consider f(), below:

```
f <- function() {
  pause(0.1)
  g()
  h()
}
g <- function() {
  pause(0.1)
  h()
}
h <- function() {
  pause(0.1)
}
```

(I use profvis::pause() instead of Sys.sleep() because Sys.sleep() does not appear in profiling outputs because as far as R can tell, it doesn't use up any computing time.)

If we profiled the execution of f(), stopping the execution of code every 0.1 s, we'd see a profile like this:

```
"pause" "f"
"pause" "f" "g"
"pause" "f" "g" "h"
"pause" "f" "h"
```

Each line represents one "tick" of the profiler (0.1 s in this case), and function calls are recorded from right to left: the first line shows f() calling pause().

It shows that the code spends 0.1 s running `f()`, then 0.2 s running `g()`, then 0.1 s running `h()`.

If we actually profile `f()`, using `utils::Rprof()` as in the code below, we're unlikely to get such a clear result.

```
tmp <- tempfile()
Rprof(tmp, interval = 0.1)
f()
Rprof(NULL)
writeLines(readLines(tmp))
#> sample.interval=100000
#> "pause" "g" "f"
#> "pause" "h" "g" "f"
#> "pause" "h" "f"
```

That's because all profilers must make a fundamental trade-off between accuracy and performance. The compromise that makes, using a sampling profiler, only has minimal impact on performance, but is fundamentally stochastic because there's some variability in both the accuracy of the timer and in the time taken by each operation. That means each time that you profile you'll get a slightly different answer. Fortunately, the variability most affects functions that take very little time to run, which are also the functions of least interest.

23.2.1 Visualising profiles

The default profiling resolution is quite small, so if your function takes even a few seconds it will generate hundreds of samples. That quickly grows beyond our ability to look at directly, so instead of using `utils::Rprof()` we'll use the profvis package to visualise aggregates. profvis also connects profiling data back to the underlying source code, making it easier to build up a mental model of what you need to change. If you find profvis doesn't help for your code, you might try one of the other options like `utils::summaryRprof()` or the proftools package [Tierney and Jarjour, 2016].

There are two ways to use profvis:

- From the Profile menu in RStudio.
- With `profvis::profvis()`. I recommend storing your code in a separate file and `source()`ing it in; this will ensure you get the best connection between profiling data and source code.

```
source("profiling-example.R")
profvis(f())
```

After profiling is complete, profvis will open an interactive HTML document that allows you to explore the results. There are two panes, as shown in Figure 23.1.

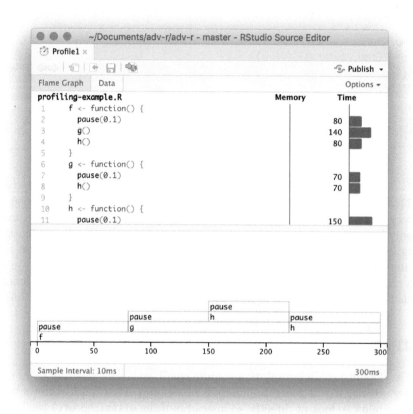

Figure 23.1 profvis output showing source on top and flame graph below.

The top pane shows the source code, overlaid with bar graphs for memory and execution time for each line of code. Here I'll focus on time, and we'll come back to memory shortly. This display gives you a good overall feel for the bottlenecks but doesn't always help you precisely identify the cause. Here, for example, you can see that h() takes 150 ms, twice as long as g(); that's not because the function is slower, but because it's called twice as often.

The bottom pane displays a **flame graph** showing the full call stack. This allows you to see the full sequence of calls leading to each function, allowing you to see that h() is called from two different places. In this display you can mouse over individual calls to get more information, and see the corresponding line of source code, as in Figure 23.2.

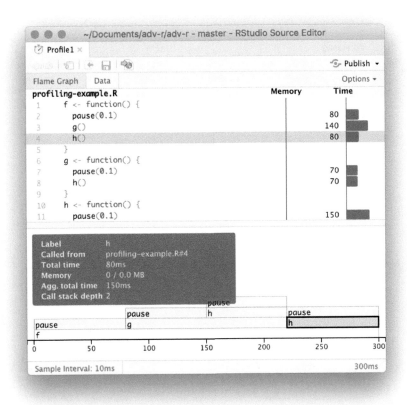

Figure 23.2 Hovering over a call in the flamegraph highlights the corresponding line of code, and displays additional information about performance.

Alternatively, you can use the **data tab**, Figure 23.3 lets you interactively dive into the tree of performance data. This is basically the same display as the flame graph (rotated 90 degrees), but it's more useful when you have very large or deeply nested call stacks because you can choose to interactively zoom into only selected components.

23.2.2 Memory profiling

There is a special entry in the flame graph that doesn't correspond to your code: <GC>, which indicates that the garbage collector is running. If <GC> is taking a lot of time, it's usually an indication that you're creating many short-lived objects. For example, take this small snippet of code:

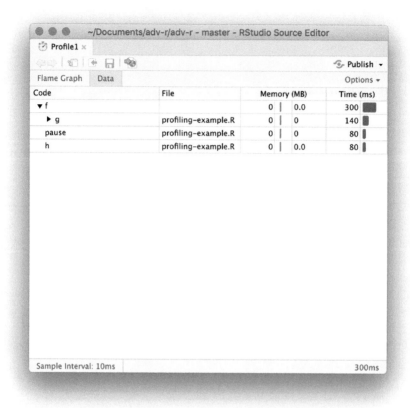

Figure 23.3 The data gives an interactive tree that allows you to selectively zoom into key components

```
x <- integer()
for (i in 1:1e4) {
  x <- c(x, i)
}
```

If you profile it, you'll see that most of the time is spent in the garbage collector, Figure 23.4.

When you see the garbage collector taking up a lot of time in your own code, you can often figure out the source of the problem by looking at the memory column: you'll see a line where large amounts of memory are being allocated (the bar on the right) and freed (the bar on the left). Here the problem arises because of copy-on-modify (Section 2.3): each iteration of the loop creates

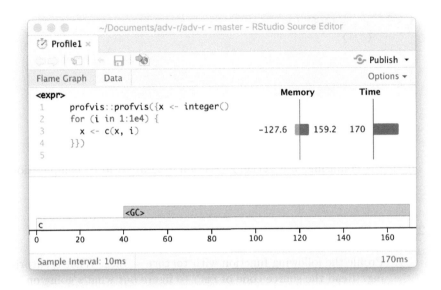

Figure 23.4 Profiling a loop that modifies an existing variable reveals that most time is spent in the garbage collector (<GC>).

another copy of x. You'll learn strategies to resolve this type of problem in Section 24.6.

23.2.3 Limitations

There are some other limitations to profiling:

- Profiling does not extend to C code. You can see if your R code calls C/C++ code but not what functions are called inside of your C/C++ code. Unfortunately, tools for profiling compiled code are beyond the scope of this book; start by looking at `https://github.com/r-prof/jointprof`.

- If you're doing a lot of functional programming with anonymous functions, it can be hard to figure out exactly which function is being called. The easiest way to work around this is to name your functions.

- Lazy evaluation means that arguments are often evaluated inside another function, and this complicates the call stack (Section 7.5.2). Unfortunately R's profiler doesn't store enough information to disentangle lazy evaluation so that in the following code, profiling would make it seem like `i()` was called by `j()` because the argument isn't evaluated until it's needed by `j()`.

```
i <- function() {
  pause(0.1)
  10
}
j <- function(x) {
  x + 10
}
j(i())
```

If this is confusing, use `force()` (Section 10.2.3) to force computation to happen earlier.

23.2.4   Exercises

1. Profile the following function with `torture = TRUE`. What is surprising? Read the source code of `rm()` to figure out what's going on.

   ```
   f <- function(n = 1e5) {
     x <- rep(1, n)
     rm(x)
   }
   ```

23.3   Microbenchmarking

A **microbenchmark** is a measurement of the performance of a very small piece of code, something that might take milliseconds (ms), microseconds (μs), or nanoseconds (ns) to run. Microbenchmarks are useful for comparing small snippets of code for specific tasks. Be very wary of generalising the results of microbenchmarks to real code: the observed differences in microbenchmarks will typically be dominated by higher-order effects in real code; a deep understanding of subatomic physics is not very helpful when baking.

A great tool for microbenchmarking in R is the bench package [Hester, 2018]. The bench package uses a a high precision timer, making it possible to compare operations that only take a tiny amount of time. For example, the following code compares the speed of two approaches to computing a square root.

```
x <- runif(100)
(lb <- bench::mark(
  sqrt(x),
  x ^ 0.5
))
#> # A tibble: 2 x 10
#>   expression      min    mean  median      max `itr/sec` mem_alloc
#>   <chr>        <bch:tm> <bch:> <bch:t>  <bch:t>     <dbl> <bch:byt>
#> 1 sqrt(x)         442ns 1.29µs   665ns  45.6µs   776952.      848B
#> 2 x^0.5          3.05µs 3.65µs   3.2µs 100.4µs   274024.      848B
#> # ... with 3 more variables: n_gc <dbl>, n_itr <int>,
#> #   total_time <bch:tm>
```

By default, bench::mark() runs each expression at least once (min_iterations = 1), and at most enough times to take 0.5 s (min_time = 0.5). It checks that each run returns the same value which is typically what you want microbenchmarking; if you want to compare the speed of expressions that return different values, set check = FALSE.

23.3.1 bench::mark() **results**

bench::mark() returns the results as a tibble, with one row for each input expression, and the following columns:

- min, mean, median, max, and itr/sec summarise the time taken by the expression. Focus on the minimum (the best possible running time) and the median (the typical time). In this example, you can see that using the special purpose sqrt() function is faster than the general exponentiation operator.

You can visualise the distribution of the individual timings with plot():

```
plot(lb)
```

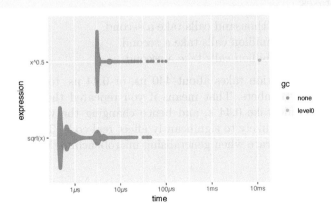

The distribution tends to be heavily right-skewed (note that the x-axis is already on a log scale!), which is why you should avoid comparing means. You'll also often see multimodality because your computer is running something else in the background.

- mem_alloc tells you the amount of memory allocated by the first run, and n_gc() tells you the total number of garbage collections over all runs. These are useful for assessing the memory usage of the expression.

- n_itr and total_time tells you how many times the expression was evaluated and how long that took in total. n_itr will always be greater than the min_iteration parameter, and total_time will always be greater than the min_time parameter.

- result, memory, time, and gc are list-columns that store the raw underlying data.

Because the result is a special type of tibble, you can use [to select just the most important columns. I'll do that frequently in the next chapter.

```
lb[c("expression", "min", "median", "itr/sec", "n_gc")]
#> # A tibble: 2 x 5
#>   expression      min    median 'itr/sec' n_gc
#>   <chr>      <bch:tm> <bch:tm>     <dbl> <dbl>
#> 1 sqrt(x)       442ns    665ns   776952.     0
#> 2 x^0.5        3.05µs    3.2µs   274024.     1
```

23.3.2 Interpreting results

As with all microbenchmarks, pay careful attention to the units: here, each computation takes about 440 ns, 440 billionths of a second. To help calibrate the impact of a microbenchmark on run time, it's useful to think about how many times a function needs to run before it takes a second. If a microbenchmark takes:

- 1 ms, then one thousand calls take a second.
- 1 µs, then one million calls take a second.
- 1 ns, then one billion calls take a second.

The sqrt() function takes about 440 ns, or 0.44 µs, to compute the square roots of 100 numbers. That means if you repeated the operation a million times, it would take 0.44 s, and hence changing the way you compute the square root is unlikely to significantly affect real code. This is the reason you need to exercise care when generalising microbenchmarking results.

23.3.3 Exercises

1. Instead of using bench::mark(), you could use the built-in function
   system.time(). But system.time() is much less precise, so you'll
   need to repeat each operation many times with a loop, and then
   divide to find the average time of each operation, as in the code
   below.

   ```
   n <- 1e6
   system.time(for (i in 1:n) sqrt(x)) / n
   system.time(for (i in 1:n) x ^ 0.5) / n
   ```

   How do the estimates from system.time() compare to those from
   bench::mark()? Why are they different?

2. Here are two other ways to compute the square root of a vector.
   Which do you think will be fastest? Which will be slowest? Use
   microbenchmarking to test your answers.

   ```
   x ^ (1 / 2)
   exp(log(x) / 2)
   ```

24

Improving performance

24.1 Introduction

> *We should forget about small efficiencies, say about 97% of the time: premature optimization is the root of all evil. Yet we should not pass up our opportunities in that critical 3%. A good programmer will not be lulled into complacency by such reasoning, he will be wise to look carefully at the critical code; but only after that code has been identified.*
>
> — *Donald Knuth*

Once you've used profiling to identify a bottleneck, you need to make it faster. It's difficult to provide general advice on improving performance, but I try my best with four techniques that can be applied in many situations. I'll also suggest a general strategy for performance optimisation that helps ensure that your faster code is still correct.

It's easy to get caught up in trying to remove all bottlenecks. Don't! Your time is valuable and is better spent analysing your data, not eliminating possible inefficiencies in your code. Be pragmatic: don't spend hours of your time to save seconds of computer time. To enforce this advice, you should set a goal time for your code and optimise only up to that goal. This means you will not eliminate all bottlenecks. Some you will not get to because you've met your goal. Others you may need to pass over and accept either because there is no quick and easy solution or because the code is already well optimised and no significant improvement is possible. Accept these possibilities and move on to the next candidate.

If you'd like to learn more about the performance characteristics of the R language, I'd highly recommend *Evaluating the Design of the R Language* [Morandat et al., 2012]. It draws conclusions by combining a modified R interpreter with a wide set of code found in the wild.

Outline

- Section 24.2 teaches you how to organise your code to make optimisation as easy, and bug free, as possible.

- Section 24.3 reminds you to look for existing solutions.

- Section 24.4 emphasises the importance of being lazy: often the easiest way to make a function faster is to let it to do less work.

- Section 24.5 concisely defines vectorisation, and shows you how to make the most of built-in functions.

- Section 24.6 discusses the performance perils of copying data.

- Section 24.7 pulls all the pieces together into a case study showing how to speed up repeated t-tests by about a thousand times.

- Section 24.8 finishes the chapter with pointers to more resources that will help you write fast code.

Prerequisites

We'll use bench (https://bench.r-lib.org/) to precisely compare the performance of small self-contained code chunks.

```
library(bench)
```

24.2 Code organisation

There are two traps that are easy to fall into when trying to make your code faster:

1. Writing faster but incorrect code.
2. Writing code that you think is faster, but is actually no better.

The strategy outlined below will help you avoid these pitfalls.

When tackling a bottleneck, you're likely to come up with multiple approaches. Write a function for each approach, encapsulating all relevant behaviour. This makes it easier to check that each approach returns the correct result and to time how long it takes to run. To demonstrate the strategy, I'll compare two approaches for computing the mean:

```
mean1 <- function(x) mean(x)
mean2 <- function(x) sum(x) / length(x)
```

I recommend that you keep a record of everything you try, even the failures. If a similar problem occurs in the future, it'll be useful to see everything you've tried. To do this I recommend RMarkdown, which makes it easy to intermingle code with detailed comments and notes.

Next, generate a representative test case. The case should be big enough to capture the essence of your problem but small enough that it only takes a few seconds at most. You don't want it to take too long because you'll need to run the test case many times to compare approaches. On the other hand, you don't want the case to be too small because then results might not scale up to the real problem. Here I'm going to use 100,000 numbers:

```
x <- runif(1e5)
```

Now use `bench::mark()` to precisely compare the variations. `bench::mark()` automatically checks that all calls return the same values. This doesn't guarantee that the function behaves the same for all inputs, so in an ideal world you'll also have unit tests to make sure you don't accidentally change the behaviour of the function.

```
bench::mark(
  mean1(x),
  mean2(x)
)[c("expression", "min", "median", "itr/sec", "n_gc")]
#> # A tibble: 2 x 5
#>   expression      min   median `itr/sec` n_gc
#>   <chr>      <bch:tm> <bch:tm>     <dbl> <dbl>
#> 1 mean1(x)      191µs    219µs     4558.     0
#> 2 mean2(x)      101µs    113µs     8832.     0
```

(You might be surprised by the results: `mean(x)` is considerably slower than `sum(x) / length(x)`. This is because, among other reasons, `mean(x)` makes two passes over the vector to be more numerically accurate.)

If you'd like to see this strategy in action, I've used it a few times on stack-overflow:

- http://stackoverflow.com/questions/22515525#22518603
- http://stackoverflow.com/questions/22515175#22515856
- http://stackoverflow.com/questions/3476015#22511936

24.3   Checking for existing solutions

Once you've organised your code and captured all the variations you can
think of, it's natural to see what others have done. You are part of a large
community, and it's quite possible that someone has already tackled the same
problem. Two good places to start are:

- CRAN task views (http://cran.rstudio.com/web/views/). If there's a CRAN
  task view related to your problem domain, it's worth looking at the packages
  listed there.

- Reverse dependencies of Rcpp, as listed on its CRAN page (http://cran.
  r-project.org/web/packages/Rcpp). Since these packages use C++, they're
  likely to be fast.

Otherwise, the challenge is describing your bottleneck in a way that helps
you find related problems and solutions. Knowing the name of the problem or
its synonyms will make this search much easier. But because you don't know
what it's called, it's hard to search for it! The best way to solve this problem
is to read widely so that you can build up your own vocabulary over time.
Alternatively, ask others. Talk to your colleagues and brainstorm some possible
names, then search on Google and StackOverflow. It's often helpful to restrict
your search to R related pages. For Google, try rseek (http://www.rseek.org/).
For stackoverflow, restrict your search by including the R tag, [R], in your
search.

Record all solutions that you find, not just those that immediately appear to
be faster. Some solutions might be slower initially, but end up being faster
because they're easier to optimise. You may also be able to combine the fastest
parts from different approaches. If you've found a solution that's fast enough,
congratulations! Otherwise, read on.

24.3.1   Exercises

1. What are faster alternatives to lm()? Which are specifically designed
   to work with larger datasets?

2. What package implements a version of match() that's faster for
   repeated lookups? How much faster is it?

3. List four functions (not just those in base R) that convert a string
   into a date time object. What are their strengths and weaknesses?

4. Which packages provide the ability to compute a rolling mean?

5. What are the alternatives to optim()?

24.4 Doing as little as possible

The easiest way to make a function faster is to let it do less work. One way to do that is use a function tailored to a more specific type of input or output, or to a more specific problem. For example:

- `rowSums()`, `colSums()`, `rowMeans()`, and `colMeans()` are faster than equivalent invocations that use `apply()` because they are vectorised (Section 24.5).

- `vapply()` is faster than `sapply()` because it pre-specifies the output type.

- If you want to see if a vector contains a single value, `any(x == 10)` is much faster than `10 %in% x` because testing equality is simpler than testing set inclusion.

Having this knowledge at your fingertips requires knowing that alternative functions exist: you need to have a good vocabulary. Expand your vocab by regularly reading R code. Good places to read code are the R-help mailing list (`https://stat.ethz.ch/mailman/listinfo/r-help`) and StackOverflow (`http://stackoverflow.com/questions/tagged/r`).

Some functions coerce their inputs into a specific type. If your input is not the right type, the function has to do extra work. Instead, look for a function that works with your data as it is, or consider changing the way you store your data. The most common example of this problem is using `apply()` on a data frame. `apply()` always turns its input into a matrix. Not only is this error prone (because a data frame is more general than a matrix), it is also slower.

Other functions will do less work if you give them more information about the problem. It's always worthwhile to carefully read the documentation and experiment with different arguments. Some examples that I've discovered in the past include:

- `read.csv()`: specify known column types with `colClasses`. (Also consider switching to `readr::read_csv()` or `data.table::fread()` which are considerably faster than `read.csv()`.)

- `factor()`: specify known levels with `levels`.

- `cut()`: don't generate labels with `labels = FALSE` if you don't need them, or, even better, use `findInterval()` as mentioned in the "see also" section of the documentation.

- `unlist(x, use.names = FALSE)` is much faster than `unlist(x)`.

- `interaction()`: if you only need combinations that exist in the data, use `drop = TRUE`.

Below, I explore how you might improve apply this strategy to improve the performance of `mean()` and `as.data.frame()`.

24.4.1 mean()

Sometimes you can make a function faster by avoiding method dispatch. If you're calling a method in a tight loop, you can avoid some of the costs by doing the method lookup only once:

- For S3, you can do this by calling generic.class() instead of generic().

- For S4, you can do this by using getMethod() to find the method, saving it to a variable, and then calling that function.

For example, calling mean.default() is quite a bit faster than calling mean() for small vectors:

```
x <- runif(1e2)

bench::mark(
  mean(x),
  mean.default(x)
)[c("expression", "min", "median", "itr/sec", "n_gc")]
#> # A tibble: 2 x 5
#>   expression            min   median `itr/sec`  n_gc
#>   <chr>            <bch:tm> <bch:tm>     <dbl> <dbl>
#> 1 mean(x)            2.48µs   2.81µs   325109.     0
#> 2 mean.default(x)    1.18µs   1.29µs   703275.     1
```

This optimisation is a little risky. While mean.default() is almost twice as fast for 100 values, it will fail in surprising ways if x is not a numeric vector.

An even riskier optimisation is to directly call the underlying .Internal function. This is faster because it doesn't do any input checking or handle NA's, so you are buying speed at the cost of safety.

```
x <- runif(1e2)
bench::mark(
  mean(x),
  mean.default(x),
  .Internal(mean(x))
)[c("expression", "min", "median", "itr/sec", "n_gc")]
#> # A tibble: 3 x 5
#>   expression             min   median `itr/sec`   n_gc
#>   <chr>             <bch:tm> <bch:tm>     <dbl>  <dbl>
#> 1 mean(x)             2.47µs   2.79µs   344523.      1
#> 2 mean.default(x)     1.16µs   1.37µs   697365.      0
#> 3 .Internal(mean(x))   325ns    343ns  2813476.      0
```

NB: most of these differences arise because x is small. If you increase the size the differences basically disappear, because most of the time is now spent computing the mean, not finding the underlying implementation. This is a good reminder that the size of the input matters, and you should motivate your optimisations based on realistic data.

```
x <- runif(1e4)
bench::mark(
  mean(x),
  mean.default(x),
  .Internal(mean(x))
)[c("expression", "min", "median", "itr/sec", "n_gc")]
#> # A tibble: 3 x 5
#>   expression               min    median `itr/sec`  n_gc
#>   <chr>               <bch:tm> <bch:tm>     <dbl> <dbl>
#> 1 mean(x)                 21µs    24.5µs    40569.     1
#> 2 mean.default(x)       19.7µs    19.8µs    48892.     0
#> 3 .Internal(mean(x))    18.9µs    20.2µs    48834.     0
```

24.4.2 as.data.frame()

Knowing that you're dealing with a specific type of input can be another way to write faster code. For example, as.data.frame() is quite slow because it coerces each element into a data frame and then rbind()s them together. If you have a named list with vectors of equal length, you can directly transform it into a data frame. In this case, if you can make strong assumptions about your input, you can write a method that's considerably faster than the default.

```
quickdf <- function(l) {
  class(l) <- "data.frame"
  attr(l, "row.names") <- .set_row_names(length(l[[1]]))
  l
}

l <- lapply(1:26, function(i) runif(1e3))
names(l) <- letters

bench::mark(
  as.data.frame = as.data.frame(l),
  quick_df      = quickdf(l)
)[c("expression", "min", "median", "itr/sec", "n_gc")]
#> # A tibble: 2 x 5
```

```
#>    expression              min    median `itr/sec`  n_gc
#>    <chr>              <bch:tm> <bch:tm>      <dbl> <dbl>
#> 1 as.data.frame        1.01ms   1.12ms       875.     9
#> 2 quick_df             6.34µs   7.16µs    125452.     2
```

Again, note the trade-off. This method is fast because it's dangerous. If you give it bad inputs, you'll get a corrupt data frame:

```
quickdf(list(x = 1, y = 1:2))
#> Warning in format.data.frame(if (omit) x[seq_len(n0), , drop = FALSE]
#> else x, : corrupt data frame: columns will be truncated or padded
#> with NAs
#>   x y
#> 1 1 1
```

To come up with this minimal method, I carefully read through and then rewrote the source code for as.data.frame.list() and data.frame(). I made many small changes, each time checking that I hadn't broken existing behaviour. After several hours work, I was able to isolate the minimal code shown above. This is a very useful technique. Most base R functions are written for flexibility and functionality, not performance. Thus, rewriting for your specific need can often yield substantial improvements. To do this, you'll need to read the source code. It can be complex and confusing, but don't give up!

24.4.3 Exercises

1. What's the difference between rowSums() and .rowSums()?

2. Make a faster version of chisq.test() that only computes the chi-square test statistic when the input is two numeric vectors with no missing values. You can try simplifying chisq.test() or by coding from the mathematical definition (http://en.wikipedia.org/wiki/Pearson%27s_chi-squared_test).

3. Can you make a faster version of table() for the case of an input of two integer vectors with no missing values? Can you use it to speed up your chi-square test?

24.5 Vectorise

If you've used R for any length of time, you've probably heard the admonishment to "vectorise your code". But what does that actually mean? Vectorising your code is not just about avoiding for loops, although that's often a step. Vectorising is about taking a whole-object approach to a problem, thinking about vectors, not scalars. There are two key attributes of a vectorised function:

- It makes many problems simpler. Instead of having to think about the components of a vector, you only think about entire vectors.

- The loops in a vectorised function are written in C instead of R. Loops in C are much faster because they have much less overhead.

Chapter 9 stressed the importance of vectorised code as a higher level abstraction. Vectorisation is also important for writing fast R code. This doesn't mean simply using map() or lapply(). Instead, vectorisation means finding the existing R function that is implemented in C and most closely applies to your problem.

Vectorised functions that apply to many common performance bottlenecks include:

- rowSums(), colSums(), rowMeans(), and colMeans(). These vectorised matrix functions will always be faster than using apply(). You can sometimes use these functions to build other vectorised functions.

```
rowAny <- function(x) rowSums(x) > 0
rowAll <- function(x) rowSums(x) == ncol(x)
```

- Vectorised subsetting can lead to big improvements in speed. Remember the techniques behind lookup tables (Section 4.5.1) and matching and merging by hand (Section 4.5.2). Also remember that you can use subsetting assignment to replace multiple values in a single step. If x is a vector, matrix or data frame then x[is.na(x)] <- 0 will replace all missing values with 0.

- If you're extracting or replacing values in scattered locations in a matrix or data frame, subset with an integer matrix. See Section 4.2.3 for more details.

- If you're converting continuous values to categorical make sure you know how to use cut() and findInterval().

- Be aware of vectorised functions like cumsum() and diff().

Matrix algebra is a general example of vectorisation. There loops are executed by highly tuned external libraries like BLAS. If you can figure out a way to use matrix algebra to solve your problem, you'll often get a very fast solution.

The ability to solve problems with matrix algebra is a product of experience. A good place to start is to ask people with experience in your domain.

Vectorisation has a downside: it is harder to predict how operations will scale. The following example measures how long it takes to use character subsetting to look up 1, 10, and 100 elements from a list. You might expect that looking up 10 elements would take 10 times as long as looking up 1, and that looking up 100 elements would take 10 times longer again. In fact, the following example shows that it only takes about ~10x longer to look up 100 elements than it does to look up 1. That happens because once you get to a certain size, the internal implementation switches to a strategy that has a higher set up cost, but scales better.

```
lookup <- setNames(as.list(sample(100, 26)), letters)

x1 <- "j"
x10 <- sample(letters, 10)
x100 <- sample(letters, 100, replace = TRUE)

bench::mark(
  lookup[x1],
  lookup[x10],
  lookup[x100],
  check = FALSE
)[c("expression", "min", "median", "itr/sec", "n_gc")]
#> # A tibble: 3 x 5
#>   expression       min    median `itr/sec` n_gc
#>   <chr>        <bch:tm>  <bch:tm>     <dbl> <dbl>
#> 1 lookup[x1]      508ns     545ns  1571545.     1
#> 2 lookup[x10]    1.55µs    1.64µs   527835.     0
#> 3 lookup[x100]   4.93µs    7.53µs   127306.     0
```

Vectorisation won't solve every problem, and rather than torturing an existing algorithm into one that uses a vectorised approach, you're often better off writing your own vectorised function in C++. You'll learn how to do so in Chapter 25.

24.5.1 Exercises

1. The density functions, e.g., `dnorm()`, have a common interface. Which arguments are vectorised over? What does `rnorm(10, mean = 10:1)` do?

2. Compare the speed of `apply(x, 1, sum)` with `rowSums(x)` for varying sizes of x.

3. How can you use `crossprod()` to compute a weighted sum? How much faster is it than the naive `sum(x * w)`?

24.6 Avoiding copies

A pernicious source of slow R code is growing an object with a loop. Whenever you use `c()`, `append()`, `cbind()`, `rbind()`, or `paste()` to create a bigger object, R must first allocate space for the new object and then copy the old object to its new home. If you're repeating this many times, like in a for loop, this can be quite expensive. You've entered Circle 2 of the *R inferno* (http://www.burns-stat.com/pages/Tutor/R_inferno.pdf).

You saw one example of this type of problem in Section 23.2.2, so here I'll show a slightly more complex example of the same basic issue. We first generate some random strings, and then combine them either iteratively with a loop using `collapse()`, or in a single pass using `paste()`. Note that the performance of `collapse()` gets relatively worse as the number of strings grows: combining 100 strings takes almost 30 times longer than combining 10 strings.

```r
random_string <- function() {
  paste(sample(letters, 50, replace = TRUE), collapse = "")
}
strings10 <- replicate(10, random_string())
strings100 <- replicate(100, random_string())

collapse <- function(xs) {
  out <- ""
  for (x in xs) {
    out <- paste0(out, x)
  }
  out
}

bench::mark(
  loop10  = collapse(strings10),
  loop100 = collapse(strings100),
  vec10   = paste(strings10, collapse = ""),
  vec100  = paste(strings100, collapse = ""),
  check = FALSE
)[c("expression", "min", "median", "itr/sec", "n_gc")]
#> # A tibble: 4 x 5
```

```
#>    expression        min    median `itr/sec`  n_gc
#>    <chr>        <bch:tm> <bch:tm>      <dbl> <dbl>
#> 1 loop10         16.87µs   19.8µs     49742.     3
#> 2 loop100       578.8µs   618.2µs      1575.     3
#> 3 vec10          4.53µs      5µs     196192.     0
#> 4 vec100        31.65µs    32.8µs     29970.     0
```

Modifying an object in a loop, e.g., x[i] <- y, can also create a copy, depending on the class of x. Section 2.5.1 discusses this issue in more depth and gives you some tools to determine when you're making copies.

24.7 Case study: t-test

The following case study shows how to make t-tests faster using some of the techniques described above. It's based on an example in *Computing thousands of test statistics simultaneously in R* (http://stat-computing.org/newsletter/ issues/scgn-18-1.pdf) by Holger Schwender and Tina Müller. I thoroughly recommend reading the paper in full to see the same idea applied to other tests.

Imagine we have run 1000 experiments (rows), each of which collects data on 50 individuals (columns). The first 25 individuals in each experiment are assigned to group 1 and the rest to group 2. We'll first generate some random data to represent this problem:

```
m <- 1000
n <- 50
X <- matrix(rnorm(m * n, mean = 10, sd = 3), nrow = m)
grp <- rep(1:2, each = n / 2)
```

For data in this form, there are two ways to use t.test(). We can either use the formula interface or provide two vectors, one for each group. Timing reveals that the formula interface is considerably slower.

```
system.time(
  for (i in 1:m) {
    t.test(X[i, ] ~ grp)$statistic
  }
)
#>    user  system elapsed
```

```
#>    0.707   0.003   0.712
system.time(
  for (i in 1:m) {
    t.test(X[i, grp == 1], X[i, grp == 2])$statistic
  }
)
#>    user  system elapsed
#>   0.186   0.002   0.189
```

Of course, a for loop computes, but doesn't save the values. We can map_dbl()
(Section 9.2.1) to do that. This adds a little overhead:

```
compT <- function(i){
  t.test(X[i, grp == 1], X[i, grp == 2])$statistic
}
system.time(t1 <- purrr::map_dbl(1:m, compT))
#>    user  system elapsed
#>   0.186   0.001   0.187
```

How can we make this faster? First, we could try doing less work. If you look at
the source code of stats:::t.test.default(), you'll see that it does a lot more
than just compute the t-statistic. It also computes the p-value and formats
the output for printing. We can try to make our code faster by stripping out
those pieces.

```
my_t <- function(x, grp) {
  t_stat <- function(x) {
    m <- mean(x)
    n <- length(x)
    var <- sum((x - m) ^ 2) / (n - 1)

    list(m = m, n = n, var = var)
  }

  g1 <- t_stat(x[grp == 1])
  g2 <- t_stat(x[grp == 2])

  se_total <- sqrt(g1$var / g1$n + g2$var / g2$n)
  (g1$m - g2$m) / se_total
}

system.time(t2 <- purrr::map_dbl(1:m, ~ my_t(X[.,], grp)))
#>    user  system elapsed
```

```
#>    0.028    0.000    0.028
stopifnot(all.equal(t1, t2))
```

This gives us about a six-fold speed improvement.

Now that we have a fairly simple function, we can make it faster still by vectorising it. Instead of looping over the array outside the function, we will modify t_stat() to work with a matrix of values. Thus, mean() becomes rowMeans(), length() becomes ncol(), and sum() becomes rowSums(). The rest of the code stays the same.

```
rowtstat <- function(X, grp){
  t_stat <- function(X) {
    m <- rowMeans(X)
    n <- ncol(X)
    var <- rowSums((X - m) ^ 2) / (n - 1)

    list(m = m, n = n, var = var)
  }

  g1 <- t_stat(X[, grp == 1])
  g2 <- t_stat(X[, grp == 2])

  se_total <- sqrt(g1$var / g1$n + g2$var / g2$n)
  (g1$m - g2$m) / se_total
}
system.time(t3 <- rowtstat(X, grp))
#>     user  system elapsed
#>    0.011   0.000    0.011
stopifnot(all.equal(t1, t3))
```

That's much faster! It's at least 40 times faster than our previous effort, and around 1000 times faster than where we started.

24.8   Other techniques

Being able to write fast R code is part of being a good R programmer. Beyond the specific hints in this chapter, if you want to write fast R code, you'll need to improve your general programming skills. Some ways to do this are to:

- Read R blogs (`http://www.r-bloggers.com/`) to see what performance problems other people have struggled with, and how they have made their code faster.

- Read other R programming books, like *The Art of R Programming* [Matloff, 2011] or Patrick Burns' *R Inferno* (`http://www.burns-stat.com/documents/books/the-r-inferno/`) to learn about common traps.

- Take an algorithms and data structure course to learn some well known ways of tackling certain classes of problems. I have heard good things about Princeton's Algorithms course (`https://www.coursera.org/course/algs4partI`) offered on Coursera.

- Learn how to parallelise your code. Two places to start are *Parallel R* [McCallum and Weston, 2011] and *Parallel Computing for Data Science* [Matloff, 2015].

- Read general books about optimisation like *Mature optimisation* [Bueno, 2013] or the *Pragmatic Programmer* [Hunt and Thomas, 1990].

You can also reach out to the community for help. StackOverflow can be a useful resource. You'll need to put some effort into creating an easily digestible example that also captures the salient features of your problem. If your example is too complex, few people will have the time and motivation to attempt a solution. If it's too simple, you'll get answers that solve the toy problem but not the real problem. If you also try to answer questions on StackOverflow, you'll quickly get a feel for what makes a good question.

25

Rewriting R code in C++

25.1 Introduction

Sometimes R code just isn't fast enough. You've used profiling to figure out where your bottlenecks are, and you've done everything you can in R, but your code still isn't fast enough. In this chapter you'll learn how to improve performance by rewriting key functions in C++. This magic comes by way of the Rcpp (http://www.rcpp.org/) package [Eddelbuettel and François, 2011] (with key contributions by Doug Bates, John Chambers, and JJ Allaire).

Rcpp makes it very simple to connect C++ to R. While it is *possible* to write C or Fortran code for use in R, it will be painful by comparison. Rcpp provides a clean, approachable API that lets you write high-performance code, insulated from R's complex C API.

Typical bottlenecks that C++ can address include:

- Loops that can't be easily vectorised because subsequent iterations depend on previous ones.

- Recursive functions, or problems which involve calling functions millions of times. The overhead of calling a function in C++ is much lower than in R.

- Problems that require advanced data structures and algorithms that R doesn't provide. Through the standard template library (STL), C++ has efficient implementations of many important data structures, from ordered maps to double-ended queues.

The aim of this chapter is to discuss only those aspects of C++ and Rcpp that are absolutely necessary to help you eliminate bottlenecks in your code. We won't spend much time on advanced features like object-oriented programming or templates because the focus is on writing small, self-contained functions, not big programs. A working knowledge of C++ is helpful, but not essential. Many good tutorials and references are freely available, including http://www.learncpp.com/ and https://en.cppreference.com/w/cpp. For more advanced topics, the *Effective C++* series by Scott Meyers is a popular choice.

Outline

- Section 25.2 teaches you how to write C++ by converting simple R functions to their C++ equivalents. You'll learn how C++ differs from R, and what the key scalar, vector, and matrix classes are called.

- Section 25.2.5 shows you how to use `sourceCpp()` to load a C++ file from disk in the same way you use `source()` to load a file of R code.

- Section 25.3 discusses how to modify attributes from Rcpp, and mentions some of the other important classes.

- Section 25.4 teaches you how to work with R's missing values in C++.

- Section 25.5 shows you how to use some of the most important data structures and algorithms from the standard template library, or STL, built-in to C++.

- Section 25.6 shows two real case studies where Rcpp was used to get considerable performance improvements.

- Section 25.7 teaches you how to add C++ code to a package.

- Section 25.8 concludes the chapter with pointers to more resources to help you learn Rcpp and C++.

Prerequisites

We'll use Rcpp (`http://www.rcpp.org/`) to call C++ from R:

```
library(Rcpp)
```

You'll also need a working C++ compiler. To get it:

- On Windows, install Rtools (`http://cran.r-project.org/bin/windows/Rtools/`).
- On Mac, install Xcode from the app store.
- On Linux, `sudo apt-get install r-base-dev` or similar.

25.2 Getting started with C++

`cppFunction()` allows you to write C++ functions in R:

```
cppFunction('int add(int x, int y, int z) {
  int sum = x + y + z;
  return sum;
}')
# add works like a regular R function
add
#> function (x, y, z)
#> .Call(<pointer: 0x109eeac90>, x, y, z)
add(1, 2, 3)
#> [1] 6
```

When you run this code, Rcpp will compile the C++ code and construct an R function that connects to the compiled C++ function. There's a lot going on underneath the hood but Rcpp takes care of all the details so you don't need to worry about them.

The following sections will teach you the basics by translating simple R functions to their C++ equivalents. We'll start simple with a function that has no inputs and a scalar output, and then make it progressively more complicated:

- Scalar input and scalar output
- Vector input and scalar output
- Vector input and vector output
- Matrix input and vector output

25.2.1 No inputs, scalar output

Let's start with a very simple function. It has no arguments and always returns the integer 1:

```
one <- function() 1L
```

The equivalent C++ function is:

```
int one() {
  return 1;
}
```

We can compile and use this from R with `cppFunction()`

```
cppFunction('int one() {
  return 1;
}')
```

This small function illustrates a number of important differences between R and C++:

- The syntax to create a function looks like the syntax to call a function; you don't use assignment to create functions as you do in R.

- You must declare the type of output the function returns. This function returns an `int` (a scalar integer). The classes for the most common types of R vectors are: `NumericVector`, `IntegerVector`, `CharacterVector`, and `LogicalVector`.

- Scalars and vectors are different. The scalar equivalents of numeric, integer, character, and logical vectors are: `double`, `int`, `String`, and `bool`.

- You must use an explicit `return` statement to return a value from a function.

- Every statement is terminated by a `;`.

25.2.2 Scalar input, scalar output

The next example function implements a scalar version of the `sign()` function which returns 1 if the input is positive, and -1 if it's negative:

```
signR <- function(x) {
  if (x > 0) {
    1
  } else if (x == 0) {
    0
  } else {
    -1
  }
}

cppFunction('int signC(int x) {
  if (x > 0) {
    return 1;
  } else if (x == 0) {
    return 0;
  } else {
    return -1;
  }
}')
```

In the C++ version:

- We declare the type of each input in the same way we declare the type of the output. While this makes the code a little more verbose, it also makes clear the type of input the function needs.

- The `if` syntax is identical — while there are some big differences between R and C++, there are also lots of similarities! C++ also has a `while` statement that works the same way as R's. As in R you can use `break` to exit the loop, but to skip one iteration you need to use `continue` instead of `next`.

25.2.3 Vector input, scalar output

One big difference between R and C++ is that the cost of loops is much lower in C++. For example, we could implement the `sum` function in R using a loop. If you've been programming in R a while, you'll probably have a visceral reaction to this function!

```r
sumR <- function(x) {
  total <- 0
  for (i in seq_along(x)) {
    total <- total + x[i]
  }
  total
}
```

In C++, loops have very little overhead, so it's fine to use them. In Section 25.5, you'll see alternatives to `for` loops that more clearly express your intent; they're not faster, but they can make your code easier to understand.

```cpp
cppFunction('double sumC(NumericVector x) {
  int n = x.size();
  double total = 0;
  for(int i = 0; i < n; ++i) {
    total += x[i];
  }
  return total;
}')
```

The C++ version is similar, but:

- To find the length of the vector, we use the `.size()` method, which returns an integer. C++ methods are called with `.` (i.e., a full stop).

- The `for` statement has a different syntax: `for(init; check; increment)`. This loop is initialised by creating a new variable called `i` with value 0. Before each iteration we check that `i < n`, and terminate the loop if it's not.

After each iteration, we increment the value of i by one, using the special prefix operator ++ which increases the value of i by 1.

- In C++, vector indices start at 0, which means that the last element is at position n - 1. I'll say this again because it's so important: **IN C++, VECTOR INDICES START AT 0**! This is a very common source of bugs when converting R functions to C++.

- Use = for assignment, not <-.

- C++ provides operators that modify in-place: total += x[i] is equivalent to total = total + x[i]. Similar in-place operators are -=, *=, and /=.

This is a good example of where C++ is much more efficient than R. As shown by the following microbenchmark, sumC() is competitive with the built-in (and highly optimised) sum(), while sumR() is several orders of magnitude slower.

```
x <- runif(1e3)
bench::mark(
  sum(x),
  sumC(x),
  sumR(x)
)[1:6]
#> # A tibble: 3 x 6
#>   expression       min     mean   median       max `itr/sec`
#>   <chr>        <bch:tm> <bch:tm> <bch:tm> <bch:tm>      <dbl>
#> 1 sum(x)         1.13µs   1.21µs   1.17µs    17.2µs   823472.
#> 2 sumC(x)        2.52µs   4.53µs   4.99µs   775.4µs   220921.
#> 3 sumR(x)       42.41µs  46.03µs   43.2µs   137.9µs    21723.
```

25.2.4   Vector input, vector output

Next we'll create a function that computes the Euclidean distance between a value and a vector of values:

```
pdistR <- function(x, ys) {
  sqrt((x - ys) ^ 2)
}
```

In R, it's not obvious that we want x to be a scalar from the function definition, and we'd need to make that clear in the documentation. That's not a problem in the C++ version because we have to be explicit about types:

```
cppFunction('NumericVector pdistC(double x, NumericVector ys) {
  int n = ys.size();
  NumericVector out(n);

  for(int i = 0; i < n; ++i) {
    out[i] = sqrt(pow(ys[i] - x, 2.0));
  }
  return out;
}')
```

This function introduces only a few new concepts:

- We create a new numeric vector of length n with a constructor: `NumericVector out(n)`. Another useful way of making a vector is to copy an existing one: `NumericVector zs = clone(ys)`.

- C++ uses `pow()`, not `^`, for exponentiation.

Note that because the R version is fully vectorised, it's already going to be fast.

```
y <- runif(1e6)
bench::mark(
  pdistR(0.5, y),
  pdistC(0.5, y)
)[1:6]
#> # A tibble: 2 x 6
#>   expression           min     mean   median      max `itr/sec`
#>   <chr>            <bch:tm> <bch:tm> <bch:tm> <bch:tm>     <dbl>
#> 1 pdistR(0.5, y)    5.21ms   5.57ms   5.24ms  10.89ms      180.
#> 2 pdistC(0.5, y)    2.31ms   2.48ms   2.39ms   3.39ms      404.
```

On my computer, it takes around 5 ms with a 1 million element y vector. The C++ function is about 2.5 times faster, ~2 ms, but assuming it took you 10 minutes to write the C++ function, you'd need to run it ~200,000 times to make rewriting worthwhile. The reason why the C++ function is faster is subtle, and relates to memory management. The R version needs to create an intermediate vector the same length as y (x - ys), and allocating memory is an expensive operation. The C++ function avoids this overhead because it uses an intermediate scalar.

25.2.5 Using sourceCpp

So far, we've used inline C++ with `cppFunction()`. This makes presentation simpler, but for real problems, it's usually easier to use stand-alone C++ files and then source them into R using `sourceCpp()`. This lets you take advantage of text editor support for C++ files (e.g., syntax highlighting) as well as making it easier to identify the line numbers in compilation errors.

Your stand-alone C++ file should have extension .cpp, and needs to start with:

```
#include <Rcpp.h>
using namespace Rcpp;
```

And for each function that you want available within R, you need to prefix it with:

```
// [[Rcpp::export]]
```

If you're familiar with roxygen2, you might wonder how this relates to @export. Rcpp::export controls whether a function is exported from C++ to R; @export controls whether a function is exported from a package and made available to the user.

You can embed R code in special C++ comment blocks. This is really convenient if you want to run some test code:

```
/*** R
# This is R code
*/
```

The R code is run with `source(echo = TRUE)` so you don't need to explicitly print output.

To compile the C++ code, use `sourceCpp("path/to/file.cpp")`. This will create the matching R functions and add them to your current session. Note that these functions can not be saved in a .Rdata file and reloaded in a later session; they must be recreated each time you restart R.

For example, running `sourceCpp()` on the following file implements mean in C++ and then compares it to the built-in `mean()`:

```
#include <Rcpp.h>
using namespace Rcpp;

// [[Rcpp::export]]
double meanC(NumericVector x) {
  int n = x.size();
```

```
  double total = 0;

  for(int i = 0; i < n; ++i) {
    total += x[i];
  }
  return total / n;
}

/*** R
x <- runif(1e5)
bench::mark(
  mean(x),
  meanC(x)
)
*/
```

NB: if you run this code, you'll notice that meanC() is much faster than the built-in mean(). This is because it trades numerical accuracy for speed.

For the remainder of this chapter C++ code will be presented stand-alone rather than wrapped in a call to cppFunction. If you want to try compiling and/or modifying the examples you should paste them into a C++ source file that includes the elements described above. This is easy to do in RMarkdown: all you need to do is specify engine = "Rcpp".

25.2.6   Exercises

1. With the basics of C++ in hand, it's now a great time to practice by reading and writing some simple C++ functions. For each of the following functions, read the code and figure out what the corresponding base R function is. You might not understand every part of the code yet, but you should be able to figure out the basics of what the function does.

    ```
    double f1(NumericVector x) {
      int n = x.size();
      double y = 0;

      for(int i = 0; i < n; ++i) {
        y += x[i] / n;
      }
      return y;
    }
    ```

```
NumericVector f2(NumericVector x) {
  int n = x.size();
  NumericVector out(n);

  out[0] = x[0];
  for(int i = 1; i < n; ++i) {
    out[i] = out[i - 1] + x[i];
  }
  return out;
}


bool f3(LogicalVector x) {
  int n = x.size();

  for(int i = 0; i < n; ++i) {
    if (x[i]) return true;
  }
  return false;
}

int f4(Function pred, List x) {
  int n = x.size();

  for(int i = 0; i < n; ++i) {
    LogicalVector res = pred(x[i]);
    if (res[0]) return i + 1;
  }
  return 0;
}

NumericVector f5(NumericVector x, NumericVector y) {
  int n = std::max(x.size(), y.size());
  NumericVector x1 = rep_len(x, n);
  NumericVector y1 = rep_len(y, n);

  NumericVector out(n);

  for (int i = 0; i < n; ++i) {
    out[i] = std::min(x1[i], y1[i]);
  }

  return out;
}
```

2. To practice your function writing skills, convert the following functions into C++. For now, assume the inputs have no missing values.

   1. `all()`.

   2. `cumprod()`, `cummin()`, `cummax()`.

   3. `diff()`. Start by assuming lag 1, and then generalise for lag n.

   4. `range()`.

   5. `var()`. Read about the approaches you can take on Wikipedia (`http://en.wikipedia.org/wiki/Algorithms_for_calculating_variance`). Whenever implementing a numerical algorithm, it's always good to check what is already known about the problem.

25.3   Other classes

You've already seen the basic vector classes (`IntegerVector`, `NumericVector`, `LogicalVector`, `CharacterVector`) and their scalar (`int`, `double`, `bool`, `String`) equivalents. Rcpp also provides wrappers for all other base data types. The most important are for lists and data frames, functions, and attributes, as described below. Rcpp also provides classes for more types like `Environment`, `DottedPair`, `Language`, `Symbol`, etc, but these are beyond the scope of this chapter.

25.3.1   Lists and data frames

Rcpp also provides `List` and `DataFrame` classes, but they are more useful for output than input. This is because lists and data frames can contain arbitrary classes but C++ needs to know their classes in advance. If the list has known structure (e.g., it's an S3 object), you can extract the components and manually convert them to their C++ equivalents with `as()`. For example, the object created by `lm()`, the function that fits a linear model, is a list whose components are always of the same type. The following code illustrates how you might extract the mean percentage error (`mpe()`) of a linear model. This isn't a good example of when to use C++, because it's so easily implemented in R, but it shows how to work with an important S3 class. Note the use of `.inherits()` and the `stop()` to check that the object really is a linear model.

```cpp
#include <Rcpp.h>
using namespace Rcpp;

// [[Rcpp::export]]
double mpe(List mod) {
  if (!mod.inherits("lm")) stop("Input must be a linear model");

  NumericVector resid = as<NumericVector>(mod["residuals"]);
  NumericVector fitted = as<NumericVector>(mod["fitted.values"]);

  int n = resid.size();
  double err = 0;
  for(int i = 0; i < n; ++i) {
    err += resid[i] / (fitted[i] + resid[i]);
  }
  return err / n;
}
```

```r
mod <- lm(mpg ~ wt, data = mtcars)
mpe(mod)
#> [1] -0.0154
```

25.3.2 Functions

You can put R functions in an object of type Function. This makes calling an R function from C++ straightforward. The only challenge is that we don't know what type of output the function will return, so we use the catchall type RObject.

```cpp
#include <Rcpp.h>
using namespace Rcpp;

// [[Rcpp::export]]
RObject callWithOne(Function f) {
  return f(1);
}
```

```r
callWithOne(function(x) x + 1)
#> [1] 2
```

```
callWithOne(paste)
#> [1] "1"
```

Calling R functions with positional arguments is obvious:

```
f("y", 1);
```

But you need a special syntax for named arguments:

```
f(_["x"] = "y", _["value"] = 1);
```

25.3.3 Attributes

All R objects have attributes, which can be queried and modified with .attr().
Rcpp also provides .names() as an alias for the name attribute. The following
code snippet illustrates these methods. Note the use of ::create(), a *class*
method. This allows you to create an R vector from C++ scalar values:

```
#include <Rcpp.h>
using namespace Rcpp;

// [[Rcpp::export]]
NumericVector attribs() {
  NumericVector out = NumericVector::create(1, 2, 3);

  out.names() = CharacterVector::create("a", "b", "c");
  out.attr("my-attr") = "my-value";
  out.attr("class") = "my-class";

  return out;
}
```

For S4 objects, .slot() plays a similar role to .attr().

25.4 Missing values

If you're working with missing values, you need to know two things:

- How R's missing values behave in C++'s scalars (e.g., double).
- How to get and set missing values in vectors (e.g., NumericVector).

25.4.1 Scalars

The following code explores what happens when you take one of R's missing values, coerce it into a scalar, and then coerce back to an R vector. Note that this kind of experimentation is a useful way to figure out what any operation does.

```
#include <Rcpp.h>
using namespace Rcpp;

// [[Rcpp::export]]
List scalar_missings() {
  int int_s = NA_INTEGER;
  String chr_s = NA_STRING;
  bool lgl_s = NA_LOGICAL;
  double num_s = NA_REAL;

  return List::create(int_s, chr_s, lgl_s, num_s);
}
```

```
str(scalar_missings())
#> List of 4
#>  $ : int NA
#>  $ : chr NA
#>  $ : logi TRUE
#>  $ : num NA
```

With the exception of bool, things look pretty good here: all of the missing values have been preserved. However, as we'll see in the following sections, things are not quite as straightforward as they seem.

25.4.1.1 Integers

With integers, missing values are stored as the smallest integer. If you don't do anything to them, they'll be preserved. But, since C++ doesn't know that the smallest integer has this special behaviour, if you do anything to it you're likely to get an incorrect value: for example, evalCpp('NA_INTEGER + 1') gives -2147483647.

So if you want to work with missing values in integers, either use a length 1 `IntegerVector` or be very careful with your code.

25.4.1.2 Doubles

With doubles, you may be able to get away with ignoring missing values and working with NaNs (not a number). This is because R's NA is a special type of IEEE 754 floating point number NaN. So any logical expression that involves a NaN (or in C++, NAN) always evaluates as FALSE:

```
evalCpp("NAN == 1")
#> [1] FALSE
evalCpp("NAN < 1")
#> [1] FALSE
evalCpp("NAN > 1")
#> [1] FALSE
evalCpp("NAN == NAN")
#> [1] FALSE
```

(Here I'm using `evalCpp()` which allows you to see the result of running a single C++ expression, making it excellent for this sort of interactive experimentation.)

But be careful when combining them with Boolean values:

```
evalCpp("NAN && TRUE")
#> [1] TRUE
evalCpp("NAN || FALSE")
#> [1] TRUE
```

However, in numeric contexts NaNs will propagate NAs:

```
evalCpp("NAN + 1")
#> [1] NaN
evalCpp("NAN - 1")
#> [1] NaN
evalCpp("NAN / 1")
#> [1] NaN
evalCpp("NAN * 1")
#> [1] NaN
```

25.4.2 Strings

String is a scalar string class introduced by Rcpp, so it knows how to deal with missing values.

25.4.3 Boolean

While C++'s bool has two possible values (true or false), a logical vector in R has three (TRUE, FALSE, and NA). If you coerce a length 1 logical vector, make sure it doesn't contain any missing values; otherwise they will be converted to TRUE. An easy fix is to use int instead, as this can represent TRUE, FALSE, and NA.

25.4.4 Vectors

With vectors, you need to use a missing value specific to the type of vector, NA_REAL, NA_INTEGER, NA_LOGICAL, NA_STRING:

```cpp
#include <Rcpp.h>
using namespace Rcpp;

// [[Rcpp::export]]
List missing_sampler() {
  return List::create(
    NumericVector::create(NA_REAL),
    IntegerVector::create(NA_INTEGER),
    LogicalVector::create(NA_LOGICAL),
    CharacterVector::create(NA_STRING)
  );
}
```

```r
str(missing_sampler())
#> List of 4
#>  $ : num NA
#>  $ : int NA
#>  $ : logi NA
#>  $ : chr NA
```

25.4.5 Exercises

1. Rewrite any of the functions from the first exercise of Section 25.2.6 to deal with missing values. If na.rm is true, ignore the missing values. If na.rm is false, return a missing value if the input contains any missing values. Some good functions to practice with are min(), max(), range(), mean(), and var().

2. Rewrite cumsum() and diff() so they can handle missing values. Note that these functions have slightly more complicated behaviour.

25.5 Standard Template Library

The real strength of C++ is revealed when you need to implement more complex algorithms. The standard template library (STL) provides a set of extremely useful data structures and algorithms. This section will explain some of the most important algorithms and data structures and point you in the right direction to learn more. I can't teach you everything you need to know about the STL, but hopefully the examples will show you the power of the STL, and persuade you that it's useful to learn more.

If you need an algorithm or data structure that isn't implemented in STL, a good place to look is boost (http://www.boost.org/doc/). Installing boost on your computer is beyond the scope of this chapter, but once you have it installed, you can use boost data structures and algorithms by including the appropriate header file with (e.g.) #include <boost/array.hpp>.

25.5.1 Using iterators

Iterators are used extensively in the STL: many functions either accept or return iterators. They are the next step up from basic loops, abstracting away the details of the underlying data structure. Iterators have three main operators:

1. Advance with ++.
2. Get the value they refer to, or **dereference**, with *.
3. Compare with ==.

For example we could re-write our sum function using iterators:

```
#include <Rcpp.h>
using namespace Rcpp;

// [[Rcpp::export]]
double sum3(NumericVector x) {
  double total = 0;

  NumericVector::iterator it;
  for(it = x.begin(); it != x.end(); ++it) {
    total += *it;
  }
  return total;
}
```

The main changes are in the for loop:

- We start at x.begin() and loop until we get to x.end(). A small optimization is to store the value of the end iterator so we don't need to look it up each time. This only saves about 2 ns per iteration, so it's only important when the calculations in the loop are very simple.

- Instead of indexing into x, we use the dereference operator to get its current value: *it.

- Notice the type of the iterator: NumericVector::iterator. Each vector type has its own iterator type: LogicalVector::iterator, CharacterVector::iterator, etc.

This code can be simplified still further through the use of a C++11 feature: range-based for loops. C++11 is widely available, and can easily be activated for use with Rcpp by adding [[Rcpp::plugins(cpp11)]].

```
// [[Rcpp::plugins(cpp11)]]
#include <Rcpp.h>
using namespace Rcpp;

// [[Rcpp::export]]
double sum4(NumericVector xs) {
  double total = 0;

  for(const auto &x : xs) {
    total += x;
  }
  return total;
}
```

Iterators also allow us to use the C++ equivalents of the apply family of functions. For example, we could again rewrite sum() to use the accumulate() function, which takes a starting and an ending iterator, and adds up all the values in the vector. The third argument to accumulate gives the initial value: it's particularly important because this also determines the data type that accumulate uses (so we use 0.0 and not 0 so that accumulate uses a double, not an int.). To use accumulate() we need to include the <numeric> header.

```
#include <numeric>
#include <Rcpp.h>
using namespace Rcpp;

// [[Rcpp::export]]
double sum5(NumericVector x) {
  return std::accumulate(x.begin(), x.end(), 0.0);
}
```

25.5.2 Algorithms

The <algorithm> header provides a large number of algorithms that work with iterators. A good reference is available at https://en.cppreference.com/w/cpp/algorithm. For example, we could write a basic Rcpp version of findInterval() that takes two arguments a vector of values and a vector of breaks, and locates the bin that each x falls into. This shows off a few more advanced iterator features. Read the code below and see if you can figure out how it works.

```
#include <algorithm>
#include <Rcpp.h>
using namespace Rcpp;

// [[Rcpp::export]]
IntegerVector findInterval2(NumericVector x, NumericVector breaks) {
  IntegerVector out(x.size());

  NumericVector::iterator it, pos;
  IntegerVector::iterator out_it;

  for(it = x.begin(), out_it = out.begin(); it != x.end();
      ++it, ++out_it) {
    pos = std::upper_bound(breaks.begin(), breaks.end(), *it);
    *out_it = std::distance(breaks.begin(), pos);
  }
```

```
  return out;
}
```

The key points are:

- We step through two iterators (input and output) simultaneously.

- We can assign into an dereferenced iterator (`out_it`) to change the values in out.

- `upper_bound()` returns an iterator. If we wanted the value of the `upper_bound()` we could dereference it; to figure out its location, we use the `distance()` function.

- Small note: if we want this function to be as fast as `findInterval()` in R (which uses handwritten C code), we need to compute the calls to `.begin()` and `.end()` once and save the results. This is easy, but it distracts from this example so it has been omitted. Making this change yields a function that's slightly faster than R's `findInterval()` function, but is about 1/10 of the code.

It's generally better to use algorithms from the STL than hand rolled loops. In *Effective STL*, Scott Meyers gives three reasons: efficiency, correctness, and maintainability. Algorithms from the STL are written by C++ experts to be extremely efficient, and they have been around for a long time so they are well tested. Using standard algorithms also makes the intent of your code more clear, helping to make it more readable and more maintainable.

25.5.3 Data structures

The STL provides a large set of data structures: `array`, `bitset`, `list`, `forward_list`, `map`, `multimap`, `multiset`, `priority_queue`, `queue`, `deque`, `set`, `stack`, `unordered_map`, `unordered_set`, `unordered_multimap`, `unordered_multiset`, and `vector`. The most important of these data structures are the `vector`, the `unordered_set`, and the `unordered_map`. We'll focus on these three in this section, but using the others is similar: they just have different performance trade-offs. For example, the deque (pronounced "deck") has a very similar interface to vectors but a different underlying implementation that has different performance trade-offs. You may want to try it for your problem. A good reference for STL data structures is `https://en.cppreference.com/w/cpp/container` — I recommend you keep it open while working with the STL.

Rcpp knows how to convert from many STL data structures to their R equivalents, so you can return them from your functions without explicitly converting to R data structures.

25.5.4 Vectors

An STL vector is very similar to an R vector, except that it grows efficiently. This makes vectors appropriate to use when you don't know in advance how big the output will be. Vectors are templated, which means that you need to specify the type of object the vector will contain when you create it: vector<int>, vector<bool>, vector<double>, vector<String>. You can access individual elements of a vector using the standard [] notation, and you can add a new element to the end of the vector using .push_back(). If you have some idea in advance how big the vector will be, you can use .reserve() to allocate sufficient storage.

The following code implements run length encoding (rle()). It produces two vectors of output: a vector of values, and a vector lengths giving how many times each element is repeated. It works by looping through the input vector x comparing each value to the previous: if it's the same, then it increments the last value in lengths; if it's different, it adds the value to the end of values, and sets the corresponding length to 1.

```
#include <Rcpp.h>
using namespace Rcpp;

// [[Rcpp::export]]
List rleC(NumericVector x) {
  std::vector<int> lengths;
  std::vector<double> values;

  // Initialise first value
  int i = 0;
  double prev = x[0];
  values.push_back(prev);
  lengths.push_back(1);

  NumericVector::iterator it;
  for(it = x.begin() + 1; it != x.end(); ++it) {
    if (prev == *it) {
      lengths[i]++;
    } else {
      values.push_back(*it);
      lengths.push_back(1);
```

```
    i++;
    prev = *it;
  }
}

return List::create(
  _["lengths"] = lengths,
  _["values"] = values
);
}
```

(An alternative implementation would be to replace i with the iterator lengths.rbegin() which always points to the last element of the vector. You might want to try implementing that.)

Other methods of a vector are described at https://en.cppreference.com/w/cpp/container/vector.

25.5.5 Sets

Sets maintain a unique set of values, and can efficiently tell if you've seen a value before. They are useful for problems that involve duplicates or unique values (like unique, duplicated, or in). C++ provides both ordered (std::set) and unordered sets (std::unordered_set), depending on whether or not order matters for you. Unordered sets tend to be much faster (because they use a hash table internally rather than a tree), so even if you need an ordered set, you should consider using an unordered set and then sorting the output. Like vectors, sets are templated, so you need to request the appropriate type of set for your purpose: unordered_set<int>, unordered_set<bool>, etc. More details are available at https://en.cppreference.com/w/cpp/container/set and https://en.cppreference.com/w/cpp/container/unordered_set.

The following function uses an unordered set to implement an equivalent to duplicated() for integer vectors. Note the use of seen.insert(x[i]).second. insert() returns a pair, the .first value is an iterator that points to element and the .second value is a Boolean that's true if the value was a new addition to the set.

```
// [[Rcpp::plugins(cpp11)]]
#include <Rcpp.h>
#include <unordered_set>
using namespace Rcpp;
```

```
// [[Rcpp::export]]
LogicalVector duplicatedC(IntegerVector x) {
  std::unordered_set<int> seen;
  int n = x.size();
  LogicalVector out(n);

  for (int i = 0; i < n; ++i) {
    out[i] = !seen.insert(x[i]).second;
  }

  return out;
}
```

25.5.6 Map

A map is similar to a set, but instead of storing presence or absence, it can store additional data. It's useful for functions like table() or match() that need to look up a value. As with sets, there are ordered (std::map) and unordered (std::unordered_map) versions. Since maps have a value and a key, you need to specify both types when initialising a map: map<double, int>, unordered_map<int, double>, and so on. The following example shows how you could use a map to implement table() for numeric vectors:

```
#include <Rcpp.h>
using namespace Rcpp;

// [[Rcpp::export]]
std::map<double, int> tableC(NumericVector x) {
  std::map<double, int> counts;

  int n = x.size();
  for (int i = 0; i < n; i++) {
    counts[x[i]]++;
  }

  return counts;
}
```

25.5.7   Exercises

To practice using the STL algorithms and data structures, implement the
following using R functions in C++, using the hints provided:

1. `median.default()` using `partial_sort`.

2. `%in%` using `unordered_set` and the `find()` or `count()` methods.

3. `unique()` using an `unordered_set` (challenge: do it in one line!).

4. `min()` using `std::min()`, or `max()` using `std::max()`.

5. `which.min()` using `min_element`, or `which.max()` using `max_element`.

6. `setdiff()`, `union()`, and `intersect()` for integers using sorted ranges
   and `set_union`, `set_intersection` and `set_difference`.

25.6   Case studies

The following case studies illustrate some real life uses of C++ to replace slow
R code.

25.6.1   Gibbs sampler

The following case study updates an example blogged about (http://dirk.
eddelbuettel.com/blog/2011/07/14/) by Dirk Eddelbuettel, illustrating the
conversion of a Gibbs sampler in R to C++. The R and C++ code shown
below is very similar (it only took a few minutes to convert the R version to
the C++ version), but runs about 20 times faster on my computer. Dirk's
blog post also shows another way to make it even faster: using the faster random number generator functions in GSL (easily accessible from R through the
RcppGSL package) can make it another two to three times faster.

The R code is as follows:

```
gibbs_r <- function(N, thin) {
  mat <- matrix(nrow = N, ncol = 2)
  x <- y <- 0

  for (i in 1:N) {
    for (j in 1:thin) {
      x <- rgamma(1, 3, y * y + 4)
```

```
      y <- rnorm(1, 1 / (x + 1), 1 / sqrt(2 * (x + 1)))
    }
    mat[i, ] <- c(x, y)
  }
  mat
}
```

This is straightforward to convert to C++. We:

- Add type declarations to all variables.

- Use (instead of [to index into the matrix.

- Subscript the results of rgamma and rnorm to convert from a vector into a scalar.

```
#include <Rcpp.h>
using namespace Rcpp;

// [[Rcpp::export]]
NumericMatrix gibbs_cpp(int N, int thin) {
  NumericMatrix mat(N, 2);
  double x = 0, y = 0;

  for(int i = 0; i < N; i++) {
    for(int j = 0; j < thin; j++) {
      x = rgamma(1, 3, 1 / (y * y + 4))[0];
      y = rnorm(1, 1 / (x + 1), 1 / sqrt(2 * (x + 1)))[0];
    }
    mat(i, 0) = x;
    mat(i, 1) = y;
  }

  return(mat);
}
```

Benchmarking the two implementations yields:

```
bench::mark(
  gibbs_r(100, 10),
  gibbs_cpp(100, 10),
  check = FALSE
)
#> # A tibble: 2 x 10
#>   expression        min    mean   median     max `itr/sec` mem_alloc
```

```
#>     <chr>        <bch:tm> <bch:tm> <bch:tm> <bch:>      <dbl> <bch:byt>
#> 1 gibbs_r(1...    4.25ms   5.53ms   4.96ms 9.38ms      181.    4.97MB
#> 2 gibbs_cpp... 223.76µs 259.25µs 255.99µs 1.17ms      3857.    4.1KB
#> # ... with 3 more variables: n_gc <dbl>, n_itr <int>,
#> #   total_time <bch:tm>
```

25.6.2   R vectorisation versus C++ vectorisation

This example is adapted from "Rcpp is smoking fast for agent-based models in data frames" (https://gweissman.github.io/babelgraph/blog/2017/06/15/rcpp-is-smoking-fast-for-agent-based-models-in-data-frames.html). The challenge is to predict a model response from three inputs. The basic R version of the predictor looks like:

```r
vacc1a <- function(age, female, ily) {
  p <- 0.25 + 0.3 * 1 / (1 - exp(0.04 * age)) + 0.1 * ily
  p <- p * if (female) 1.25 else 0.75
  p <- max(0, p)
  p <- min(1, p)
  p
}
```

We want to be able to apply this function to many inputs, so we might write a vector-input version using a for loop.

```r
vacc1 <- function(age, female, ily) {
  n <- length(age)
  out <- numeric(n)
  for (i in seq_len(n)) {
    out[i] <- vacc1a(age[i], female[i], ily[i])
  }
  out
}
```

If you're familiar with R, you'll have a gut feeling that this will be slow, and indeed it is. There are two ways we could attack this problem. If you have a good R vocabulary, you might immediately see how to vectorise the function (using ifelse(), pmin(), and pmax()). Alternatively, we could rewrite vacc1a() and vacc1() in C++, using our knowledge that loops and function calls have much lower overhead in C++.

Either approach is fairly straightforward. In R:

```
vacc2 <- function(age, female, ily) {
  p <- 0.25 + 0.3 * 1 / (1 - exp(0.04 * age)) + 0.1 * ily
  p <- p * ifelse(female, 1.25, 0.75)
  p <- pmax(0, p)
  p <- pmin(1, p)
  p
}
```

(If you've worked R a lot you might recognise some potential bottlenecks in this code: ifelse, pmin, and pmax are known to be slow, and could be replaced with p * 0.75 + p * 0.5 * female, p[p < 0] <- 0, p[p > 1] <- 1. You might want to try timing those variations.)

Or in C++:

```cpp
#include <Rcpp.h>
using namespace Rcpp;

double vacc3a(double age, bool female, bool ily){
  double p = 0.25 + 0.3 * 1 / (1 - exp(0.04 * age)) + 0.1 * ily;
  p = p * (female ? 1.25 : 0.75);
  p = std::max(p, 0.0);
  p = std::min(p, 1.0);
  return p;
}

// [[Rcpp::export]]
NumericVector vacc3(NumericVector age, LogicalVector female,
                    LogicalVector ily) {
  int n = age.size();
  NumericVector out(n);

  for(int i = 0; i < n; ++i) {
    out[i] = vacc3a(age[i], female[i], ily[i]);
  }

  return out;
}
```

We next generate some sample data, and check that all three versions return the same values:

```
n <- 1000
age <- rnorm(n, mean = 50, sd = 10)
```

```
female <- sample(c(T, F), n, rep = TRUE)
ily <- sample(c(T, F), n, prob = c(0.8, 0.2), rep = TRUE)

stopifnot(
  all.equal(vacc1(age, female, ily), vacc2(age, female, ily)),
  all.equal(vacc1(age, female, ily), vacc3(age, female, ily))
)
```

The original blog post forgot to do this, and introduced a bug in the C++
version: it used `0.004` instead of `0.04`. Finally, we can benchmark our three
approaches:

```
bench::mark(
  vacc1 = vacc1(age, female, ily),
  vacc2 = vacc2(age, female, ily),
  vacc3 = vacc3(age, female, ily)
)
#> # A tibble: 3 x 10
#>   expression     min     mean   median     max `itr/sec` mem_alloc
#>   <chr>      <bch:t> <bch:tm> <bch:tm> <bch:t>     <dbl> <bch:byt>
#> 1 vacc1       1.62ms   1.81ms   1.79ms   2.6ms      552.    7.86KB
#> 2 vacc2      84.73µs 115.71µs 100.89µs 569.1µs     8642.     224KB
#> 3 vacc3      13.38µs  15.63µs   15.6µs  75.4µs    63988.   14.48KB
#> # ... with 3 more variables: n_gc <dbl>, n_itr <int>,
#> #    total_time <bch:tm>
```

Not surprisingly, our original approach with loops is very slow. Vectorising in
R gives a huge speedup, and we can eke out even more performance (about
ten times) with the C++ loop. I was a little surprised that the C++ was so
much faster, but it is because the R version has to create 11 vectors to store
intermediate results, where the C++ code only needs to create 1.

25.7 Using Rcpp in a package

The same C++ code that is used with `sourceCpp()` can also be bundled into a
package. There are several benefits of moving code from a stand-alone C++
source file to a package:

1. Your code can be made available to users without C++ development
    tools.

2. Multiple source files and their dependencies are handled automatically by the R package build system.

3. Packages provide additional infrastructure for testing, documentation, and consistency.

To add `Rcpp` to an existing package, you put your C++ files in the `src/` directory and create or modify the following configuration files:

- In `DESCRIPTION` add

  ```
  LinkingTo: Rcpp
  Imports: Rcpp
  ```

- Make sure your `NAMESPACE` includes:

  ```
  useDynLib(mypackage)
  importFrom(Rcpp, sourceCpp)
  ```

  We need to import something (anything) from Rcpp so that internal Rcpp code is properly loaded. This is a bug in R and hopefully will be fixed in the future.

The easiest way to set this up automatically is to call `usethis::use_rcpp()`.

Before building the package, you'll need to run `Rcpp::compileAttributes()`. This function scans the C++ files for `Rcpp::export` attributes and generates the code required to make the functions available in R. Re-run `compileAttributes()` whenever functions are added, removed, or have their signatures changed. This is done automatically by the devtools package and by Rstudio.

For more details see the Rcpp package vignette, `vignette("Rcpp-package")`.

25.8 Learning more

This chapter has only touched on a small part of Rcpp, giving you the basic tools to rewrite poorly performing R code in C++. As noted, Rcpp has many other capabilities that make it easy to interface R to existing C++ code, including:

- Additional features of attributes including specifying default arguments, linking in external C++ dependencies, and exporting C++ interfaces from packages. These features and more are covered in the Rcpp attributes vignette, `vignette("Rcpp-attributes")`.

- Automatically creating wrappers between C++ data structures and R data structures, including mapping C++ classes to reference classes.

A good introduction to this topic is the Rcpp modules vignette, `vignette("Rcpp-modules")`.

- The Rcpp quick reference guide, `vignette("Rcpp-quickref")`, contains a useful summary of Rcpp classes and common programming idioms.

I strongly recommend keeping an eye on the Rcpp homepage (`http://www.rcpp.org`) and signing up for the Rcpp mailing list (`http://lists.r-forge.r-project.org/cgi-bin/mailman/listinfo/rcpp-devel`).

Other resources I've found helpful in learning C++ are:

- *Effective C++* [Meyers, 2005] and *Effective STL* [Meyers, 2001].

- *C++ Annotations* (`http://www.icce.rug.nl/documents/cplusplus/cplusplus.html`), aimed at knowledgeable users of C (or any other language using a C-like grammar, like Perl or Java) who would like to know more about, or make the transition to, C++.

- *Algorithm Libraries* (`http://www.cs.helsinki.fi/u/tpkarkka/alglib/k06/`), which provides a more technical, but still concise, description of important STL concepts. (Follow the links under notes.)

Writing performance code may also require you to rethink your basic approach: a solid understanding of basic data structures and algorithms is very helpful here. That's beyond the scope of this book, but I'd suggest the *Algorithm Design Manual* [Skiena, 1998], MIT's *Introduction to Algorithms* (`http://ocw.mit.edu/courses/electrical-engineering-and-computer-science/6-046j-introduction-to-algorithms-sma-5503-fall-2005/`), *Algorithms* by Robert Sedgewick and Kevin Wayne which has a free online textbook (`http://algs4.cs.princeton.edu/home/`) and a matching Coursera course (`https://www.coursera.org/learn/algorithms-part1`).

25.9   Acknowledgments

I'd like to thank the Rcpp-mailing list for many helpful conversations, particularly Romain Francois and Dirk Eddelbuettel who have not only provided detailed answers to many of my questions, but have been incredibly responsive at improving Rcpp. This chapter would not have been possible without JJ Allaire; he encouraged me to learn C++ and then answered many of my dumb questions along the way.

Bibliography

Harold Abelson, Gerald Jay Sussman, and Julie Sussman. *Structure and Interpretation of Computer Programs.* MIT Press, 1996.

Stefan Milton Bache and Hadley Wickham. *magrittr: A forward-pipe operator for R*, 2014. URL `http://magrittr.tidyverse.org/`.

James Balamuta. *errorist: Automatically Search Errors or Warnings*, 2018a. URL `https://github.com/coatless/errorist`.

James Balamuta. *searcher: Query Search Interfaces*, 2018b. URL `https://github.com/coatless/searcher`.

Douglas Bates and Martin Maechler. Matrix: Sparse and dense matrix classes and methods, 2018. URL `https://CRAN.R-project.org/package=Matrix`.

Alan Bawden. Quasiquotation in Lisp. In *PEPM '99*, pages 4–12, 1999. URL `http://citeseerx.ist.psu.edu/viewdoc/summary?doi=10.1.1.309.227`.

Henrik Bengtsson. The R.oo package - object-oriented programming with references using standard R code. In Kurt Hornik, Friedrich Leisch, and Achim Zeileis, editors, *Proceedings of the 3rd International Workshop on Distributed Statistical Computing (DSC 2003)*, Vienna, Austria, March 2003. URL `https://www.r-project.org/conferences/DSC-2003/Proceedings/Bengtsson.pdf`.

Christopher Brown. *hash: Full feature implementation of hash/associated arrays/dictionaries*, 2013. URL `https://CRAN.R-project.org/package=hash`.

Carlos Bueno. *Mature Optimization Handbook.* 2013. URL `http://carlos.bueno.org/optimization/`.

John M Chambers. *Programming with Data: A Guide to the S Language.* Springer, 1998.

John M Chambers. *Software for Data Analysis: Programming with R.* Springer, 2008.

John M Chambers. Object-oriented programming, functional programming and R. *Statistical Science*, 29(2):167–180, 2014. URL `https://projecteuclid.org/download/pdfview_1/euclid.ss/1408368569`.

John M Chambers. *Extending R.* CRC Press, 2016.

John M Chambers and Trevor J Hastie. *Statistical Models in S.* Wadsworth & Brooks/Cole Advanced Books & Software, 1992.

Winston Chang. *R6: Classes with reference semantics*, 2017. URL https: //r6.r-lib.org.

Dirk Eddelbuettel and Romain François. Rcpp: Seamless R and C++ integration. *Journal of Statistical Software*, 40(8):1–18, 2011. doi: 10.18637/jss. v040.i08. URL http://www.jstatsoft.org/v40/i08/.

Martin Fowler. *Domain-specific Languages.* Pearson Education, 2010. URL http://amzn.com/0321712943.

Garrett Grolemund and Hadley Wickham. Dates and times made easy with lubridate. *Journal of Statistical Software*, 40(3):1–25, 2011. URL http: //www.jstatsoft.org/v40/i03/.

Gabor Grothendieck, Louis Kates, and Thomas Petzoldt. *proto: prototype object-based programming*, 2016. URL https://CRAN.R-project.org/ package=proto.

Lionel Henry and Hadley Wickham. *purrr: functional programming tools*, 2018a. URL https://purrr.tidyverse.org.

Lionel Henry and Hadley Wickham. *rlang: tools for low-level R programming*, 2018b. URL https://rlang.r-lib.org.

Jim Hester. *bench: high precision timing of R expressions*, 2018. URL http: //bench.r-lib.org/.

Jim Hester, Kirill Müller, Kevin Ushey, Hadley Wickham, and Winston Chang. *withr: Run code with temporarily modified global state*, 2018. URL http://withr.r-lib.org.

Andrew Hunt and David Thomas. *The Pragmatic Programmer.* Addison Wesley, 1990.

Thomas Lumley. Programmer's niche: Macros in R. *R News*, 1(3):11–13, 2001. URL https://www.r-project.org/doc/Rnews/Rnews_2001-3.pdf.

Norman Matloff. *The Art of R Programming.* No Starch Press, 2011.

Norman Matloff. *Parallel Computing for Data Science.* Chapman & Hall/CRC, 2015. URL http://amzn.com/1466587016.

Q. Ethan McCallum and Steve Weston. *Parallel R.* O'Reilly, 2011. URL http://amzn.com/B005Z29QT4.

Scott Meyers. *Effective STL: 50 specific ways to improve your use of the standard template library.* Pearson Education, 2001. URL http://amzn. com/0201749629.

Scott Meyers. *Effective C++: 55 specific ways to improve your programs and designs.* Pearson Education, 2005. URL http://amzn.com/0321334876.

Floréal Morandat, Brandon Hill, Leo Osvald, and Jan Vitek. Evaluating the design of the R language. In *European Conference on Object-Oriented Programming*, pages 104–131. Springer, 2012. URL http://r.cs.purdue.edu/pub/ecoop12.pdf.

Kirill Müller and Lorenz Walthert. *styler: Non-Invasive Pretty Printing of R Code*, 2018. URL http://styler.r-lib.org.

Kirill Müller and Hadley Wickham. *tibble: simple data frames*, 2018. URL http://tibble.tidyverse.org/.

R Core Team. Writing R extensions. *R Foundation for Statistical Computing*, 2018a. URL https://cran.r-project.org/doc/manuals/r-devel/R-exts.html.

R Core Team. R internals. *R Foundation for Statistical Computing*, 2018b. URL https://cran.r-project.org/doc/manuals/r-devel/R-ints.html.

Steven S Skiena. *The Algorithm Design Manual*. Springer, 1998. URL http://amzn.com/0387948600.

Nathan Teetor. *zeallot: multiple, unpacking, and destructuring assignment*, 2018. URL https://CRAN.R-project.org/package=zeallot.

Luke Tierney and Riad Jarjour. *proftools: Profile Output Processing Tools for R*, 2016. URL https://CRAN.R-project.org/package=proftools.

Peter Van-Roy and Seif Haridi. *Concepts, Techniques, and Models of Computer Programming*. MIT press, 2004.

Hadley Wickham. mutatr: mutable objects for R. *Computational Statistics*, 26(3):405—418, 2011. doi: 10.1007/s00180-011-0235-7.

Hadley Wickham. *forcats: tools for working with categorical variables*, 2018. URL http://forcats.tidyverse.org.

Hadley Wickham and Yihui Xie. *evaluate: Parsing and Evaluation Tools that Provide More Details than the Default*, 2018. URL https://github.com/r-lib/evaluate.

Hadley Wickham, Jim Hester, Kirill Müller, and Daniel Cook. *memoise: Memoisation of Functions*, 2018. URL https://github.com/r-lib/memoise.

Index